Shigella

Molecular and Cellular Biology

Edited by:

William D. Picking and Wendy L. Picking
Department of Pharmaceutical Chemistry
University of Kansas
Lawrence
KS, USA
picking@ku.edu

Caister Academic Press www.caister.com

ISBN: 978-1-910190-19-7 (paperback)
ISBN: 978-1-910190-20-3 (ebook)

Ebooks

Contents

Preface 5

1 *Shigella* and Shigellosis 7
Philip R. Adam and William D. Picking

2 *Shigella* Iron Transport Systems 27
A. R. Mey, E. E. Wyckoff and S. M. Payne

3 *Shigella* and Antivirulence: the Dark Side of Bacterial Evolution 49
Kimberly A. Bliven and Anthony T. Maurelli

4 The Genetic Organization and Transcriptional Regulation
of *Shigella* Virulence Genes 65
Natasha Weatherspoon-Griffin, Michael A. Picker and Helen J. Wing

5 Ribo-regulation in *Shigella* 109
Erin R. Murphy, William H. Broach, and Andrew B. Kouse

6 IpaB Ion Channels Induce Pyroptosis in Macrophages 139
Lidija Senerovic, Anna Brotcke Zumsteg and Michael Kolbe

7 The Cellular Microbiology of *Shigella* Invasion of Epithelial Cells............151
Guy Tran Van Nhieu and Philippe Sansonetti

8 The Molecular Role of the Polar Outer Membrane Protein IcsA (VirG)
in *Shigella* Actin-based Motility 169
Anuradha Janakiraman

9 Trans-epithelial Communication in the Context of *Shigella* Infection........193
Ana Maldonado and Beth A. McCormick

10 Imaging Techniques to Track *Shigella* Infection at the Cellular Level 211
Nora Mellouk and Jost Enninga

11 Immune Responses Against *Shigella* During Natural Infection
and Induced by Vaccination 229
Francisco Martinez-Becerra

12 Dissecting *Shigella flexneri* Type III Secretion System Tip Complex Maturation 243
Olivia Arizmendi and Nicholas E. Dickenson

13 Molecular Approaches to Identifying Type 3 Secreted Effectors:
a *Shigella* Case Study 259
Analise Z. Reeves, Sonia C. Costa and Cammie F. Lesser

Index 277

Current books of interest

- Aquatic Biofilms: Ecology, Water Quality and Wastewater Treatment 2016
- Alphaviruses: Current Biology 2016
- Thermophilic Microorganisms 2015
- Flow Cytometry in Microbiology: Technology and Applications 2015
- Probiotics and Prebiotics: Current Research and Future Trends 2015
- Epigenetics: Current Research and Emerging Trends 2015
- *Corynebacterium glutamicum*: From Systems Biology to Biotechnological Applications 2015
- Advanced Vaccine Research Methods for the Decade of Vaccines 2015
- Antifungals: From Genomics to Resistance and the Development of Novel Agents 2015
- Bacteria-Plant Interactions: Advanced Research and Future Trends 2015
- *Aeromonas* *2015*
- Antibiotics: Current Innovations and Future Trends 2015
- *Leishmania*: Current Biology and Control 2015
- *Acanthamoeba*: Biology and Pathogenesis (2nd edition) 2015
- Microarrays: Current Technology, Innovations and Applications 2014
- Metagenomics of the Microbial Nitrogen Cycle: Theory, Methods and Applications 2014
- Pathogenic *Neisseria*: Genomics, Molecular Biology and Disease Intervention 2014
- Proteomics: Targeted Technology, Innovations and Applications 2014
- Biofuels: From Microbes to Molecules 2014
- Human Pathogenic Fungi: Molecular Biology and Pathogenic Mechanisms 2014
- Applied RNAi: From Fundamental Research to Therapeutic Applications 2014
- Halophiles: Genetics and Genomes 2014
- Molecular Diagnostics: Current Research and Applications 2014
- Phage Therapy: Current Research and Applications 2014
- Bioinformatics and Data Analysis in Microbiology 2014
- The Cell Biology of Cyanobacteria 2014
- Pathogenic *Escherichia coli*: Molecular and Cellular Microbiology 2014
- *Campylobacter* Ecology and Evolution 2014
- *Burkholderia*: From Genomes to Function 2014
- Myxobacteria: Genomics, Cellular and Molecular Biology 2014
- Next-generation Sequencing: Current Technologies and Applications 2014
- Omics in Soil Science 2014
- Applications of Molecular Microbiological Methods 2014
- *Mollicutes*: Molecular Biology and Pathogenesis 2014
- Genome Analysis: Current Procedures and Applications 2014

www.caister.com

Preface

Members of the genus *Shigella* are responsible for a moderate to severe dysenteric syndrome called shigellosis. This form of bacillary dysentery is ubiquitous around the world with greatly elevated morbidity and mortality in developing nations. Typical symptoms include diarrhea, fever, and stomach cramps starting one or two days after ingestion of contaminated food or water. In many cases the initial diarrhea then turns bloody and contains mucus. Patients in developed nations do not usually require hospitalization, however, children can develop severe symptoms that require immediate medical care to prevent dehydration and longer term effects. This is especially true for children between the ages of two and five in underdeveloped parts of the world where there are simultaneous threats of malnourishment and high incidence of other infectious diseases.

As considered here, the genus *Shigella* is considered a stand-alone group of species that specifically cause diarrhea in humans. They were discovered over 100 years ago and classified as four distinct species containing numerous O antigen serotypes. In reality, the *Shigella* species are so closely related to *Escherichia coli* that they should possibly not be classified as a genus but rather a distinctive species within the genus *Escherichia* (for example, see: Fukushima, et al. (2002) Phylogenetic Analysis of *Salmonella*, *Shigella*, and *Escherichia coli* Strains on the Basis of the *gyrB* Gene Sequence. J. Clin. Microbiol. 40:2779-85 & van den Beld and Reubsaet (2012) Differentiation between Shigella, enteroinvasive Escherichia coli (EIEC) and noninvasive Escherichia coli. Eur. J. Clin. Microbiol. 31:899-904). Nevertheless, the vast majority of cases of bacillary dysentery are attributable to what we refer to here as *Shigella* and they thus warrant special consideration as a diarrheal pathogen of global public health importance.

This book is roughly divided into three sections that bring us up to date on the molecular biology of *Shigella*. The first section will explore essential aspects of metabolism and gene regulation in this human adapted pathogen. Topics here include the acquisition of limiting iron as an essential means for survival within the host, the evolution of the *Shigella* genome to make its pathogenic lifestyle compatible with its metabolism, and the mechanisms by which virulence genes are expressed. This section explores critical aspects of the molecular and cellular biology of *Shigella* as it has diversified from its evolutionary place among the *Escherichia coli*. While there is much more to be learned about the metabolomics and genomics of *Shigella*, this section provides the reader with a flavor of often overlooked features of bacterial pathogens and the recently developed appreciation for RNA-based regulation of gene expression.

The second section of this work focuses on the important area of host-pathogen interplay. *Shigella* is among a group of organisms for which a wealth of information exists on its ability to subvert host signaling processes for the benefit of the pathogen. Humans have evolved to fight infections by a variety of innate and adaptive processes, however, the most successful human pathogens have developed exquisite means for avoiding these processes to allow them to gain a foothold to ultimately allow colonization and cause host cell damage. *Shigella* has evolved specific mechanisms to trigger pathways for the destruction of phagocytes (macrophage killing) that are portrayed differently to intestinal epithelial cells which are prompted to internalize the bacterial pathogen to provide an intracellular niche where it can replicate and spread to neighboring cells. All three of these topics are covered in excellent detail in this section of the book, however, *Shigella* interplay with the host is only beginning to be fully appreciated. The ability for this pathogen to alter signaling pathways within tissues is also a topic of interest and this is evident from its manipulation of host innate immunity and its ability to elicit signals across the intestinal epithelium. A full understanding of all of these events is prompting the development of novel methods for assessing the host cell targets that are subverted by the effector proteins that *Shigella* skillfully delivers, which is also addressed in this section. Lastly, an overview of the adaptive immunity elicited by *Shigella* infection is presented along with a comparison of this response with that induced by current *Shigella* vaccine candidates.

In the final section of this book, two aspects of what is probably *Shigella*'s most prominent virulence are discussed. The *Shigella* type III secretion system is considered one of the paradigms for type III secretion in Gram-negative bacteria. The T3SS is recognized as a highly evolved nanomachine for promoting cross-species communications between prokaryotes and eukaryotes in what are often pathogenic relationships, but which can be symbiotic in some cases. While we have not gone into tremendous detail on the overall structure of the type III secretion apparatus (T3SA), two chapters are dedicated to dynamic aspects of the *Shigella* T3SS. One is the nature of the needle tip complex of the T3SA which is essential for sensing contact with host cells and ultimately creating the translocon pore through which effector proteins are delivered into host cells. The second chapter describes new methods that are allowing the identification of newly recognized effector proteins that are the tools by which *Shigella* alters target cell behavior. It should be noted, however, that previous chapters also touch upon the topics of type III secretion and the manipulation of host cell signaling by effector proteins. Thus, all the sections of this book should be viewed as an integrated overview of the molecular and cellular biology of *Shigella* and its human host.

There are many important facets of *Shigella* that are not covered in this book, but these are by no means to be slighted by their omission. The overall goal of this compendium is to provide the researcher, especially those new to the field, with a flavor of some of the topics that are important as we move forward in understanding *Shigella* the pathogen and the delicate balance it has with its primary host – human beings. Just a few of the topics not covered here would include the importance of lipopolysaccharide as a virulence factor and the effect of newly developed antibiotic resistances on the changing epidemiology of shigellosis. We hope it is thus clear to the reader that much remains to be learned about this organism and the public health problem it poses.

We would like to express our gratitude to the authors who contributed to this book and we thank them for their important contributions to our current understanding of *Shigella* and shigellosis. We also thank the publisher for realizing the importance of providing this overview of the research being done on the cellular and molecular biology of this pathogen and how they contribute to bacillary dysentery.

William D. Picking
Wendy L. Picking

1

Shigella and Shigellosis

Philip R. Adam[1] **and William D. Picking**[2]

[1]Department of Microbiology and Molecular Genetics, Oklahoma State University, Stillwater, Oklahoma, 74078 USA
[2]Department of Pharmaceutical Chemistry, 320B MRB, University of Kansas, Lawrence, KS 66045 USA, picking@ku.edu

Abstract
Shigella spp. are Gram-negative, nonmotile bacilli that are important gastrointestinal pathogens responsible for significant morbidity and mortality, especially in developing nations. It is particularly devastating to children ages 2 to 5 who suffer most from the dehydration and complications of shigellosis. The survivors often exhibit impaired growth due to malnutrition and this is exacerbated by repeated diarrheal episodes. The difficulties posed by shigellosis are further complicated by increased antibiotic resistance and the possibility of post-infection sequelae such as episodes of reactive arthritis and inflammatory bowel disease in genetically susceptible individuals. Advances in understanding the pathogenic mechanisms have revealed novel forms of crosstalk between *Shigella* and its human host. New insights into the type III secretion system of this pathogen, the subversion of normal mammalian cell functions for the benefit of the pathogen, and the use of toxins that enhance infection and cause targeted tissue damage are all part of the *Shigella* pathogenic arsenal. Nevertheless, shigellosis continues to be a devastating public health problem. This collection of works will illuminate our current understanding of important aspects of *Shigella* molecular biology and its crosstalk with human cells, and provides the foundation for future research on this important agent of infectious diarrhea.

Shigella species
Bacteria of the genus *Shigella* cause a severe disease in humans that, in the absence of supportive care, can lead to death. *Shigella* infections are spread via the fecal-oral route, usually through the consumption of contaminated drinking water. Flies have an affinity for fecal matter and are known to land on human food, thus flies have also been implicated in the spread of this and other diarrheal diseases (Cohen et al., 1991; Lindsay et al., 2012). A study by the Israel Defense Force found that population-control measures against the common housefly (*Musca domestica*) reduced the incidence of shigellosis by 85% (Cohen et al., 1991). In developing countries, individual cases of shigellosis can quickly turn into outbreaks due to the absence of public sanitation and a lack of clean drinking water. Members of the genus *Shigella* (considered a stand-alone genus here) are non-motile, Gram-negative bacilli. There are four species within the genus *Shigella*: *S. dysenteriae* (serogroup A), *S. flexneri* (serogroup B), *S. boydii* (serogroup C), and *S.*

sonnei (serogroup D) (Hale TL, 1996). Serogroup A contains 15 serotypes, B contains 8 serotypes, C contains 19 serotypes, and D contains only a single serotype (CDC., 2011; Hale TL, 1996).

Shigella has a remarkably low infectious dose of 10-200 organisms due to its acid-tolerance systems (DuPont et al., 1989). Once in the large intestine, *Shigella* crosses the epithelial cell barrier, evades the innate immune system and induces its uptake into colonic epithelial cells from the basolateral side (Mounier et al., 1992). Once inside an epithelial cell the bacterium escapes its vacuole, begins to multiply and spreads directly to neighboring epithelial cells without entering the extracellular environment (Bernardini et al., 1989). The symptoms of shigellosis occur as a result of inflammation at the site of infection and include fever, severe abdominal cramping, and watery diarrhea (Sansonetti, 1998). Within two days, the patient will develop bright-red bloody, mucus-laden low volume stools and tenesmus (Koneman, 1997). Disease severity ranges from relatively minor watery diarrhea to complicated, life-threatening dysentery. Dehydration is a life-threatening complication that can result in renal failure and death, however, most healthy adults can recover from the infection. Hemolytic uremic syndrome can develop if an individual is infected with *S. dysenteriae* due to Shiga toxin production by this species (Bergan et al., 2012).

Clinical laboratories typically use MacConkey agar, Hektoen enteric agar, and xylose lysine deoxycholate agar to isolate *Shigella* (Luis M de la Maza, 2004). Serological assays are the most reliable means of distinguishing between *Shigella* species. Humans appear to be the natural reservoir for *Shigella*, although intriguing evidence suggests that *Shigella* can live within the amoeba *Acanthamoeba castellanii,* which could serve as an environmental reservoir (Saeed et al., 2009). It should be noted that *Shigella* has not been found within amoebae in the environment and this potential association remains unconfirmed. *Shigella* outbreaks have occurred in non-human primate colonies (Nizeyi et al., 2001). Individuals can asymptomatically carry *Shigella*, though the reason for carriage versus disease is not clear. Nevertheless, this phenomenon may explain why *Shigella* remains endemic in some areas of the world (Bovee et al., 2012; Hossain et al., 1994). The pathology of *Shigella* in primates is similar to humans and they have been used as laboratory animal models. Outside of non-human primates, an entirely reliable animal model for dysenteric disease is not available.

Discovery of Shigella

Kiyoshi Shiga isolated *Bacillus dysenteriae* in 1898, which distinguished bacillary dysentery from amoebic dysentery (Hale, 1991). *Shigella* was initially thought to be a member of a human's natural flora, and that disease occurred when the equilibrium between the host, *Shigella*, and the natural intestinal flora was disrupted (Watkins, 1960). This theory is similar to today's understanding of *Clostridium difficile* induction of disease (Knight and Surawicz, 2013). The disease produced by *Shigella* was thought to be a result of vascular damage caused by excessive endotoxin release by the pathogen (Watkins, 1960). It was not until the 1950's that the Congress of the International Association of Microbiologists *Shigella* Commission extensively reviewed epidemiological, physiological, and serological reports of bacilli causing dysentery and recommended that the genus be named *Shigella* and established the species subgroups as we know them today (Hale, 1991). There are over 40 serotypes of *Shigella* and infection provides only homologous serotype protection. Thus, the need for a vaccine is urgent (Hale, 1991; Kweon, 2008).

Epidemiology

The World Health Organization estimates that about 1.7 billion cases of diarrheal disease occur globally each year (2013a). Approximately 1.9 million people die each year as a result of diarrheal disease, with about 800,000 of these deaths being in children under five years of age (2013a). *Shigella* causes about 125 million cases of diarrhea and ~14,000 deaths annually (Bardhan et al., 2010). 99% of all cases of shigellosis occur in the developing world where *S. flexneri* is the species most commonly isolated, while *S. sonnei* is most common in developed countries (Kotloff et al., 1999). In a real-time demonstration of this phenomenon in China, Thailand, South Korea and Taiwan, the prevalence of *S. flexneri* has decreased and *S. sonnei* has increased during the recent economic growth in these countries (Chang et al., 2012; Seol et al., 2006; von Seidlein et al., 2006; Wei et al., 2007). *S. flexneri* 2a is the most commonly isolated strain, however, a recently emerged strain, *S. flexneri* 4c, has been the second most-common strain in China in the past ten years (Chang et al., 2012).

Deaths from shigellosis are rare in developed countries due to greater access to medical care. In the developed world, outbreaks can occur among children in daycare and among inmates at correctional institutions (Mahoney et al., 1993; Pickering et al., 1981). *Shigella* spp. are also known to cause traveler's diarrhea in people who have visited endemic regions of the world (Sharp et al., 1995). Those returning from North Africa, South and Southeast Asia, and Oceania are at the highest risk (Ross et al., 2013).

Recent studies challenge the 1.1 million deaths attributable to *Shigella* reported by Kotloff, et al. in 1999. One such study was limited to shigellosis in Asia, and reports that approximately 125 million cases of shigellosis occur and only 14,000 of these cases are fatal annually (Bardhan et al., 2010). The reason cited for the recent decrease in the number of *Shigella*-related deaths is improved health of children upon the onset of infection (Bardhan et al., 2010). The availability of the measles vaccine may have played a role in the decreased number of deaths, because dysentery and measles were often comorbid (Bardhan et al., 2010). Measles exacerbates the diarrhea caused by *Shigella* and other enteric pathogens, leading to increased mortality (Bardhan et al., 2010; Koster et al., 1981; Mathur et al., 1989).

The global enteric multicenter study (GEMS) paired children experiencing diarrhea with a healthy control individual, and found that *Shigella* spp. were the fourth, second and first leading causes of moderate to severe diarrhea in children aged < 12 months, 12-23 months, and 24-59 months (Kotloff et al., 2013). This study, however, found that *Shigella* spp. were not associated with an increased risk of death relative to the healthy control group (Kotloff et al., 2013). This fact can be misleading, as case-fatality rates of up to 52% have been reported during *S. dysenteriae* outbreaks (Butler, 2012). The mortality from diarrheal disease is declining at a rate of ~4% per year, but the prevalence of disease in general has remained high (Fischer Walker et al., 2012; Kotloff et al., 2013; Modell et al., 2012).

Diarrheal disease is only one aspect of a complicated public health scenario in developing countries. It has an impact on long-term nutritional status that is associated with growth stunting (Checkley et al., 2008). The risk of pneumonia increases by 1.06% for every day of diarrheal disease in the 4 weeks prior to the onset of pneumonia (Ashraf et al., 2013; Schmidt et al., 2009). The risk of secondary infections can be decreased by promoting hand washing and other personal sanitation techniques (Ashraf et al., 2013). The impact of drinking water quality on diarrheal disease burden is only evident when other factors are satisfied beforehand, namely community and personal sanitation (VanDerslice and Briscoe, 1995). In very poor countries the sanitary disposal of human excrement is not always possible. Providing the infrastructure to dispose of waste in a more sanitary way is expected to decrease the diarrheal disease burden by 30-40% (VanDerslice and Briscoe, 1995). Decreasing the burden of fecal coliforms in drinking water by two orders of magnitude would then decrease diarrheal disease incidence by another 40% once community sanitation is improved (VanDerslice and Briscoe, 1995). Building sewer systems is expensive and not a practical solution for poorly developed countries. Thus, the need for a low-cost vaccine that can be provided by philanthropic organizations would be beneficial.

Emergence of antibiotic resistance

Multi-drug resistance is a problem for many important human pathogens, and *Shigella* is no exception. One study examined 81 *S. flexneri* isolates from India, and found resistance to 21 different antibiotics including many that are considered first line therapy for shigellosis (Bhattacharya et al., 2013). Almost half of these isolates were resistant to ten or more drugs (Bhattacharya et al., 2013). Over 70% of the isolates were resistant to second-generation fluoroquinolones (ciprofloxacin, nofloxacin), and 17% were resistant to third generation cephalosporins (cefixime, ceftriaxone) (Bhattacharya et al., 2013). Resistance to tetracycline, chloramphenicols, sulfonamides, and gentamycin was also noted (Bhattacharya et al., 2013). Multidrug efflux pumps and mutations in DNA gyrase and topoisomerase were noted as major contributors to the appearance of multidrug resistant strains (Bhattacharya et al., 2013).

A review published in 2012 revealed that antibiotic resistance is much more pervasive in Asia and Africa than in Europe and America (Gu et al., 2012). For *S. flexneri* and *S. sonnei* in Asia and Africa, resistance to nalidixic acid is has risen from 4% in 2000 to 84% in 2009, whereas the rate has remained close to 3% in Europe and America over the same time span (Gu et al., 2012). *S. flexneri* isolates are also rapidly acquiring resistance to ciprofloxacin in Asia and Africa, increasing from 0.3% in 2000 to 13% in 2006 (Gu et al., 2012).

Shigella isolates acquire their antibiotic resistance genes from mobile genetic elements identified within islands in the bacterial genome such as R plasmids, transposons, and integrons (Ke et al., 2011). Class 1 and class 2 integrons have been found within *Shigella* isolates and contain the genes required for resistance to ampicillin, streptomycin, trimethoprim, and streptothricin (DeLappe et al., 2003; Ke et al., 2011; Mammina et al., 2005). Isolates of *S dysenteriae* type 1 are typically susceptible to nalidixic acid, ciprofloxacin, and ceftriaxone, whereas they are resistant to ampicillin, chloramphenicol, tetracycline, and sulfa drugs (Butler, 2012).

Recent documented outbreaks in developed world

A clonal outbreak of *S. sonnei* was reported in Los Angeles, California in May 2012 (2013b). Forty-three cases of shigellosis were reported among members of a private club aged from 54-98 years (2013b). No deaths occurred despite the severity of the illness with disease severity attributed to the advanced age of the patients (2013b). Most notably, this was the first reported outbreak of *S. sonnei* with decreased susceptibility to azithromycin in the United States (2013b).

In North London in 2011 an outbreak of *S. sonnei* was reported among members of the Orthodox Jewish community (McDonnell et al., 2013). The outbreak was caused by a single strain and lasted a total of eight weeks, causing 82 illnesses with no associated mortality (McDonnell et al., 2013). The cases occurred in both children and the parents of young children with daycare centers and schools serving as the conduit for transmission (McDonnell et al., 2013). Genome-level examination of the isolates in this outbreak revealed that the isolates were all very closely related, but the strain accumulated up to seven single nucleotide polymorphisms (SNPs) during the outbreak, though no phenotypic differences were evident (McDonnell et al., 2013). The length of this outbreak was attributed to poor communication between public health officials and members of the Orthodox Jewish community (McDonnell et al., 2013).

In the United Kingdom, there is an ongoing outbreak of *S. flexneri* 3a in men who have sex with men (Borg et al., 2012). This outbreak was detected in 2011 but started in 2009, causing 381 confirmed cases (Borg et al., 2012). This epidemic is slow moving and causing relatively few cases, only 1-7 per month, but the continued low-level transmission is due to poor hygiene during and after sex (Borg et al., 2012). A similar outbreak of *S. sonnei* occurred in Tokyo, Japan in late 2011 encompassing five patients, however, an epidemiologic investigation was not conducted so the number of cases may be underreported in this case (Okame et al., 2012).

Complications

Hemolytic uremic syndrome

S. dysenteriae type 1 is associated with a high incidence of severe gastrointestinal symptoms and hemolytic uremic syndrome (HUS) (Butler, 2012; Khan et al., 2013). Disease progression to HUS results in mortality rates of up to 52% (Butler, 2012). HUS caused by *S. dysenteriae* type 1 nearly always occurs in young children and is primarily due to the massive production of Shiga toxin (Tzipori et al., 2004). Lipopolysaccharide (LPS) is another potential mediator of HUS as it was detected in the blood of several patients with *Shigella*-mediated HUS and can induce similar damage to the kidneys of rabbits during experimental infections (Butler, 2012). *Escherichia coli* O157:H7 is also noted for causing HUS, but it does not occur exclusively within one age group (Gould et al., 2009). Three standard clinical/laboratory signs are typical of HUS development including hemolytic anemia, thrombocytopenia (low platelet count), and elevated serum creatinine concentrations (Butler, 2012). Classically, HUS is considered a blood clotting disorder in which clots form preferentially in the glomeruli of the kidneys (Butler, 2012). Decreased serum sodium concentrations cause the serum to become hypotonic which may exacerbate hemolysis of red blood cells, especially in young children (Butler, 2012). Development of HUS is strongly linked to the start of antibiotic therapy four or more days after the onset of symptoms (Butler, 2012). Furthermore, the choice of antibiotic is important when treating *S. dysenteriae* Type I infections (Al-Qarawi et al., 1995; Bin Saeed et al., 1995). Clinical outcomes of *S. dysenteriae* type 1 treatment with either ampicillin or nalidixic acid were compared, and it was found that HUS occurred at a much higher frequency in the ampicillin-treated patients than the nalidixic acid-treated patients (Al-Qarawi et al., 1995; Bin Saeed et al., 1995). Anti-motility drugs are contraindicated in the treatment of dysentery and increase the risk of HUS (Butler, 2012). The use of appropriate antibiotics early in the onset of disease symptoms is recommended (Butler, 2012). Resistance to nalidixic acid has led to its replacement with ciprofloxacin as the recommended treatment for shigellosis (Vinh et al., 2011).

Reactive Arthritis

Reactive arthritis (ReA) is an inflammatory disease of the joints that can be triggered by gastrointestinal pathogens including *Shigella*, *Salmonella*, *Campylobacter*, *Escherichia coli*, *Yersinia* spp. and other gastrointestinal parasites (Connor and Riddle, 2013; Hannu et al., 2005). The disease onset is 1-5 weeks post-infection, is usually polyarticular and tends to occur in joints of the legs, although it can also manifest itself outside of the joints as a mucosal, urethral, cutaneous, or ocular inflammation (Connor and Riddle, 2013). ReA symptoms have a broad range of severity, ranging from a relatively minor and temporary discomfort to chronic, debilitating arthritis (Connor and Riddle, 2013). Incidences of ReA due to *Shigella* infection are typically associated with *S. flexneri*, but *S. sonnei* has also been implicated (Chen et al., 2002; Lauhio et al., 1988; Morse et al., 1980; Simon et al., 1981). ReA may occur in up to 7% of individuals who recover from infection with *Shigella*, but is uncommon in children (Hannu et al., 2005). Other reported complications include tendonitis, enthesopathy, and bursitis (Hannu et al., 2005). Most individuals that develop ReA after enteric infection are HLA-B27 positive (Hannu et al., 2005; Schiellerup et al., 2008). Development of irritable bowel syndrome has been linked to previous infection with an enteric pathogen, with an incidence ranging anywhere from 4-32% (Borgaonkar et al., 2006; Connor, 2005; McKendrick and Read, 1994; Spiller and Garsed, 2009). Risk factors for post-infection IBS include young age, stress, high production of TNF-alpha and low production of IL-10 (Bashashati et al., 2012; Gwee et al., 1996). Clearly, the development of post-infection complications is the result of a complex interplay between the pathogen, host, and a multitude of host-specific factors and is the subject of continuing research (Bashashati et al., 2012; Connor and Riddle, 2013; Hannu et al., 2005).

Virulence Factors

Type Three Secretion System

S. flexneri possesses a type three secretion system (T3SS) that is encoded on a 200 kb virulence plasmid composed of ~100 genes (Buchrieser et al., 2000; Sansonetti et al., 1982; Venkatesan et al., 2001). T3SS's are used by a diverse array of Gram-negative pathogens to deliver proteins into the extracellular environment or directly into the target cell's cytoplasm. For *Shigella*, the T3SS is an essential virulence factor that is used for macrophage killing and to promote entry into colonic epithelial cells. More detailed descriptions of the structure and function of the *Shigella* T3SS will appear in later chapters within this compilation.

The ability to invade human intestinal cells is encoded by the "entry region" of the large *Shigella* virulence plasmid. The entry region is a conserved 31-kb segment that harbors 34 genes that encode the complete T3SS including the type III secretion apparatus (T3SA) and the various effector proteins that are required for invasion of host epithelial cells, immune evasion, and intracellular survival (Maurelli et al., 1985; Sasakawa et al., 1988; Sasakawa et al., 1993; Schroeder and Hilbi, 2008). In general terms, the entry region contains four distinct functional groups of genes: structural components of the T3SA (including the so called translocator proteins), proteins that affect host cell function, transcriptional activators, and chaperones (Schroeder and Hilbi, 2008). These genes of the entry region are only expressed at 37°C, but some require additional environmental signals for their expression (Schroeder and Hilbi, 2008; Tobe et al., 1991). Transcription of the *mxi-spa* locus and the first set of effectors (the early effectors) is activated by exposure to 37°C, thus priming the bacterium to invade epithelial cells following ingestion by a human host. Initial activation of secretion results in the formation of a translocation pore (the translocon) in the host cell membrane comprised of the invasion plasmid antigens IpaB and IpaC. These remain anchored to the T3SA by IpaD which is located at the distal tip of the T3SA needle. Following secretion of the translocator proteins along with injection of early effectors into the host cell cytoplasm, the chaperone for IpaB and

IpaC (IpgC) is freed in the *Shigella* cytoplasm (Mavris et al., 2002). IpgC then binds to MxiE and induces expression of the late effector proteins that are delivered to the host cell in a coordinated manner (Mavris et al., 2002).

Invasion plasmid antigens (Ipa) B, C and D are essential virulence factors secreted by the T3SS. IpaB is necessary for macrophage killing and formation of the translocon pore through which effector proteins are injected into the host cell. IpaC induces major actin rearrangements by activating small GTPases and is also thought to participate in translocon pore formation (Blocker et al., 1999; Tran Van Nhieu et al., 1999). IpaD and IpaB control secretion from the T3SA needle tip, and IpaD serves as an environmental sensor by binding to bile salts and recruiting IpaB to the needle tip (Menard et al., 1994; Menard et al., 1993; Olive et al., 2007; Stensrud et al., 2008). IpaD's effector function is unclear. Meanwhilie, structural components of the T3SS are encoded by two groups of genes: membrane expression of *ipa* (*mxi*) and surface presentation of *ipa* antigen (*spa*) genes (Hromockyj and Maurelli, 1989; Venkatesan et al., 1992). Genes of the *mxi-spa* locus are absolutely essential for secretion of all T3SS effector proteins (Blocker et al., 1999).

Two T3SS-transcriptional activators, VirB and MxiE, comprise another distinct group of genes within the entry region (Adler et al., 1989; Kane et al., 2002). These proteins regulate various genes in the entry region and in other regions of the virulence plasmid. This regulation ensures proper coordination of the events leading to host cell invasion and growth within the subsequent intracellular niche.

The four T3SS chaperones comprise the next group of genes within the entry region and these include IpgA, IpgC, IpgE, and Spa15 (Page et al., 1999; Page et al., 2002; Parsot et al., 2003). IpgC and Spa15 are also involved as co-factors in transcriptional regulation (Parsot et al., 2005; Pilonieta and Munson, 2008). T3SS effectors that are encoded outside of the entry region, such as the outer *Shigella* protein (Osp) and IpaH families of proteins, help shape the immune response by altering host cell transcription (Schroeder and Hilbi, 2008).

Intracellular motility via actin polymerization is provided by a set of genes located outside of the entry region and, with the exception of VirA, these proteins are not secreted by the T3SS (Schroeder and Hilbi, 2008). A brief description of intracellular motility and intercellular spread is provided below with a more detailed one provided later in this compilation.

T3SS's Role in Invasion

Shigella initially interacts with host cells by binding to CD44 and $\alpha_5\beta_1$ integrins within cholesterol-rich membrane raft microdomains and this may be mediated by surface-exposed Ipa proteins (Skoudy et al., 2000; Watarai et al., 1996). Adherence of *Shigella* to the host cell triggers aggregation of membrane rafts creating a locally high concentration of receptors and their signaling components, and induces the preliminary cytoskeletal rearrangements necessary for invasion (Lafont et al., 2002). The T3SS effector proteins IpaC, IpgB1, IpgB2, VirA, IpgD, and IpaA work cooperatively to alter the host cell's cytoskeleton to promote bacterial uptake. IpaC interacts with small GTPases to form membrane extensions around the bacterium (Tran Van Nhieu et al., 1999). IpgB1 and IpgB2 destabilize cytoskeletal components by mimicking the actions of small GTPases, whereas VirA may destabilize these structures by virtue of a protease activity (Alto et al., 2006; Handa et al., 2007; Yoshida et al., 2002). IpgD is a phophoinositide-4-phosphatase that hydrolyzes membrane components to ultimately separate the membrane from the cytoskeleton (Niebuhr et al., 2002). IpaA then increases vinculin's affinity for actin, leading to vinculin capping of actin filaments and further disassociation of the cytoskeleton (Bourdet-Sicard et al., 1999; Ramarao et al., 2007). Further details on the cell biology of host cell invasion is provided later in this book.

T3SS's Role in Immune Evasion

Before *Shigella* can invade the host's epithelium, it must first escape the innate immune system. In short, *Shigella* is ingested and passes through the gastrointestinal tract to the large intestine, where it crosses the epithelium via M-cells and is phagocytosed by a macrophage (Sansonetti et al., 1996; Wassef et al., 1989). Once inside the macrophage *Shigella* lyses the resulting phagosome (Chen et al., 1996) and then induces macrophage apoptosis via the T3SS effector IpaB which directly binds to and activates caspase-1 (Chen et al., 1996; Hilbi et al., 1998; Zychlinsky et al., 1994). Macrophage cell death leads to the release of proinflammatory cytokines that trigger the inflammatory response which will ultimately lead to the overt symptoms of dysentery (Sansonetti et al., 2000). Nonetheless, *Shigella* is now free to use its T3SS to induce its own uptake into host epithelial cells from the basal side (Mounier et al., 1992).

Shigella uses its T3SS to lyse its phagosome within 15 minutes of bacterial uptake into an epithelial cell (Clerc et al., 1987). IpaB, IpaC, and IpaD are all important for phagosome escape and while IpaC may play the most pivotal role here (Barzu et al., 1997; High et al., 1992; Picking et al., 2005), the precise mechanism of vacuolar escape remains unknown. The late effector IpaH7.8 also appears to be required for phagosomal escape, however, its role remains unclear (Fernandez-Prada et al., 2000). Once in the host epithelial cytoplasm, *Shigella* secretes numerous T3SS effector proteins that subvert normal host cell transcription and signaling, thus allowing the bacterium to persist in the epithelial cell cytoplasm and avoid detection by the immune system. IpgD promotes host cell survival by setting off a signaling cascade that activates the serine-threonine kinase Akt (Pendaries et al., 2006). Another unknown T3SS effector also inhibits apoptosis (Clark and Maurelli, 2007). OspF interferes with intracellular pathogen detection systems by dephosphorylating MAPK, thus inhibiting MAPK-mediated signaling (Kramer et al., 2007; Li et al., 2007). In the absence of MAPK signaling the transcription of pro-inflammatory genes is greatly reduced (Arbibe et al., 2007). OspG then promotes the proteolytic degradation of the transcription factor NF-κB to further down-regulate the inflammatory response (Kim et al., 2005). IpaH9.8 may also contribute to this process by acting as an ubiquitin ligase, thereby destroying downstream components of the MAPK kinase-dependent pathway (Rohde et al., 2007).

In the cytoplasm, *Shigella* grows, divides and spreads cell-to-cell using actin-based motility (Schroeder and Hilbi, 2008). VirA degrades microtubules in front of the bacterium, while the intercellular spread protein IcsA (also referred to as VirG) localizes to one pole of the bacterium and nucleates actin to create an actin tail that pushes the bacterium through the cytosol (Goldberg et al., 1993; Goldberg and Theriot, 1995; Yoshida et al., 2006). The bacterium creates membrane bulges into neighboring cells at tight junctions. The neighboring cell then pinches off the invagination which leaves the bacterium surrounded by a double membrane (Rathman et al., 2000). The double membrane is then lysed in a T3SS-dependent manner and the bacterium perpetuates the cycle of intracellular growth and intercellular spread all over again (Picking et al., 2005; Schuch et al., 1999).

Toxins

Shiga Toxin

Shiga toxin was first described in *S. dysenteriae* and is the "original" member of the Shiga toxin family (Bergan et al., 2012). Several homologous toxins, referred to as Shiga toxins 1 and 2, are present in various strains of *E. coli*, most notably strain O157:H7 (Bergan et al., 2012; Calderwood et al., 1987; Takao et al., 1988). Shiga toxins are AB_5-type proteinaceous toxins, where the toxic A subunit is enzymatically active and the non-toxic B subunits bind to the target receptors (Bergan et al., 2012). The C-terminus of the A subunit is inserted into the center of the pore formed by the five B subunits (Fraser et al., 1994; Stein et al., 1992). The B subunit binds to the glycosphingolipid Gb3 resulting in uptake of the toxin complex into target cells (Lindberg et al., 1987). Following uptake, the enzymatically active, 27.5 kDa A1 fragment is released upon cleavage by the intracellular protease furin (Garred et al., 1995). Eventually the A1 fragment enters the cytosol where it displays single-site RNA *N*-glycosidase activity, thereby inactivating ribosomes through the removal of a single adenine residue from the 28S rRNA of 60S ribosomal subunits (Saxena et al., 1989). Inhibition of protein synthesis leads to apoptosis of the intoxicated cells (Tesh, 2010). How Shiga toxin reaches the kidney where it is known to cause damage of the renal endothelium is poorly understood. The most likely scenario is that the secreted toxin enters the bloodstream from the intestine via the bloody lesions that are the result of shigellosis (Butler, 2012). Free Shiga toxin has never been found in the blood, but it has been observed to bind with low affinity to polymorphonuclear leukocytes via the A subunit and could reach the kidney in this way (Brigotti et al., 2011).

Enterotoxins

Four enterotoxins are encoded by *S. flexneri*: ShET1 (*setAB*), ShET2 (*sen*), Pic, and SepA (Faherty et al., 2012). Collectively, these enterotoxins are thought to cause electrolyte secretion into the small intestine, contributing to the watery diarrhea that often accompanies and is seen at the onset of shigellosis (Faherty et al., 2012). What advantage *Shigella* would gain by increased fluid level in the small intestine is unclear, but it has been proposed that this may expedite transit to the colon where *Shigella* actually causes disease (Schroeder and Hilbi, 2008). ShET1 and Pic are encoded on the bacterial chromosome, whereas ShET2 and SepA are encoded on the *Shigella* virulence plasmid (Faherty et al., 2012). ShET1 is only found in *S. flexneri* 2a and 2b, whereas ShET2 is found in most serotypes of *S. flexneri* as well as in *S. boydii*, *S. sonnei*, *S. dysenteriae*, and enteroinvasive *Escherichia coli* (Fasano et al., 1995; Niyogi et al., 2004; Yavzori et al., 2002). ShET1 is an A-B type toxin, but the stoichiometry remains undefined (Niyogi et al., 2004). ShET2, also known as OspD3, is secreted via the T3SS (Nataro et al., 1995). ShET2 is found in 80-90% of all *Shigella* isolates tested that also harbor the virulence plasmid (Roy et al., 2006). ShET2 increases the amount of the pro-inflammatory cytokine interleukin-8 (IL-8) secreted post-invasion by cultured epithelial cells (Farfan et al., 2011). The prevalence of these toxins does not appear to follow any geographical pattern. The molecular mechanisms governing the action of ShET1, ShET2 and SepA are poorly understood, however, Pic has been studied in some detail.

Pic (Protease involved in colonization) is a member of the serine protease autotransporters of Enterobacteriaceae (SPATE) family, class 2. Pic binds to and cleaves proteins modified with Sialyl Lewis-X (SLeX)-O-glycans (Ruiz-Perez et al., 2011). SepA is also a SPATE that contributes to the enterotoxicity of *Shigella*, but its target remains unknown (Faherty et al., 2012). Pic is involved in colonizing the mucosal epithelium by cleaving mucin and inducing mucin release (Harrington et al., 2009; Ruiz-Perez et al., 2011). This toxin is only expressed in *S. flexneri* 2a (Ruiz-Perez et al., 2011). Pic may be involved in immunomodulation in the host because it cleaves many cell-surface O-linked glycans (*e.g.* CD43) that are involved in leukocyte trafficking (Ruiz-Perez et al., 2011).

Deletion of ShET1, ShET2, and Pic allowed for the development of a live attenuated vaccine that does not cause watery diarrhea, thus solving a problem that has plagued vaccine development for some time (Kotloff et al., 2004). Live attenuated vaccines are effective at inducing homologous serotype protection, however, they are noted for their lack of cross-protection against distinct serotypes, which compromises their use as broadly protective vaccines. The development of a universal and broadly protective vaccine candidate remains a priority (Kotloff et al., 2004). An intriguing finding was that this live attenuated vaccine also generated very high titers to antigens conserved across all *Shigella* species, namely IpaB, IpaC, and IpaD (Kotloff et al., 2004).

Future considerations

Bacillary dysentery (now known as shigellosis) has been a recognized public health problem since before the discovery of the causative agent more than a century ago. While *Shigella* species continue to be responsible for outbreaks, especially in developing countries, great strides have been made in understanding the mechanisms that govern the virulence of these human-adapted pathogens. In the past three decades a deep appreciation has developed for the elegant interplay that occurs between *Shigella* and its human hosts. In the chapters that follow there is a wealth of information on the molecular biology of *Shigella* and how it regulates the genes that govern its virulence, how it has evolved metabolically to become the pathogen we know today, how it has learned to communicate with cells of the human body and how the human body in turn responds to the pathogen.

From a public health perspective, the fight against shigellosis has a long distance to go. The emergence of new multidrug resistant strains and the rapidity in which such strains appear present an ever-changing target with regard to treatment. Pockets of global political and ethnic upheaval further ensure that shigellosis will continue to plague humankind. On the upside, our ever increasing understanding of the virulence factors that are responsible for *Shigella*'s ability to cause infection are providing the seeds for future anti-infective agents that may replace currently used antibiotics. Furthermore, new findings in the areas of genomics, proteomics, metabolomics and intestinal microbiomics are showing promise in defining new ways to prevent and treat infections caused by *Shigella*. One area that has been especially promising is the development of new vaccines that might be used to prevent shigellosis among young children in the developing world, who are especially vulnerable to the ill effects of dysentery. Several types of vaccines have been or are being explored for their potential in preventing *Shigella* infection. These vaccines range from live attenuated strains to whole-killed bacteria to conjugate vaccines to subunit vaccines using conserved protein antigens or combinations of these. The development of these and other vaccines that target diarrheal disease in the developing world is being helped by charitable organizations that promote improved public health on a global scale. In the end, however, the basic science needed to better understand *Shigella* as a pathogen is revealing just how exquisite the interaction is between this disease causing agent and the human hosts with which it has co-evolved. Some of that information will be explored in the chapters that follow.

References.

(2013a). Diarrhoeal disease (World Health Organization).

(2013b). Notes from the field: Outbreak of infections caused by Shigella sonnei with decreased susceptibility to azithromycin--Los Angeles, California, 2012. MMWR Morbidity and mortality weekly report *62*, 171.

Adler, B., Sasakawa, C., Tobe, T., Makino, S., Komatsu, K., and Yoshikawa, M. (1989). A dual transcriptional activation system for the 230 kb plasmid genes coding for virulence-associated antigens of *Shigella flexneri*. Mol Microbiol *3*, 627-635.

Al-Qarawi, S., Fontaine, R.E., and Al-Qahtani, M.S. (1995). An outbreak of hemolytic uremic syndrome associated with antibiotic treatment of hospital inpatients for dysentery. Emerg Infect Dis *1*, 138-140.

Alto, N.M., Shao, F., Lazar, C.S., Brost, R.L., Chua, G., Mattoo, S., McMahon, S.A., Ghosh, P., Hughes, T.R., Boone, C., *et al.* (2006). Identification of a bacterial type III effector family with G protein mimicry functions. Cell *124*, 133-145.

Arbibe, L., Kim, D.W., Batsche, E., Pedron, T., Mateescu, B., Muchardt, C., Parsot, C., and Sansonetti, P.J. (2007). An injected bacterial effector targets chromatin access for transcription factor NF-kappaB to alter transcription of host genes involved in immune responses. Nat Immunol *8*, 47-56.

Ashraf, S., Huque, M.H., Kenah, E., Agboatwalla, M., and Luby, S.P. (2013). Effect of recent diarrhoeal episodes on risk of pneumonia in children under the age of 5 years in Karachi, Pakistan. International journal of epidemiology *42*, 194-200.

Bardhan, P., Faruque, A.S., Naheed, A., and Sack, D.A. (2010). Decrease in shigellosis-related deaths without Shigella spp.-specific interventions, Asia. Emerg Infect Dis *16*, 1718-1723.

Barzu, S., Benjelloun-Touimi, Z., Phalipon, A., Sansonetti, P., and Parsot, C. (1997). Functional analysis of the *Shigella flexneri* IpaC invasin by insertional mutagenesis. Infect Immun *65*, 1599-1605.

Bashashati, M., Rezaei, N., Bashashati, H., Shafieyoun, A., Daryani, N.E., Sharkey, K.A., and Storr, M. (2012). Cytokine gene polymorphisms are associated with irritable bowel syndrome: a systematic review and meta-analysis. Neurogastroenterology and motility : the official journal of the European Gastrointestinal Motility Society *24*, 1102-e1566.

Bergan, J., Dyve Lingelem, A.B., Simm, R., Skotland, T., and Sandvig, K. (2012). Shiga toxins. Toxicon : official journal of the International Society on Toxinology *60*, 1085-1107.

Bernardini, M.L., Mounier, J., d'Hauteville, H., Coquis-Rondon, M., and Sansonetti, P.J. (1989). Identification of icsA, a plasmid locus of *Shigella flexneri* that governs bacterial intra- and intercellular spread through interaction with F-actin. Proc Natl Acad Sci U S A *86*, 3867-3871.

Bhattacharya, D., Bhattacharya, H., Thamizhmani, R., Sayi, D.S., Reesu, R., Anwesh, M., Kartick, C., Bharadwaj, A.P., Singhania, M., Sugunan, A.P., *et al.* (2013). Shigellosis in Bay of Bengal Islands, India: clinical and seasonal patterns, surveillance of antibiotic susceptibility patterns, and molecular characterization of multidrug-resistant Shigella strains isolated during a 6-year period from 2006 to 2011. European journal of clinical microbiology and infectious diseases : official publication of the European Society of Clinical Microbiology.

Bin Saeed, A.A., El Bushra, H.E., and Al-Hamdan, N.A. (1995). Does treatment of bloody diarrhea due to Shigella dysenteriae type 1 with ampicillin precipitate hemolytic uremic syndrome? Emerg Infect Dis *1*, 134-137.

Blocker, A., Gounon, P., Larquet, E., Niebuhr, K., Cabiaux, V., Parsot, C., and Sansonetti, P. (1999). The tripartite type III secreton of *Shigella flexneri* inserts IpaB and IpaC into host membranes. J Cell Biol *147*, 683-693.

Borg, M.L., Modi, A., Tostmann, A., Gobin, M., Cartwright, J., Quigley, C., Crook, P., Boxall, N., Paul, J., Cheasty, T., *et al.* (2012). Ongoing outbreak of Shigella flexneri serotype 3a in men who have sex with men in England and Wales, data from 2009-2011. Euro surveillance : bulletin Europeen sur les maladies transmissibles = European communicable disease bulletin *17*.

Borgaonkar, M.R., Ford, D.C., Marshall, J.K., Churchill, E., and Collins, S.M. (2006). The incidence of irritable bowel syndrome among community subjects with previous acute enteric infection. Digestive diseases and sciences *51*, 1026-1032.

Bourdet-Sicard, R., Rudiger, M., Jockusch, B.M., Gounon, P., Sansonetti, P.J., and Nhieu, G.T. (1999). Binding of the Shigella protein IpaA to vinculin induces F-actin depolymerization. Embo J *18*, 5853-5862.

Bovee, L., Whelan, J., Sonder, G.J., van Dam, A.P., and van den Hoek, A. (2012). Risk factors for secondary transmission of Shigella infection within households: implications for current prevention policy. BMC infectious diseases *12*, 347.

Brigotti, M., Tazzari, P.L., Ravanelli, E., Carnicelli, D., Rocchi, L., Arfilli, V., Scavia, G., Minelli, F., Ricci, F., Pagliaro, P., *et al.* (2011). Clinical relevance of shiga toxin concentrations in the blood of patients with hemolytic uremic syndrome. The Pediatric infectious disease journal *30*, 486-490.

Buchrieser, C., Glaser, P., Rusniok, C., Nedjari, H., D'Hauteville, H., Kunst, F., Sansonetti, P., and Parsot, C. (2000). The virulence plasmid pWR100 and the repertoire of proteins secreted by the type III secretion apparatus of Shigella flexneri. Mol Microbiol *38*, 760-771.

Butler, T. (2012). Haemolytic uraemic syndrome during shigellosis. Transactions of the Royal Society of Tropical Medicine and Hygiene *106*, 395-399.

Calderwood, S.B., Auclair, F., Donohue-Rolfe, A., Keusch, G.T., and Mekalanos, J.J. (1987). Nucleotide sequence of the Shiga-like toxin genes of Escherichia coli. Proc Natl Acad Sci U S A *84*, 4364-4368.

CDC., C.f.D.C.a.P. (2011). National Shigella Surveillance Overview, C. US Department of Health and Human Services, ed.

Chang, Z., Lu, S., Chen, L., Jin, Q., and Yang, J. (2012). Causative species and serotypes of shigellosis in mainland China: systematic review and meta-analysis. PLoS One *7*, e52515.

Checkley, W., Buckley, G., Gilman, R.H., Assis, A.M., Guerrant, R.L., Morris, S.S., Molbak, K., Valentiner-Branth, P., Lanata, C.F., and Black, R.E. (2008). Multi-country analysis of the effects of diarrhoea on childhood stunting. International journal of epidemiology *37*, 816-830.

Chen, M., Delpech, V., O'Sullivan, B., and Donovan, B. (2002). Shigella sonnei: another cause of sexually acquired reactive arthritis. International journal of STD and AIDS *13*, 135-136.

Chen, Y., Smith, M.R., Thirumalai, K., and Zychlinsky, A. (1996). A bacterial invasin induces macrophage apoptosis by binding directly to ICE. EMBO J *15*, 3853-3860.

Clark, C.S., and Maurelli, A.T. (2007). Shigella flexneri inhibits staurosporine-induced apoptosis in epithelial cells. Infect Immun *75*, 2531-2539.

Clerc, P.L., Ryter, A., Mounier, J., and Sansonetti, P.J. (1987). Plasmid-mediated early killing of eucaryotic cells by Shigella flexneri as studied by infection of J774 macrophages. Infect Immun *55*, 521-527.

Cohen, D., Green, M., Block, C., Slepon, R., Ambar, R., Wasserman, S.S., and Levine, M.M. (1991). Reduction of transmission of shigellosis by control of houseflies (Musca domestica). Lancet *337*, 993-997.

Connor, B.A. (2005). Sequelae of traveler's diarrhea: focus on postinfectious irritable bowel syndrome. Clin Infect Dis *41 Suppl 8*, S577-586.

Connor, B.A., and Riddle, M.S. (2013). Post-infectious sequelae of travelers' diarrhea. J Travel Med *20*, 303-312.
DeLappe, N., O'Halloran, F., Fanning, S., Corbett-Feeney, G., Cheasty, T., and Cormican, M. (2003). Antimicrobial resistance and genetic diversity of Shigella sonnei isolates from western Ireland, an area of low incidence of infection. J Clin Microbiol *41*, 1919-1924.
DuPont, H.L., Levine, M.M., Hornick, R.B., and Formal, S.B. (1989). Inoculum size in shigellosis and implications for expected mode of transmission. J Infect Dis *159*, 1126-1128.
Faherty, C.S., Harper, J.M., Shea-Donohue, T., Barry, E.M., Kaper, J.B., Fasano, A., and Nataro, J.P. (2012). Chromosomal and plasmid-encoded factors of Shigella flexneri induce secretogenic activity ex vivo. PLoS One *7*, e49980.
Farfan, M.J., Toro, C.S., Barry, E.M., and Nataro, J.P. (2011). Shigella enterotoxin-2 is a type III effector that participates in Shigella-induced interleukin 8 secretion by epithelial cells. FEMS Immunol Med Microbiol *61*, 332-339.
Fasano, A., Noriega, F.R., Maneval, D.R., Jr., Chanasongcram, S., Russell, R., Guandalini, S., and Levine, M.M. (1995). Shigella enterotoxin 1: an enterotoxin of Shigella flexneri 2a active in rabbit small intestine in vivo and in vitro. J Clin Invest *95*, 2853-2861.
Fernandez-Prada, C.M., Hoover, D.L., Tall, B.D., Hartman, A.B., Kopelowitz, J., and Venkatesan, M.M. (2000). Shigella flexneri IpaH(7.8) facilitates escape of virulent bacteria from the endocytic vacuoles of mouse and human macrophages. Infect Immun *68*, 3608-3619.
Fischer Walker, C.L., Perin, J., Aryee, M.J., Boschi-Pinto, C., and Black, R.E. (2012). Diarrhea incidence in low- and middle-income countries in 1990 and 2010: a systematic review. BMC public health *12*, 220.
Fraser, M.E., Chernaia, M.M., Kozlov, Y.V., and James, M.N. (1994). Crystal structure of the holotoxin from Shigella dysenteriae at 2.5 A resolution. Nat Struct Biol *1*, 59-64.
Garred, O., van Deurs, B., and Sandvig, K. (1995). Furin-induced cleavage and activation of Shiga toxin. J Biol Chem *270*, 10817-10821.
Goldberg, M.B., Barzu, O., Parsot, C., and Sansonetti, P.J. (1993). Unipolar localization and ATPase activity of IcsA, a *Shigella flexneri* protein involved in intracellular movement. Infect Agents Dis *2*, 210-211.
Goldberg, M.B., and Theriot, J.A. (1995). Shigella flexneri surface protein IcsA is sufficient to direct actin-based motility. Proc Natl Acad Sci U S A *92*, 6572-6576.
Gould, L.H., Demma, L., Jones, T.F., Hurd, S., Vugia, D.J., Smith, K., Shiferaw, B., Segler, S., Palmer, A., Zansky, S., *et al.* (2009). Hemolytic uremic syndrome and death in persons with Escherichia coli O157:H7 infection, foodborne diseases active surveillance network sites, 2000-2006. Clin Infect Dis *49*, 1480-1485.
Gu, B., Cao, Y., Pan, S., Zhuang, L., Yu, R., Peng, Z., Qian, H., Wei, Y., Zhao, L., Liu, G., *et al.* (2012). Comparison of the prevalence and changing resistance to nalidixic acid and ciprofloxacin of Shigella between Europe-America and Asia-Africa from 1998 to 2009. Int J Antimicrob Agents *40*, 9-17.
Gwee, K.A., Graham, J.C., McKendrick, M.W., Collins, S.M., Marshall, J.S., Walters, S.J., and Read, N.W. (1996). Psychometric scores and persistence of irritable bowel after infectious diarrhoea. Lancet *347*, 150-153.
Hale, T.L. (1991). Genetic basis of virulence in Shigella species. Microbiol Rev *55*, 206-224.
Hale TL, K.G. (1996). Shigella. In Medical Microbiology, B. S, ed. (Galveston, TX: University of Texas Medical Branch at Galveston).

Handa, Y., Suzuki, M., Ohya, K., Iwai, H., Ishijima, N., Koleske, A.J., Fukui, Y., and Sasakawa, C. (2007). *Shigella* IpgB1 promotes bacterial entry through the ELMO-Dock180 machinery. Nat Cell Biol *9*, 121-128.
Hannu, T., Mattila, L., Siitonen, A., and Leirisalo-Repo, M. (2005). Reactive arthritis attributable to Shigella infection: a clinical and epidemiological nationwide study. Annals of the rheumatic diseases *64*, 594-598.
Harrington, S.M., Sheikh, J., Henderson, I.R., Ruiz-Perez, F., Cohen, P.S., and Nataro, J.P. (2009). The Pic protease of enteroaggregative Escherichia coli promotes intestinal colonization and growth in the presence of mucin. Infect Immun *77*, 2465-2473.
High, N., Mounier, J., Prevost, M.C., and Sansonetti, P.J. (1992). IpaB of *Shigella flexneri* causes entry into epithelial cells and escape from the phagocytic vacuole. Embo J *11*, 1991-1999.
Hilbi, H., Moss, J.E., Hersh, D., Chen, Y., Arondel, J., Banerjee, S., Flavell, R.A., Yuan, J., Sansonetti, P.J., and Zychlinsky, A. (1998). Shigella-induced apoptosis is dependent on caspase-1 which binds to IpaB. J Biol Chem *273*, 32895-32900.
Hossain, M.A., Hasan, K.Z., and Albert, M.J. (1994). Shigella carriers among non-diarrhoeal children in an endemic area of shigellosis in Bangladesh. Tropical and geographical medicine *46*, 40-42.
Hromockyj, A.E., and Maurelli, A.T. (1989). Identification of Shigella invasion genes by isolation of temperature-regulated inv::lacZ operon fusions. Infect Immun *57*, 2963-2970.
Kane, C.D., Schuch, R., Day, W.A., Jr., and Maurelli, A.T. (2002). MxiE regulates intracellular expression of factors secreted by the Shigella flexneri 2a type III secretion system. J Bacteriol *184*, 4409-4419.
Ke, X., Gu, B., Pan, S., and Tong, M. (2011). Epidemiology and molecular mechanism of integron-mediated antibiotic resistance in Shigella. Archives of microbiology *193*, 767-774.
Khan, W.A., Griffiths, J.K., and Bennish, M.L. (2013). Gastrointestinal and extra-intestinal manifestations of childhood shigellosis in a region where all four species of Shigella are endemic. PLoS One *8*, e64097.
Kim, D.W., Lenzen, G., Page, A.L., Legrain, P., Sansonetti, P.J., and Parsot, C. (2005). The Shigella flexneri effector OspG interferes with innate immune responses by targeting ubiquitin-conjugating enzymes. Proc Natl Acad Sci U S A *102*, 14046-14051.
Knight, C.L., and Surawicz, C.M. (2013). Clostridium difficile Infection. The Medical clinics of North America *97*, 523-536, ix.
Koneman, E. (1997). Color Atlas and Textbook of Diagnostic Microbiology (New York: Lippincott-Raven Publishers).
Koster, F.T., Curlin, G.C., Aziz, K.M., and Haque, A. (1981). Synergistic impact of measles and diarrhoea on nutrition and mortality in Bangladesh. Bull World Health Organ *59*, 901-908.
Kotloff, K.L., Nataro, J.P., Blackwelder, W.C., Nasrin, D., Farag, T.H., Panchalingam, S., Wu, Y., Sow, S.O., Sur, D., Breiman, R.F., *et al.* (2013). Burden and aetiology of diarrhoeal disease in infants and young children in developing countries (the Global Enteric Multicenter Study, GEMS): a prospective, case-control study. Lancet *382*, 209-222.
Kotloff, K.L., Pasetti, M.F., Barry, E.M., Nataro, J.P., Wasserman, S.S., Sztein, M.B., Picking, W.D., and Levine, M.M. (2004). Deletion in the Shigella enterotoxin genes further attenuates Shigella flexneri 2a bearing guanine auxotrophy in a phase 1 trial of CVD 1204 and CVD 1208. J Infect Dis *190*, 1745-1754.
Kotloff, K.L., Winickoff, J.P., Ivanoff, B., Clemens, J.D., Swerdlow, D.L., Sansonetti, P.J., Adak, G.K., and Levine, M.M. (1999). Global burden of Shigella infections: implications

for vaccine development and implementation of control strategies. Bull World Health Organ *77*, 651-666.

Kramer, R.W., Slagowski, N.L., Eze, N.A., Giddings, K.S., Morrison, M.F., Siggers, K.A., Starnbach, M.N., and Lesser, C.F. (2007). Yeast functional genomic screens lead to identification of a role for a bacterial effector in innate immunity regulation. PLoS pathogens *3*, e21.

Kweon, M.N. (2008). Shigellosis: the current status of vaccine development. Current opinion in infectious diseases *21*, 313-318.

Lafont, F., Tran Van Nhieu, G., Hanada, K., Sansonetti, P., and van der Goot, F.G. (2002). Initial steps of *Shigella* infection depend on the cholesterol/sphingolipid raft-mediated CD44-IpaB interaction. Embo J *21*, 4449-4457.

Lauhio, A., Lahdevirta, J., Janes, R., Kontiainen, S., and Repo, H. (1988). Reactive arthritis associated with Shigella sonnei infection. Arthritis and rheumatism *31*, 1190-1193.

Li, H., Xu, H., Zhou, Y., Zhang, J., Long, C., Li, S., Chen, S., Zhou, J.M., and Shao, F. (2007). The phosphothreonine lyase activity of a bacterial type III effector family. Science *315*, 1000-1003.

Lindberg, A.A., Brown, J.E., Stromberg, N., Westling-Ryd, M., Schultz, J.E., and Karlsson, K.A. (1987). Identification of the carbohydrate receptor for Shiga toxin produced by Shigella dysenteriae type 1. J Biol Chem *262*, 1779-1785.

Lindsay, S.W., Lindsay, T.C., Duprez, J., Hall, M.J., Kwambana, B.A., Jawara, M., Nurudeen, I.U., Sallah, N., Wyatt, N., D'Alessandro, U., *et al.* (2012). Chrysomya putoria, a putative vector of diarrheal diseases. PLoS neglected tropical diseases *6*, e1895.

Luis M de la Maza, M.T.P., Janet T Shigei, Ellena M Peterson (2004). Color Atlas of Medical Bacteriology (ASM Press).

Mahoney, F.J., Farley, T.A., Burbank, D.F., Leslie, N.H., and McFarland, L.M. (1993). Evaluation of an intervention program for the control of an outbreak of shigellosis among institutionalized persons. J Infect Dis *168*, 1177-1180.

Mammina, C., Pontello, M., Dal Vecchio, A., and Nastasi, A. (2005). Identification of Shigella sonnei biotype g isolates carrying class 2 integrons in Italy (2001 to 2003). J Clin Microbiol *43*, 2467-2470.

Mathur, R., Mathur, Y.N., and Verma, S.D. (1989). An outbreak of shigellosis in central India: higher death rate in post-measles shigellosis. Journal of diarrhoeal diseases research *7*, 28-29.

Maurelli, A.T., Baudry, B., d'Hauteville, H., Hale, T.L., and Sansonetti, P.J. (1985). Cloning of plasmid DNA sequences involved in invasion of HeLa cells by *Shigella flexneri*. Infect Immun *49*, 164-171.

Mavris, M., Page, A.L., Tournebize, R., Demers, B., Sansonetti, P., and Parsot, C. (2002). Regulation of transcription by the activity of the *Shigella flexneri* type III secretion apparatus. Mol Microbiol *43*, 1543-1553.

McDonnell, J., Dallman, T., Atkin, S., Turbitt, D.A., Connor, T.R., Grant, K.A., Thomson, N.R., and Jenkins, C. (2013). Retrospective analysis of whole genome sequencing compared to prospective typing data in further informing the epidemiological investigation of an outbreak of Shigella sonnei in the UK. Epidemiol Infect *141*, 2568-2575.

McKendrick, M.W., and Read, N.W. (1994). Irritable bowel syndrome--post salmonella infection. J Infect *29*, 1-3.

Menard, R., Sansonetti, P., and Parsot, C. (1994). The secretion of the *Shigella flexneri* Ipa invasins is activated by epithelial cells and controlled by IpaB and IpaD. Embo J *13*, 5293-5302.

Menard, R., Sansonetti, P.J., and Parsot, C. (1993). Nonpolar mutagenesis of the ipa genes defines IpaB, IpaC, and IpaD as effectors of *Shigella flexneri* entry into epithelial cells. J Bacteriol *175*, 5899-5906.
Modell, B., Berry, R.J., Boyle, C.A., Christianson, A., Darlison, M., Dolk, H., Howson, C.P., Mastroiacovo, P., Mossey, P., and Rankin, J. (2012). Global regional and national causes of child mortality. Lancet *380*, 1556; author reply 1556-1557.
Morse, H.G., Rate, R.G., Bonnell, M.D., and Kuberski, T.T. (1980). Reiter's syndrome in a five-year-old girl. Arthritis and rheumatism *23*, 960-961.
Mounier, J., Vasselon, T., Hellio, R., Lesourd, M., and Sansonetti, P.J. (1992). Shigella flexneri enters human colonic Caco-2 epithelial cells through the basolateral pole. Infect Immun *60*, 237-248.
Nataro, J.P., Seriwatana, J., Fasano, A., Maneval, D.R., Guers, L.D., Noriega, F., Dubovsky, F., Levine, M.M., and Morris, J.G., Jr. (1995). Identification and cloning of a novel plasmid-encoded enterotoxin of enteroinvasive Escherichia coli and Shigella strains. Infect Immun *63*, 4721-4728.
Niebuhr, K., Giuriato, S., Pedron, T., Philpott, D.J., Gaits, F., Sable, J., Sheetz, M.P., Parsot, C., Sansonetti, P.J., and Payrastre, B. (2002). Conversion of PtdIns(4,5)P(2) into PtdIns(5)P by the *S. flexneri* effector IpgD reorganizes host cell morphology. Embo J *21*, 5069-5078.
Niyogi, S.K., Vargas, M., and Vila, J. (2004). Prevalence of the sat, set and sen genes among diverse serotypes of Shigella flexneri strains isolated from patients with acute diarrhoea. Clin Microbiol Infect *10*, 574-576.
Nizeyi, J.B., Innocent, R.B., Erume, J., Kalema, G.R., Cranfield, M.R., and Graczyk, T.K. (2001). Campylobacteriosis, salmonellosis, and shigellosis in free-ranging human-habituated mountain gorillas of Uganda. Journal of wildlife diseases *37*, 239-244.
Okame, M., Adachi, E., Sato, H., Shimizu, S., Kikuchi, T., Miyazaki, N., Koga, M., Nakamura, H., Suzuki, M., Oyaizu, N., *et al.* (2012). Shigella sonnei outbreak among men who have sex with men in Tokyo. Jpn J Infect Dis *65*, 277-278.
Olive, A.J., Kenjale, R., Espina, M., Moore, D.S., Picking, W.L., and Picking, W.D. (2007). Bile salts stimulate recruitment of IpaB to the *Shigella flexneri* surface, where it colocalizes with IpaD at the tip of the type III secretion needle. Infect Immun *75*, 2626-2629.
Paciello, I., Silipo, A., Lembo-Fazio, L., Curcuru, L., Zumsteg, A., Noel, G., Ciancarella, V., Sturiale, L., Molinaro, A., and Bernardini, M.L. (2013). Intracellular Shigella remodels its LPS to dampen the innate immune recognition and evade inflammasome activation. Proc Natl Acad Sci U S A.
Page, A.L., Ohayon, H., Sansonetti, P.J., and Parsot, C. (1999). The secreted IpaB and IpaC invasins and their cytoplasmic chaperone IpgC are required for intercellular dissemination of *Shigella flexneri.* Cell Microbiol *1*, 183-193.
Page, A.L., Sansonetti, P., and Parsot, C. (2002). Spa15 of *Shigella flexneri*, a third type of chaperone in the type III secretion pathway. Mol Microbiol *43*, 1533-1542.
Parsot, C., Ageron, E., Penno, C., Mavris, M., Jamoussi, K., d'Hauteville, H., Sansonetti, P., and Demers, B. (2005). A secreted anti-activator, OspD1, and its chaperone, Spa15, are involved in the control of transcription by the type III secretion apparatus activity in Shigella flexneri. Mol Microbiol *56*, 1627-1635.
Parsot, C., Hamiaux, C., and Page, A.L. (2003). The various and varying roles of specific chaperones in type III secretion systems. Curr Opin Microbiol *6*, 7-14.
Pendaries, C., Tronchere, H., Arbibe, L., Mounier, J., Gozani, O., Cantley, L., Fry, M.J., Gaits-Iacovoni, F., Sansonetti, P.J., and Payrastre, B. (2006). PtdIns5P activates the host cell PI3-kinase/Akt pathway during Shigella flexneri infection. EMBO J *25*, 1024-1034.

Pickering, L.K., Evans, D.G., DuPont, H.L., Vollet, J.J., 3rd, and Evans, D.J., Jr. (1981). Diarrhea caused by Shigella, rotavirus, and Giardia in day-care centers: prospective study. The Journal of pediatrics *99*, 51-56.

Picking, W.L., Nishioka, H., Hearn, P.D., Baxter, M.A., Harrington, A.T., Blocker, A., and Picking, W.D. (2005). IpaD of *Shigella flexneri* is independently required for regulation of Ipa protein secretion and efficient insertion of IpaB and IpaC into host membranes. Infect Immun *73*, 1432-1440.

Pilonieta, M.C., and Munson, G.P. (2008). The chaperone IpgC copurifies with the virulence regulator MxiE. J Bacteriol *190*, 2249-2251.

Ramarao, N., Le Clainche, C., Izard, T., Bourdet-Sicard, R., Ageron, E., Sansonetti, P.J., Carlier, M.F., and Tran Van Nhieu, G. (2007). Capping of actin filaments by vinculin activated by the Shigella IpaA carboxyl-terminal domain. FEBS Lett *581*, 853-857.

Rathman, M., de Lanerolle, P., Ohayon, H., Gounon, P., and Sansonetti, P. (2000). Myosin light chain kinase plays an essential role in S. flexneri dissemination. J Cell Sci *113 Pt 19*, 3375-3386.

Rohde, J.R., Breitkreutz, A., Chenal, A., Sansonetti, P.J., and Parsot, C. (2007). Type III secretion effectors of the IpaH family are E3 ubiquitin ligases. Cell host and microbe *1*, 77-83.

Ross, A.G.P., Olds, G.R., Cripps, A.W., Farrar, J.J., and McManus, D.P. (2013). Enteropathogens and Chronic Illness in Returning Travelers. New England Journal of Medicine *368*, 1817-1825.

Roy, S., Thanasekaran, K., Dutta Roy, A.R., and Sehgal, S.C. (2006). Distribution of Shigella enterotoxin genes and secreted autotransporter toxin gene among diverse species and serotypes of shigella isolated from Andaman Islands, India. Tropical medicine and international health : TM and IH *11*, 1694-1698.

Ruiz-Perez, F., Wahid, R., Faherty, C.S., Kolappaswamy, K., Rodriguez, L., Santiago, A., Murphy, E., Cross, A., Sztein, M.B., and Nataro, J.P. (2011). Serine protease autotransporters from Shigella flexneri and pathogenic Escherichia coli target a broad range of leukocyte glycoproteins. Proc Natl Acad Sci U S A *108*, 12881-12886.

Saeed, A., Abd, H., Edvinsson, B., and Sandstrom, G. (2009). Acanthamoeba castellanii an environmental host for Shigella dysenteriae and Shigella sonnei. Archives of microbiology *191*, 83-88.

Sansonetti, P.J. (1998). Pathogenesis of shigellosis: from molecular and cellular biology of epithelial cell invasion to tissue inflammation and vaccine development. Jpn J Med Sci Biol *51 Suppl*, S69-80.

Sansonetti, P.J., Arondel, J., Cantey, J.R., Prevost, M.C., and Huerre, M. (1996). Infection of rabbit Peyer's patches by Shigella flexneri: effect of adhesive or invasive bacterial phenotypes on follicle-associated epithelium. Infect Immun *64*, 2752-2764.

Sansonetti, P.J., Kopecko, D.J., and Formal, S.B. (1982). Involvement of a plasmid in the invasive ability of *Shigella flexneri*. Infect Immun *35*, 852-860.

Sansonetti, P.J., Phalipon, A., Arondel, J., Thirumalai, K., Banerjee, S., Akira, S., Takeda, K., and Zychlinsky, A. (2000). Caspase-1 activation of IL-1beta and IL-18 are essential for *Shigella flexneri*-induced inflammation. Immunity *12*, 581-590.

Sasakawa, C., Kamata, K., Sakai, T., Makino, S., Yamada, M., Okada, N., and Yoshikawa, M. (1988). Virulence-associated genetic regions comprising 31 kilobases of the 230-kilobase plasmid in *Shigella flexneri 2a*. J Bacteriol *170*, 2480-2484.

Sasakawa, C., Komatsu, K., Tobe, T., Suzuki, T., and Yoshikawa, M. (1993). Eight genes in region 5 that form an operon are essential for invasion of epithelial cells by *Shigella flexneri 2a*. J Bacteriol *175*, 2334-2346.

Saxena, S.K., O'Brien, A.D., and Ackerman, E.J. (1989). Shiga toxin, Shiga-like toxin II variant, and ricin are all single-site RNA N-glycosidases of 28 S RNA when microinjected into Xenopus oocytes. J Biol Chem *264*, 596-601.

Schiellerup, P., Krogfelt, K.A., and Locht, H. (2008). A comparison of self-reported joint symptoms following infection with different enteric pathogens: effect of HLA-B27. J Rheumatol *35*, 480-487.

Schmidt, W.P., Cairncross, S., Barreto, M.L., Clasen, T., and Genser, B. (2009). Recent diarrhoeal illness and risk of lower respiratory infections in children under the age of 5 years. International journal of epidemiology *38*, 766-772.

Schroeder, G.N., and Hilbi, H. (2008). Molecular pathogenesis of *Shigella* spp.: controlling host cell signaling, invasion, and death by type III secretion. Clin Microbiol Rev *21*, 134-156.

Schuch, R., Sandlin, R.C., and Maurelli, A.T. (1999). A system for identifying post-invasion functions of invasion genes: requirements for the Mxi-Spa type III secretion pathway of Shigella flexneri in intercellular dissemination. Mol Microbiol *34*, 675-689.

Seol, S.Y., Kim, Y.T., Jeong, Y.S., Oh, J.Y., Kang, H.Y., Moon, D.C., Kim, J., Lee, Y.C., Cho, D.T., and Lee, J.C. (2006). Molecular characterization of antimicrobial resistance in Shigella sonnei isolates in Korea. J Med Microbiol *55*, 871-877.

Sharp, T.W., Thornton, S.A., Wallace, M.R., Defraites, R.F., Sanchez, J.L., Batchelor, R.A., Rozmajzl, P.J., Hanson, R.K., Echeverria, P., Kapikian, A.Z., *et al.* (1995). Diarrheal disease among military personnel during Operation Restore Hope, Somalia, 1992-1993. The American journal of tropical medicine and hygiene *52*, 188-193.

Simon, D.G., Kaslow, R.A., Rosenbaum, J., Kaye, R.L., and Calin, A. (1981). Reiter's syndrome following epidemic shigellosis. J Rheumatol *8*, 969-973.

Skoudy, A., Mounier, J., Aruffo, A., Ohayon, H., Gounon, P., Sansonetti, P., and Tran Van Nhieu, G. (2000). CD44 binds to the *Shigella* IpaB protein and participates in bacterial invasion of epithelial cells. Cell Microbiol 2, 19-33.

Spiller, R., and Garsed, K. (2009). Postinfectious irritable bowel syndrome. Gastroenterology *136*, 1979-1988.

Stein, P.E., Boodhoo, A., Tyrrell, G.J., Brunton, J.L., and Read, R.J. (1992). Crystal structure of the cell-binding B oligomer of verotoxin-1 from E. coli. Nature *355*, 748-750.

Stensrud, K.F., Adam, P.R., La Mar, C.D., Olive, A.J., Lushington, G.H., Sudharsan, R., Shelton, N.L., Givens, R.S., Picking, W.L., and Picking, W.D. (2008). Deoxycholate interacts with IpaD of Shigella flexneri in inducing the recruitment of IpaB to the type III secretion apparatus needle tip. J Biol Chem *283*, 18646-18654.

Takao, T., Tanabe, T., Hong, Y.M., Shimonishi, Y., Kurazono, H., Yutsudo, T., Sasakawa, C., Yoshikawa, M., and Takeda, Y. (1988). Identity of molecular structure of Shiga-like toxin I (VT1) from *Escherichia coli* O157:H7 with that of Shiga toxin. Microb Pathog *5*, 57-69.

Tesh, V.L. (2010). Induction of apoptosis by Shiga toxins. Future microbiology *5*, 431-453.

Tobe, T., Nagai, S., Okada, N., Adler, B., Yoshikawa, M., and Sasakawa, C. (1991). Temperature-regulated expression of invasion genes in Shigella flexneri is controlled through the transcriptional activation of the virB gene on the large plasmid. Mol Microbiol *5*, 887-893.

Tran Van Nhieu, G., Caron, E., Hall, A., and Sansonetti, P.J. (1999). IpaC induces actin polymerization and filopodia formation during *Shigella* entry into epithelial cells. Embo J *18*, 3249-3262.

Tzipori, S., Sheoran, A., Akiyoshi, D., Donohue-Rolfe, A., and Trachtman, H. (2004). Antibody therapy in the management of shiga toxin-induced hemolytic uremic syndrome. Clin Microbiol Rev *17*, 926-941, table of contents.

VanDerslice, J., and Briscoe, J. (1995). Environmental interventions in developing countries: interactions and their implications. American journal of epidemiology *141*, 135-144.

Venkatesan, M.M., Buysse, J.M., and Oaks, E.V. (1992). Surface presentation of Shigella flexneri invasion plasmid antigens requires the products of the spa locus. J Bacteriol *174*, 1990-2001.

Venkatesan, M.M., Goldberg, M.B., Rose, D.J., Grotbeck, E.J., Burland, V., and Blattner, F.R. (2001). Complete DNA sequence and analysis of the large virulence plasmid of Shigella flexneri. Infect Immun *69*, 3271-3285.

Vinh, H., Anh, V.T., Anh, N.D., Campbell, J.I., Hoang, N.V., Nga, T.V., Nhu, N.T., Minh, P.V., Thuy, C.T., Duy, P.T., *et al.* (2011). A multi-center randomized trial to assess the efficacy of gatifloxacin versus ciprofloxacin for the treatment of shigellosis in Vietnamese children. PLoS neglected tropical diseases *5*, e1264.

von Seidlein, L., Kim, D.R., Ali, M., Lee, H., Wang, X., Thiem, V.D., Canh do, G., Chaicumpa, W., Agtini, M.D., Hossain, A., *et al.* (2006). A multicentre study of Shigella diarrhoea in six Asian countries: disease burden, clinical manifestations, and microbiology. PLoS medicine *3*, e353.

Wassef, J.S., Keren, D.F., and Mailloux, J.L. (1989). Role of M cells in initial antigen uptake and in ulcer formation in the rabbit intestinal loop model of shigellosis. Infect Immun *57*, 858-863.

Watarai, M., Funato, S., and Sasakawa, C. (1996). Interaction of Ipa proteins of Shigella flexneri with alpha5beta1 integrin promotes entry of the bacteria into mammalian cells. J Exp Med *183*, 991-999.

Watkins, H.M. (1960). Some attributes of virulence in Shigella. Ann N Y Acad Sci *88*, 1167-1186.

Wei, H.L., Wang, Y.W., Li, C.C., Tung, S.K., and Chiou, C.S. (2007). Epidemiology and evolution of genotype and antimicrobial resistance of an imported Shigella sonnei clone circulating in central Taiwan. Diagnostic microbiology and infectious disease *58*, 469-475.

Yavzori, M., Cohen, D., and Orr, N. (2002). Prevalence of the genes for shigella enterotoxins 1 and 2 among clinical isolates of shigella in Israel. Epidemiol Infect *128*, 533-535.

Yoshida, S., Handa, Y., Suzuki, T., Ogawa, M., Suzuki, M., Tamai, A., Abe, A., Katayama, E., and Sasakawa, C. (2006). Microtubule-severing activity of Shigella is pivotal for intercellular spreading. Science *314*, 985-989.

Yoshida, S., Katayama, E., Kuwae, A., Mimuro, H., Suzuki, T., and Sasakawa, C. (2002). Shigella deliver an effector protein to trigger host microtubule destabilization, which promotes Rac1 activity and efficient bacterial internalization. Embo J *21*, 2923-2935.

Zychlinsky, A., Kenny, B., Menard, R., Prevost, M.C., Holland, I.B., and Sansonetti, P.J. (1994). IpaB mediates macrophage apoptosis induced by *Shigella flexneri*. Mol Microbiol *11*, 619-627.

2

Shigella Iron Transport Systems

A. R. Mey, E. E. Wyckoff and S. M. Payne
Department of Molecular Biosciences, University of Texas at Austin, USA
smpayne@austin.utexas.edu

Abstract
Shigella species have several different transport systems for acquisition of the essential element iron. These include both ferric and ferrous iron transporters. These pathogens synthesize and secrete siderophores, high-affinity ferric iron chelators, and both catechol and hydroxamate siderophores have been identified in *Shigella*. Additionally, the *Shigella* species have transporters for free ferric (Sit) and ferrous (Feo) iron, and *Shigella dysenteriae* has an outer membrane receptor and transport system for heme. The expression of genes encoding these transporters is regulated by environmental cues including iron, oxygen and temperature. Iron regulation is accomplished by Fur, a transcriptional regulator that represses these genes in the presence of iron. FNR and ArcAB regulate expression of the genes in response to oxygen levels, while temperature controls expression of the heme transporter ShuA post-transcriptionally through formation of a FourU thermometer structure in the mRNA. Thus, *Shigella* are able to adapt to a variety of different environments and different iron sources to obtain sufficient iron for growth. In aerobic, iron-limiting conditions in vitro, expression of siderophores allows maximal growth, while the Sit system is crucial to replication in the intracellular environment of the host epithelium.

Introduction
Shigella species have an absolute requirement for iron. Iron is a component of electron transport proteins, central metabolism enzymes such as succinate dehydrogenase, and essential enzymes such as ribonucleotide reductase. Like most facultative anaerobes, which live in a variety of environments, iron acquisition is problematic for *Shigella*. Although iron is one of the most abundant elements in the Earth's crust, the bioavailability of this essential element in the environment is extremely limited due to its insolubility at neutral or alkaline pH in the presence of oxygen. In addition, iron-limitation presents a particular problem for pathogenic bacteria such as *Shigella* that occupy the mammalian host. Most of the iron present in mammals is tightly sequestered in compounds such as heme, transferrin and other iron complexes that are not readily available to bacteria. Thus, competition with the host for iron is one of the limiting factors in establishing a successful bacterial infection.

Shigella spp. have evolved several strategies to compete for sufficient iron, and there is considerable diversity among the *Shigellae*. The genes encoding iron acquisition systems and selected iron-responsive regulators in the reference strains for each of the four

Shigella species are shown together with their approximate genomic map locations in Figure 2.1, and the details of each system are discussed below. The iron transport systems include the synthesis and secretion of low molecular weight iron chelators, termed siderophores, transport systems for iron complexes, and transport systems for

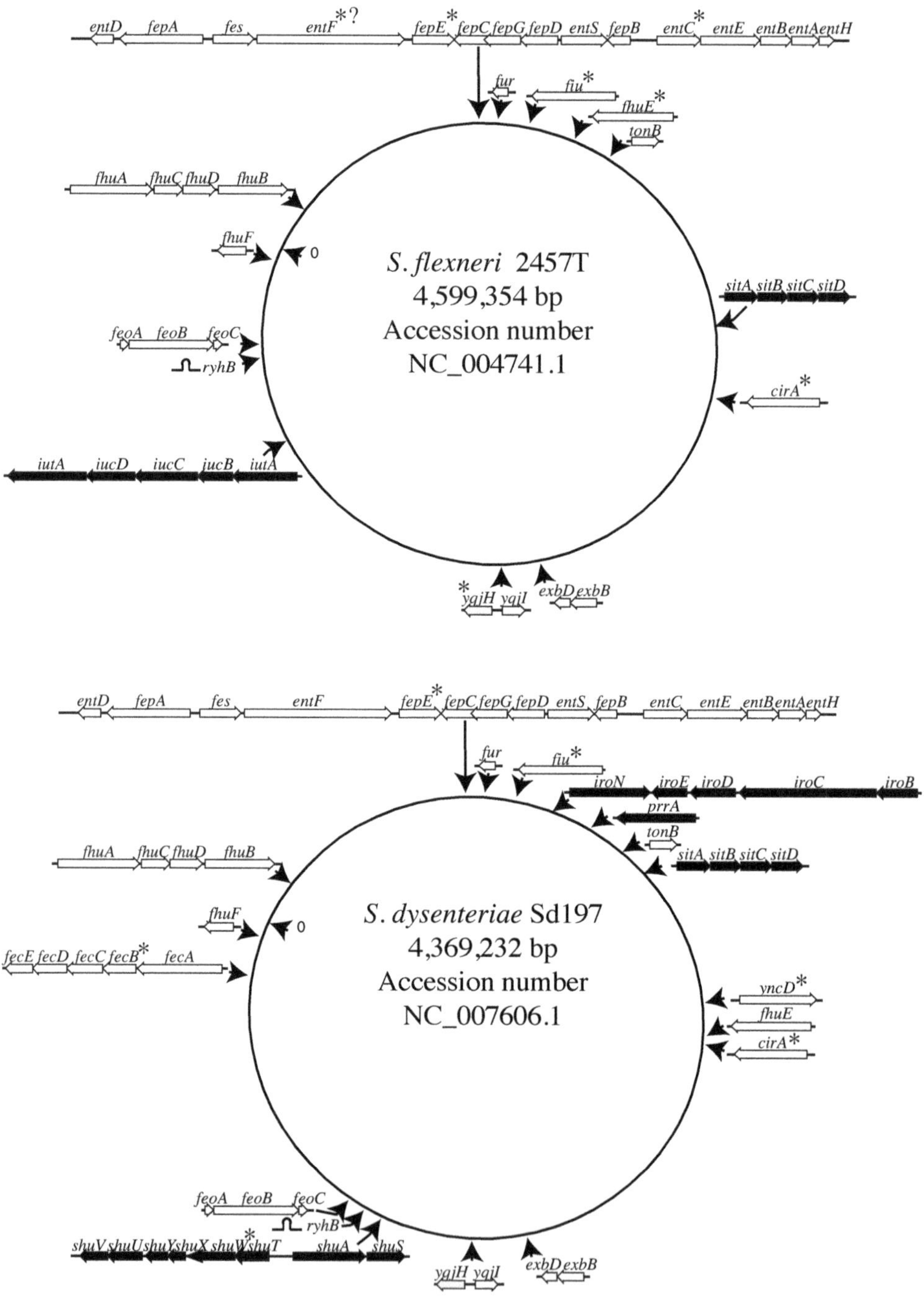

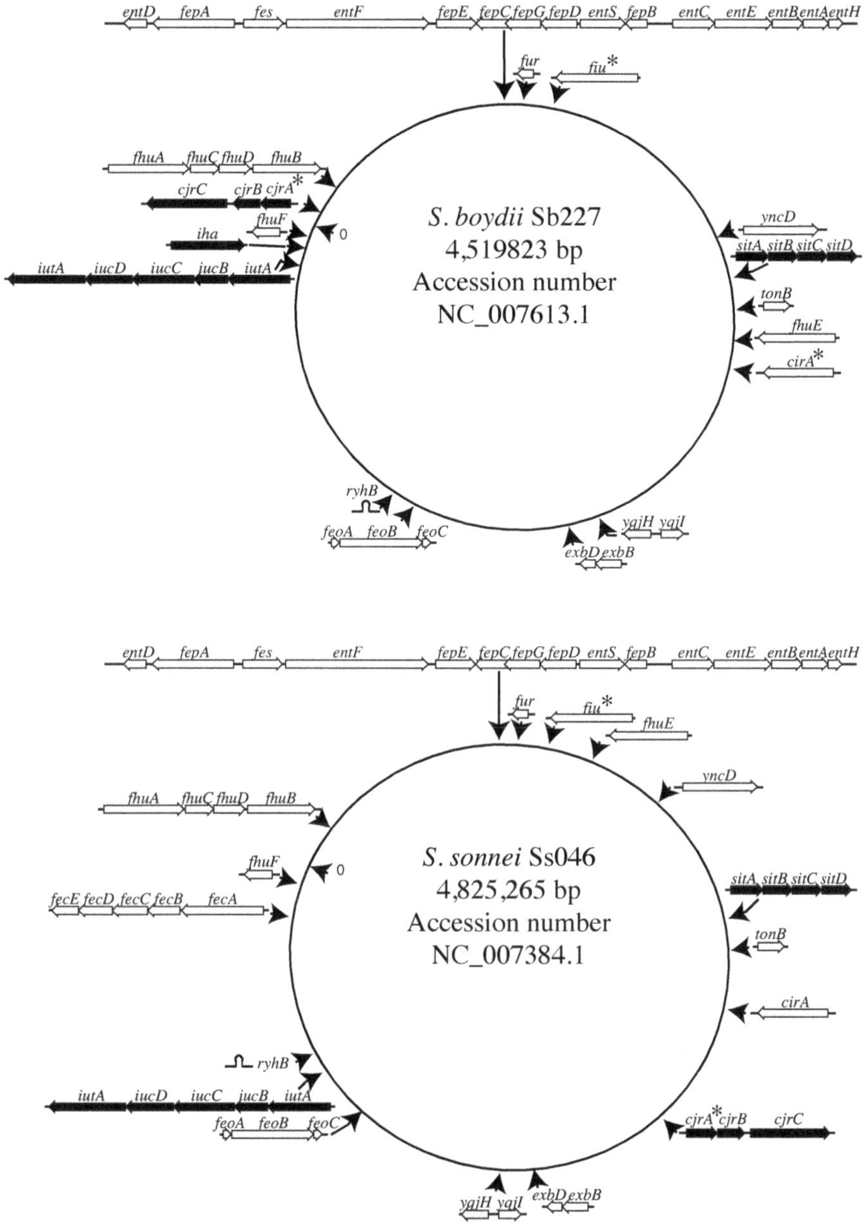

Figure 2.1. Map of iron acquisition genes for the *Shigella* species. The approximate map locations of known and putative iron acquisition genes are shown for reference genomes for each of the four *Shigella* species: *S. flexneri* 2457T (Wei et al., 2003), *S. dysenteriae* Sd197 (Yang et al., 2005), *S. boydii* Sb227 (Yang et al., 2005) and *S. sonnei* Ss046 (Yang et al., 2005). There is some variation within each of the species, particularly in genes for the synthesis and transport of enterobactin and its linear derivatives and for transport of citrate (not shown). The direction of transcription is indicated by the arrows, and open arrows represent genes that are also present in *E. coli* K12, while filled arrows indicate genes that are not shared with *E. coli* K12. The exception is *ryhB*, which encodes a small RNA. An asterisk * indicates a defective gene, including frameshifts, premature stop codons, deletions or IS element insertions. The maps are drawn using data from the PATRIC database website (Gillespie et al., 2011). None of the known iron transporters are encoded on plasmids in the strains shown.

free ferrous iron. *Shigella* isolates typically have three or more of these systems. This does not represent redundancy. Rather, each system appears to be important in a particular set of environmental conditions. There is considerable diversity in *Shigella* iron transport systems, and there is evidence for independent acquisition events for some of the systems found in more than one species.

Iron transport mechanisms in *Shigella*

Many of the iron transport mechanisms found in *Shigella* are conserved among Gram negative enteric pathogens, including species of *Escherichia coli*, *Salmonella*, and *Yersinia*. One of the most common strategies for iron acquisition among these organisms is receptor-mediated transport of iron ligands such as siderophores and heme across the outer membrane (OM) (Figure 2.2). The energy for transport across the OM is provided by an inner membrane (IM) protein complex composed of TonB and its accessory proteins ExbB and ExbD (Postle and Larsen, 2007). This complex harvests energy generated by the proton motive force in the IM and transduces it to receptors in the OM. Once in the periplasm, iron complexes are bound by a periplasmic binding protein (PBP), which shuttles the iron ligand to an IM permease complex (Chu and Vogel, 2011). The IM permease uses ATP to transport its ligand from the periplasm into the cytoplasm of the cell. The IM permease systems are classified as PBP-dependent ABC (ATP Binding Cassette) transporters (Davidson et al., 2008; Kos and Ford, 2009) (Figure 2.2). The periplasmic binding proteins and ABC transporters are generally less specific than the OM receptors, frequently transporting several related siderophores within a given siderophore class. Most Gram negative bacteria also have IM transport systems for free ferric or ferrous iron. Some, but not all, of these systems belong to the PBP-dependent

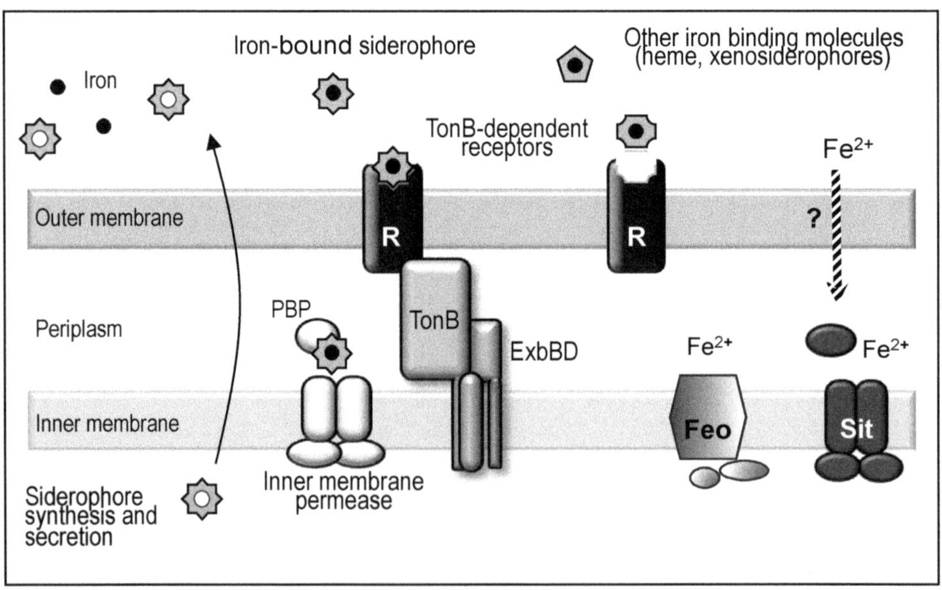

Figure 2.2. Iron transport systems in *Shigella* species. The TonB-dependent systems consist of a receptor (R) which requires the TonB, ExbB, ExbD complex, a periplasmic binding protein (PBP), and an inner membrane permease. Additional systems to transport ferrous iron (e.g. Sit, Feo) are found in the inner membrane. It is unknown how free ferrous ions cross the outer membrane.

ABC transporter class of IM permeases. It is not known how free iron becomes available in the periplasm for transport into the cell. It is possible that free iron or small iron chelates can cross the OM through non-specific porins. The fate of the transported iron substrate depends on the nature of the substrate; some iron complexes, such as heme, can be used directly by the cell. Others, such as iron-loaded siderophores, must release their bound iron inside the cell. Removal of iron from siderophores generally involves reduction of the ferric iron to reduce the affinity of the siderophore for its iron ligand, hydrolysis of the siderophore, or a combination of these activities (Andrews et al., 2003). The released iron can be used for cellular processes or stored in ferritin or bacterioferritin (Andrews, 1998) .

Siderophores

The siderophore systems allow efficient acquisition of ferric iron. Under conditions of iron starvation, bacteria secrete these compounds into the medium where they bind ferric iron with high affinity. The ferri-siderophore complex is brought back into the cell via specific transport systems, and the iron is released in the cytoplasm for use or for storage. *Shigella* species are closely related to *E. coli*, which produces the catechol siderophore enterobactin, a cyclic trimer of 2,3,di-hydroxybenzoyl serine (O'Brien and Gibson, 1970). Enterobactin is recognized by the TonB-dependent OM receptor FepA (Wookey and Rosenberg, 1978), while its linear dihydroxybenzoyl serine derivatives can be transported by CirA or Fiu (Strahsburger et al., 2005). Some *Shigella* strains have Iha, which has been reported to transport both enterobactin and its linear derivatives, and is an important virulence factor in some strains of pathogenic *E. coli* (Léveillé et al., 2006; Tarr et al., 2000). The FepBDGC PBP-dependent ABC transporter is the only known system for transport of enterobactin and its derivative across the IM. The iron is removed from the siderophore in the cytoplasm through the action of the ferric reductase YqjH (Miethke et al., 2011) and the enterobactin esterase Fes (Brickman and McIntosh, 1992). Interestingly, although *yqjH* is part of the Fur regulon (see below), it is also negatively regulated by the product of the divergently transcribed gene *yqjI*. YqjI is a nickel binding protein, and it is believed that this regulatory circuit may be involved in maintaining proper iron homeostasis in the presence of high levels of nickel (Wang et al., 2011). The synthesis and transport of enterobactin is determined by genes organized into 4 contiguous operons (Figure 2.1). Although the genes are present in all four *Shigella* species, enterobactin is not made by all strains, due to the presence of deletions and other mutations that prevent the synthesis and/or uptake of this siderophore. For example, in *S. flexneri*, there are at least two mutations within the enterobactin biosynthesis genes: an amber mutation in *entC,* and an insertion sequence at the 3' end of *entF*. Mutations in the transport genes include a deletion in the promoter of the *fepB* periplasmic binding protein gene and a mutation that reduces the amount of the receptor, FepA, localized to the outer membrane (Schmitt and Payne, 1988, 1991) (Figure 2.1). Although *S. flexneri* strains do not typically produce enterobactin, they do produce a siderophore, the secondary hydroxamate aerobactin (discussed in more detail below). Some *Shigella* isolates, particularly *S. sonnei*, produce both classes of siderophores (Figure 2.1).

In addition to enterobactin, *S. dysenteriae* type 1 produces salmochelins, modified forms of enterobactin in which the siderophore is C-glucosylated (Hantke et al., 2003). This modification was first identified in *Salmonella* and is found in several enteric pathogens. Production of salmochelin may be advantageous for pathogens in mammalian hosts, since the unmodified ferri-enterobactin can be bound by siderocalin (lipocalin-2) in the intestinal tract, rendering the siderophore unavailable to the bacteria (Flo et al., 2004).

Salmochelin is not recognized by this component of the host innate immunity. Salmochelin has a separate TonB-dependent OM receptor, IroN, but uses the remainder of the enterobactin transport proteins for uptake.

The aerobactin system is widespread among enteric pathogens, including *Shigella spp.* (Lawlor and Payne, 1984, Wyckoff 2009 Biometals). Aerobactin is a secondary hydroxamate synthesized from L-lysine and citrate by four enzymes specified by the *iucABCD* genes (Gibson and Magrath, 1969; de Lorenzo and Neilands, 1986; de Lorenzo et al., 1986). The fifth gene in the operon, *iutA*, encodes the TonB-dependent OM receptor for the siderophore (Bindereif et al., 1982) (Figure 2.2). Following transport into the periplasm via IutA, aerobactin is recognized by the FhuBCD PBP-dependent IM permease system specific for hydroxamate siderophores. The *S. boydii* aerobactin genes are part of a horizontally acquired island that maps between the *pheU* and *lysU* tRNA genes and carries with it a P4-like integrase gene (Purdy and Payne, 2001). The *S. flexneri* aerobactin locus maps to *selC* (Moss et al., 1999; Vokes et al., 1999).

Transporters for xenosiderophores

In addition to receptors for their endogenous siderophores, *Shigella* spp. also produce receptors for siderophores made by other organisms. *Shigella* strains, like *E. coli* and *Salmonella* strains, synthesize FhuA, a TonB-dependent receptor for the fungal siderophore ferrichrome (Figure 2.1). Like aerobactin, ferrichrome is a hydroxamate siderophore and is transported through the periplasm and across the IM by the FhuBCD system. Removal of iron from the ferri-ferrichrome complex requires the iron reductase FhuF (Matzanke et al., 2004). Many of the *Shigella* strains also have the *fhuE* gene, which encodes the receptor for the hydroxamate siderophores coprogen, ferrioxamine B, and rhodotorulic acid (Hantke, 1983). As with other hydroxamates, these siderophores are transported across the IM by the FhuBCD system (Figure 2.1). Many *Shigella* strains also have genes that encode apparent TonB-dependent outer membrane receptors for which the ligand has not been identified. These include: *prrA* (Okeke et al., 2004; Ye and Xu, 2001), *ycnD* and *cjrC*. YcnD is a virulence factor in *Salmonella enterica* serovar Typhi (Xiong et al., 2012). CjrC is the receptor for colicin J, and the linked gene *cjrB* appears to encode a TonB-type protein dedicated to the Cir system (Smajs and Weinstock, 2001).

Heme transport

Heme is the most abundant iron source in mammalian hosts. Thus, it is not surprising that human pathogens often have transport systems for heme, either free or bound to hemoglobin (Nobles and Maresso, 2011; Runyen-Janecky, 2013). Among the *Shigella* spp., heme transport is characteristic of *S. dysenteriae* type I strains (Mills and Payne, 1997; Wyckoff et al., 1998) (Figure 2.2). Interestingly, the heme transport system found in *S. dysenteriae* is essentially identical to that in enterohemorrhagic *E. coli* O157:H7 (Mills and Payne, 1995; Torres and Payne, 1997; Wyckoff et al., 1998). This system is not present in other *Shigella* spp. or in commensal *E. coli* (Wyckoff et al., 1998), although the ability to use heme for growth in low-iron conditions has been observed in other *Shigella* isolates (Lawlor et al., 1987). The *S. dysenteriae* heme transport system is analogous to the siderophore transporters (Wyckoff et al., 1998). There is a TonB-dependent OM receptor for heme, ShuA, which binds heme at the cell surface and delivers it to a PBP-dependent ABC transporter (Mills and Payne, 1997). ShuT is the periplasmic binding protein (Eakanunkul et al., 2005), and ShuU and ShuV comprise the IM permease and ATPase (Burkhard and Wilks, 2008).

ShuS, a soluble cytoplasmic protein, may function in heme trafficking, or as a heme storage and DNA binding protein to protect the DNA from heme-mediated oxidative damage (Wilks, 2001; Wyckoff et al., 2005). The entire heme moiety is transported into the cell, since addition of heme to the medium allows growth of a heme biosynthesis mutant. However, it is not known whether the iron can be removed from heme to satisfy the non-heme iron needs of the cell. No heme oxygenase has been identified in *S. dysenteriae*, although heme oxygenases have been found in other bacteria and have been shown to catalyze the removal of iron from heme.

Transporters for other chelates

Iron can bind to organic acids, including citrate, and these complexes promote the growth of some *Shigella*. In *S. sonnei,* the ferric citrate transport system is closely related to that found in commensal *E. coli* strains. Ferric citrate transport is not typical of *S. flexneri* strains, although one strain of *S. flexneri* has been shown to encode the Fec system comprising a TonB-dependent receptor FecA, and a PBP-dependent IM permease system FecBCD (Luck et al., 2001). In this isolate the *fec* genes are encoded within a large genomic island that also includes multiple genes for resistance to antibiotics, and was likely acquired by horizontal gene transfer.

Ferrous iron transport

Shigella species have two transport systems for ferrous iron, Feo and Sit (Figure 2.2). In contrast to transporters for ferric iron chelates and heme, the ferrous systems do not appear to require a specific outer membrane receptor. It is not known how ferrous iron crosses the outer membrane, although porins or channels for other ions may be used.

All *Shigella* species possess a Feo system that is essentially identical that of *E. coli* Feo (Kammler et al., 1993) (Figure 2.1). Feo homologs are widely distributed among bacteria, and Feo may be derived from an ancient transporter that evolved during the time when the Earth's atmosphere lacked oxygen and iron was generally found in the reduced, ferrous form. As in *E. coli*, the *feoABC* genes are arranged in an apparent operon (Figure 2.1). Interestingly, none of the Feo proteins have sequence similarity with other iron transporters, and the mechanism of iron transport by this system is not well understood (1). FeoB has an amino terminal cytoplasmic domain with GTPase activity. This domain has structural similarity with eukaryotic G-proteins, and mutational analysis suggests that GTP binding and hydrolysis are required for iron transport (Marlovits et al., 2002) (4). The carboxy-terminal domain of FeoB is embedded in the inner membrane and is thought to form the channel for iron transport, but this has not been experimentally demonstrated. As with other Feo systems, mutation of *feoB* prevents iron uptake by the Feo system in *S. flexneri* (11).

The roles of FeoA and FeoC are only now beginning to be unraveled. Examination of the amino acid sequence of these small, basic proteins led to the prediction that one or both of them may function as regulators of the *feo* operon (Cartron et al., 2006), but this has not been confirmed in subsequent experiments, and it appears more likely that they are required for Feo activity (Fetherston et al., 2012; Hung et al., 2010; Kim et al., 2012, 2013; Weaver et al., 2013). FeoA has an SH3 domain-like fold (Lau et al., 2013; Su et al., 2010). It is required for iron transport by the Feo system (Kim et al., 2012; Perry et al., 2007; Weaver et al., 2013), but it is not known how it promotes this activity. It did not stimulate GTP hydrolysis by FeoB in vitro (Lau et al., 2013). Although FeoC is required for Feo iron transport in some species, including *Vibrio cholerae* and *Salmonella enterica,* (Kim et al., 2013; Weaver et al., 2013), it is not essential for Feo function in *Yesinia pestis*

(Perry et al., 2007). In *V. cholerae, S. enterica, and Klebsiella pneumoniae*, FeoC was shown to interact directly with FeoB (Hung et al., 2012; Kim et al., 2013; Weaver et al., 2013). In *Salmonella*, FeoB was rapidly degraded by the FtsH protease in the absence of FeoC (Kim et al., 2013). It is not known whether protection of FeoB from proteolysis is the primary function of FeoC, or whether FeoC increases the activity of the Feo transporter, rendering FeoB more stable as part of an active complex.

The second ferrous iron uptake system in *Shigella* spp. is the Sit system (Runyen-Janecky et al., 2003). This is a PBP-dependent ABC transporter that has homology to the *S. enterica* Sit (Zhou et al., 1999) and *Y. pestis* Yfe (Perry et al., 2007) transporters. SitA is the periplasmic binding protein, SitC and SitD form the cytoplasmic membrane permease, and SitB is the ATPase. *S. flexneri* Sit also functions in manganese acquisition (Runyen-Janecky et al., 2006), as has been noted for the *S. enterica* Sit transporter (Kehres et al., 2002).

The *sit* operon has been found in all *Shigella* strains tested, and in a variety of pathogenic *E. coli,* particularly those with an intracellular lifestyle (Payne and Mey, 2004). The *sit* genes are well conserved in these pathogens, but their locations in the genomes and the sequences surrounding them are highly variable (Figure 2.1). The *sit* genes are generally found on genomic islands or on plasmids, and may be in different locations in isolates of the same *Shigella* species (Fisher et al., 2009). This variability in location and context suggest that the *sit* genes have been acquired more than once in the lineages that make up the *Shigella* species.

Regulation of iron transport systems

Regulation via Fur and RyhB

While iron acquisition is essential for normal growth and metabolism of *Shigella*, excess iron is toxic, and its transport into the cell is tightly regulated. The *Shigella* iron transport systems are negatively regulated by the transcriptional repressor Fur, which uses iron as a co-repressor. Fur binding and its effect on gene regulation have been extensively characterized in *E. coli* (Hantke, 2002) and functions similarly in *Shigella* spp. In the presence of ferrous iron, Fur binds as a dimer to sequences upstream of transcription start sites, usually overlapping the -10 or -35 regions, to block binding of RNA polymerase (Figure 2.3A). When iron is limiting, iron-free Fur cannot bind its DNA target sequences, and transcription proceeds (Figure 2.3B). Fur binding sites, termed Fur boxes, are found upstream of all the *Shigella* iron transport operons, linking expression of the iron transporters to the iron status of the cell.

Iron-bound Fur also represses transcription of *ryhB*, which encodes a small RNA (Massé and Gottesman, 2002). In *E. coli*, RyhB reduces the synthesis of a large number of proteins, including the iron-dependent superoxide dismutase (SodB), and other proteins that contain iron or iron-sulfur clusters, including enzymes in the TCA cycle. Thus, when iron levels in the cell are low, expression of RyhB helps maintain overall iron homeostasis by decreasing flux through cellular processes that have high iron requirements. The mechanism of regulation via RyhB typically involves base-paring with target mRNAs. In the case of SodB, RyhB, together with the RNA-binding protein Hfq, binds the *sodB* mRNA and directs its degradation by RNaseE (Masse et al., 2003a, 2003b). In *S. flexneri,* RyhB is also important for regulating acid resistance by controlling expression of the *evgAS* operon (Oglesby et al., 2005). Resistance to stomach acid is a contributing factor in the low infectious dose of *Shigella* spp. In addition, RyhB plays a direct role in virulence gene expression in *S. flexneri* (Murphy and Payne, 2007) (Figure 2.4). The mechanism of

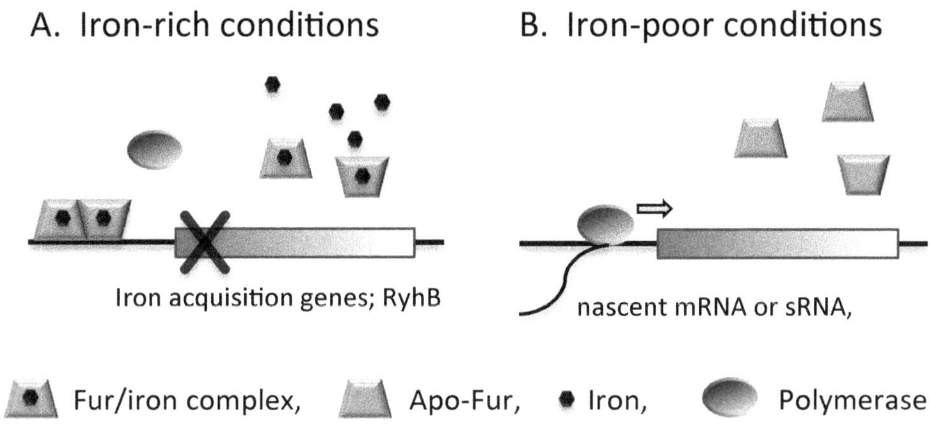

Figure 2.3. Regulation of iron acquisition genes and RyhB by iron and Fur. The expression of iron acquisition genes and RyhB is regulated by the iron-binding protein Fur, which, when complexed with iron, binds to and inhibits transcription from promoters containing Fur binding sequences (A). When iron levels in the cell are low, Fur is in the iron-free form and cannot bind its DNA targets, and transcription proceeds (B).

RyhB regulation of virulence genes is distinct from its role in destabilizing mRNAs. RyhB inhibits the synthesis of the major virulence gene regulator VirB by inhibiting *virB* transcription, rather than by directing degradation of the *virB* mRNA. RyhB-mediated regulation of *virB* expression is independent of regulation of *virB* by its upstream activator VirF (Broach et al., 2012; Murphy and Payne, 2007).

Iron also directly affects the production of *Shigella* virulence factors. Iron and Fur repress the expression of the gene encoding the *S. dysenteriae* type 1 toxin, shiga toxin (Dubos and Geiger, 1946; Svinarich and Palchaudhuri, 1992). Regulation of toxin genes by iron has been seen in other pathogens including *E. coli* and *Corynebacterium diphtheriae* (Boyd et al., 1990; Calderwood and Mekalanos, 1987; Schmitt et al., 1992). These effects on virulence gene expression suggest that iron limitation, like temperature, may be a signal to pathogens that they are within the host environment.

Regulation in response to environmental cues

Although all the characterized iron transport systems are regulated by iron and Fur, not all systems show the same levels of repression or induction in response to iron levels. Some of the difference in regulation reflects divergence in the sequences of the Fur boxes, suggesting that affinity for Fur may play a role in the levels of expression of a system. However, other environmental factors, particularly oxygen levels and temperature, also play a role (Figure 2.4). It is not surprising that the level of oxygen would influence which iron transport systems are expressed, since oxygen can determine whether ferrous or ferric iron predominates. By coupling expression of an iron transport system to environmental conditions that govern which iron sources are available, the bacteria minimize the metabolic costs of expressing transport systems in an environment where their ligand will be in low concentration. In general, ferrous iron transporters are more highly expressed in anaerobic conditions and ferric transporters in oxic environments. Oxygen regulation of iron uptake is also important in preventing damage to the cell from iron-dependent radical formation. Transcription of *fur* is activated by the redox regulators OxyR and SoxS in response to oxidative stress (Zheng et al., 1999) and is repressed under anoxic conditions by ArcA (Boulette and Payne, 2007). The increased level of Fur in the presence of oxygen ensures repression of iron transport system genes and, since Fur is an iron binding protein, it may act as a scavenger of unbound iron to prevent cell-damaging radical formation.

In bacteria, the main regulators of gene expression in response to oxygen levels are FNR and ArcA. These are global regulators whose function is to coordinate the expression of large sets of genes needed to adapt to growth under oxygen-limiting conditions. FNR is a transcription factor that can act as an activator or a repressor of gene expression. It contains an iron-sulfur cluster that is destabilized in the presence of oxygen. Only at low oxygen concentrations is the iron-sulfur cluster stable enough to allow FNR to dimerize and bind to the regulatory DNA sequences of its target genes (Kang et al., 2005; Kiley and Beinert, 1998). Like FNR, ArcA is a global regulator that can exert both positive and negative effects on gene expression. ArcA is part of a two-component system in which ArcA is the response regulator and ArcB the sensor kinase. Unlike FNR, ArcB does not sense oxygen directly through an iron cofactor, but responds instead to the redox state of ubiquinones and menaquinones, as well as to the level of reduced fermentative products in the cell, which in turn are governed by the state of aerobiosis (Bekker et al., 2010; Iuchi, 1993; Malpica et al., 2004).

In *S. flexneri*, as in *E. coli*, the *feo* genes are induced under anoxic conditions (Boulette and Payne, 2007; Kammler et al., 1993). The *S. flexneri feo* promoter has binding sites for both ArcA and FNR (Figure 2.4). In the absence of oxygen, transcription of *feo* is reduced in either an *arcA* or *fnr* mutant, and the effects are additive in the double mutant (Boulette and Payne, 2007). This indicates that these transcription factors induce *feo*

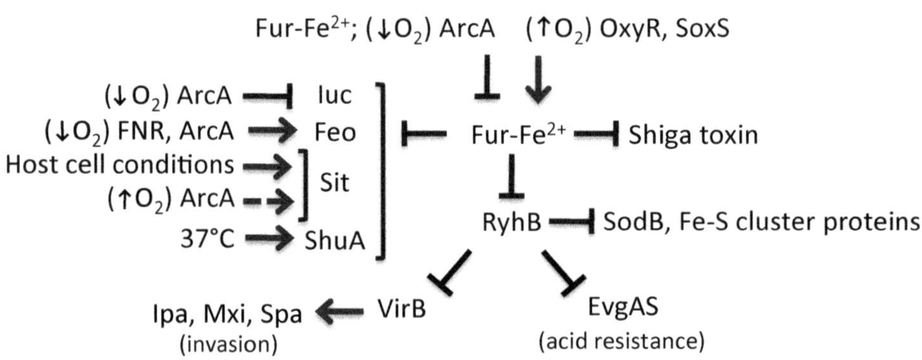

Figure 2.4. Regulation of gene expression by environmental cues in *Shigella* species. Iron uptake systems are regulated in response to iron levels through the iron-binding protein Fur. In the presence of iron, Fur acts as a transcriptional repressor by binding to specific sequences in the promoter region of its target genes. In addition to regulating iron acquisition genes, Fur represses the expression of the small RNA RyhB, which plays a role in carbon and iron metabolism, as well as in virulence gene regulation and regulation of acid resistance. Fur also autoregulates its own expression in response to iron. The genes encoding Fur and many of the iron uptake systems are affected by oxygen levels through the oxygen responsive master regulators ArcA and FNR. In addition, transcription of *fur* is activated by OxyR and SoxS in response to oxidative stress. The *sit* system genes are further induced by host cell conditions, and the heme uptake receptor *shuA* is induced at 37°C, the temperature of the mammalian host. Solid arrows indicate direct induction of transcription, whereas a dashed arrow indicates induction via an unknown, indirect mechanism.

transcription in the absence of oxygen. Both the FNR and ArcA binding sites are upstream of the Fur binding site, suggesting that Fur-occupancy of its binding site would block the ability of ArcA and FNR to induce synthesis, which is consistent with the ability of iron and Fur to inhibit iron transport under both aerobic and anaerobic conditions.

ArcA also regulates the expression of the aerobactin ferric iron transport system (Figure 2.4). In this case, ArcA represses *iuc* expression under anaerobic conditions (Boulette and Payne, 2007). The ArcA binding site in the *iuc* promoter is downstream of the transcription start site and binding of ArcA would block expression under anaerobic conditions, even in the absence of Fur binding. Therefore, ArcA and FNR work together to favor expression of the ferrous iron transporter Feo in anaerobic conditions and the ferric iron transporter aerobactin in the presence of oxygen.

An unexpected finding in the studies of oxygen and *S. flexneri* iron transport gene expression was the observation that the *sit* genes were more highly expressed in the presence than in the absence of oxygen (Boulette and Payne, 2007). Because Sit transports ferrous, and not ferric, iron, it was surprising that it would be expressed under conditions that should favor ferric iron. Additional studies provided a possible evolutionary advantage for this mode of expression. The *sit* genes are induced in *S. flexneri* during intracellular residence (Lucchini et al., 2005; Runyen-Janecky and Payne, 2002) (Figure 2.4). The iron available for bacterial growth in the epithelial cell cytoplasm is expected to be in the reduced, ferrous form; however, the eukaryotic cells would have sufficient oxygen to induce the *sit* genes while repressing the *feo* genes. Induction by oxygen would allow the Sit system to provide ferrous iron to bacteria inside eukaryotic cells under conditions where the Feo system is not expressed well.

Temperature is a regulatory signal for expression of the heme transport system of *S. dysenteriae* 1 (Figure 2.4). The TonB-dependent OM receptor ShuA is made at significantly higher levels at 37ºC than at 25ºC (Kouse et al., 2013) (Figure 2.5A). A shift in temperature from 25ºC to 37ºC corresponds to the change in temperature experienced by *Shigella* during transit to a new host, especially if transmission is not direct, but involves a period of exposure to the environment. By inducing expression of the heme transport system specifically in response to the host temperature, and not just in response to iron limitation in general, the bacteria ensure expression of this system only under conditions where heme is abundant. The mechanism of thermoregulation is post-transcriptional and involves sequences within the long 5' UTR of the *shuA* mRNA that resemble a FourU RNA thermometer (Kouse et al., 2013; Waldminghaus et al., 2007) (Figure 2.5B). In silico predictions for the folding of the *shuA* 5' UTR show that the putative FourU element may sequester the *shuA* Shine-Dalgarno sequence at the non-permissive temperature, but not at the higher, permissive, temperature, resulting in increased translation of the *shuA* mRNA at higher temperatures (Kouse et al., 2013). A FourU regulatory element is present also in the 5' UTR of the *chuA* mRNA of enteropathogenic *E. coli*, and was found to exert similar temperature control over synthesis of the ChuA heme receptor (Kouse et al., 2013).

Iron and the pathogenesis of *Shigella*

Iron acquisition is problematic for bacteria colonizing mammalian hosts (Nairz et al., 2010). Enteric pathogens may obtain some iron from host dietary sources when they are located in the lumen of the gut. However, invasive pathogens must compete with their host for the limited supply of iron within the tissues. There is essentially no free iron within the host. Most of the iron is complexed, with the largest amount in heme, which is further sequestered within hemoglobin, hemopexin, or heme-containing cellular proteins. Excess iron within the cells is stored in ferritin. Extracellular iron is tightly bound to transferrin, which serves as the major transporter of iron between cells. Transferrin is capable of binding two molecules of iron, but transferrin in the circulation is relatively unsaturated. In the normal human, circulating transferrin is only about 20-25% iron-saturated, allowing it to rapidly bind any free iron. The ability of unsaturated transferrin to withhold iron from bacteria has been shown in the laboratory. One of the first demonstrations of host proteins withholding iron from bacterial pathogens was Schade and Caroline's (Schade and Caroline, 1944) demonstration that hen egg white contained a protein, later identified as ovotransferrin, that prevented growth of *S. dysenteriae* by withholding iron. This was followed by the identification of a similar protein, transferrin, in human serum (Schade and Caroline, 1946). While unsaturated transferrin can withhold iron from pathogens, there is evidence that individuals with highly saturated transferrin have an increased incidence of infections. Hemochromatosis, which leads to iron overload, is a major predisposing factor in *V. vulnificus* septicemia (Weinberg, 2000). In a comparison of shigellosis patients with septicemic or fecal isolates, there was a two-fold increase in transferrin iron saturation in the patients with a disseminated *Shigella* infection (Struelens et al., 1990).

A second major iron-binding protein that has antibacterial properties is lactoferrin. Lactoferrin is found on mucosal surfaces, in secretions, and in the granules of phagocytic cells. It has a higher affinity for iron than transferrin and can hold the iron at a more acidic pH. Lactoferrin has both bacteriostatic and bactericidal properties. Lactoferrin not only

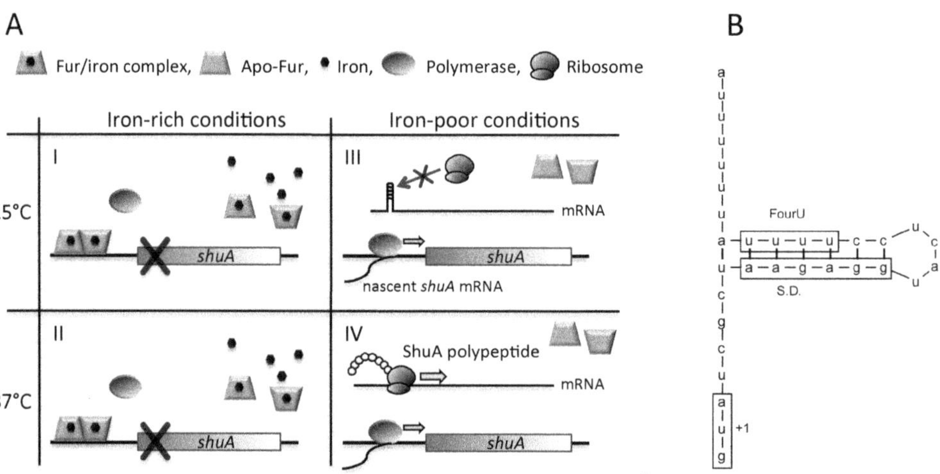

Figure 2.5. Figure 2.5. Dual regulation of *S. dysenteriae* ShuA levels by iron and temperature. (A) The expression of *shuA* is regulated by the iron-binding protein Fur, which binds to and inhibits transcription from the *shuA* promoter when iron levels are high (I and II). In addition, translation of the *shuA* mRNA is affected by temperature. The mechanism of temperature regulation involves the formation of an inhibitory FourU thermometer structure in the *shuA* mRNA at temperatures lower than 37°C (III). At 37°C, the inhibitory structure cannot form, and the message can be translated (IV). This dual regulation ensures maximal expression of the heme uptake system under conditions similar to those found in the host during infection, where heme is potentially an abundant source of iron. (B) The sequence and structure of the FourU thermometer in the *shuA* mRNA 5'UTR. The stem-loop structure of the FourU thermometer inhibits translation of the message by sequestering the Shine-Dalgarno (S.D.) sequence. *(Adapted from Kouse et al., 2013, with permission.)*

inhibits *Shigella* by withholding iron but also through specific interactions with the bacterial cell. Lactoferrin binds to the surface of *S. flexneri* (Tigyi et al., 1992), and lactoferrin binding decreases bacterial invasion by inducing the release of Ipa proteins and making IpaB and IpaC susceptible to protease degradation (Gomez et al., 2003).

Bacteria have several mechanisms for counteracting the inhibitory effects of iron-binding proteins. Some bacteria, such as *Neisseria gonorrhoeae*, have TonB-dependent receptors and associated cell surface proteins for binding transferrin or lactoferrin and removing the iron for transport into the bacterial cell (Noinaj et al., 2012). However, receptors for these proteins have not been identified in *Shigella* spp.; instead, *Shigella* rely on the production of siderophores that may allow the bacteria to remove iron from these host proteins in body fluids. Siderophore-producing *S. flexneri* can compete with transferrin or lactoferrin for iron if the proteins are relatively iron saturated, but the unsaturated proteins inhibit *Shigella* growth (Lawlor et al., 1987).

In response to bacterial infection, there is often a reduction in serum iron saturation, which serves to further restrict bacterial growth. This response has been termed the anemia of infection and is likely an innate immune response to the presence of bacterial molecules such as LPS. Weinberg (Weinberg, 1977) has described this as nutritional immunity, and there is both experimental and epidemiological evidence that it helps limit bacterial dissemination in the human. Hepcidin, an antimicrobial-like peptide hormone helps regulate the absorption and distribution of iron in human tissues, and its synthesis is regulated by innate immunity, as well as by iron (Drakesmith and Prentice, 2012).

Another component of this type of nutritional immunity is the synthesis of siderocalin (lipocalin-2, Ngal) in response to infection. Siderocalin binds unmodified catechol siderophores, such as enterobactin, preventing them from delivering iron to the bacteria (Abergel et al., 2008). This would suggest an advantage for production of the hydroxamate aerobactin, rather than enterobactin by some *Shigella* spp., but there do not appear to be any differences in aerobactin producers and enterobactin producers in their ability to cause shigellosis. As described above, *S. dysenteriae* type 1 strains also produce salmochelin, the glucosylated form of enterobactin that is not recognized by siderocalin (Reeves, 2001). Analysis of the sequenced genomes suggests that many *Shigella* strains produce at least one of the siderocalin-resistant siderophores, either aerobactin or salmochelin (E.E. Wyckoff, unpublished results). Salmochelin production has been shown to be critical for virulence of some pathogenic *E. coli* (Abergel et al., 2008; Fischbach et al., 2006; Goetz et al., 2002; Hantke et al., 2003), suggesting a role for siderocalin in the interactions between enteric pathogens and the gut epithelium.

The role of iron transport systems in *Shigella* virulence has largely been determined in cultured cells (Figure 2.6). *S. flexneri* 2a strains generally have only three iron transport systems that could supply iron within the host, the aerobactin, Feo and Sit systems. Single, double and triple mutants were constructed to see which mutations prevented growth of the bacteria within cultured epithelial cells. A triple mutant defective in aerobactin synthesis and in the Feo and Sit transporters was unable to grow in vitro in the absence of an added iron source such as aerobactin, which the mutant can transport, but not make. This mutant also failed to form plaques in the cultured cell monolayer. This indicates that *S. flexneri* does not have an additional, unidentified system that can function when the bacteria are intracellular; thus, one of the three known systems is required for intracellular growth. None of the single mutants had a plaque defect (Lawlor et al., 1987; Runyen-Janecky et al., 2003). The double mutants, each of which expressed only a single iron transport system, were tested in plaque assays using both aerobic and anaerobic culture conditions (Figure 2.6). The strain carrying only the aerobactin system made plaques aerobically, but not anaerobically, whereas the strain carrying Feo alone produced plaques only anaerobically. This is in agreement with the observation that the aerobactin synthesis genes are induced under aerobic conditions, which favor the ferric form of iron, whereas the *feo* genes are most highly expressed under anoxic conditions, which favor the ferrous form (Figure 2.4) (Boulette and Payne, 2007). The strain expressing only the Sit system grew poorly in vitro, but produced wild type plaques under both oxic and anoxic conditions (Boulette and Payne, 2007; Runyen-Janecky et al., 2003). Further, in competition assays, the Sit system conferred a growth advantage within cultured cells (Fisher et al., 2009). This points to the importance of the Sit system for intracellular growth, and is consistent with increased *sit* expression in response to host cell conditions (Figure 2.4) (Boulette and Payne, 2007). In a mouse lung model of *Shigella* infection, the Sit system was required for virulence. Mice inoculated with a *sitA*

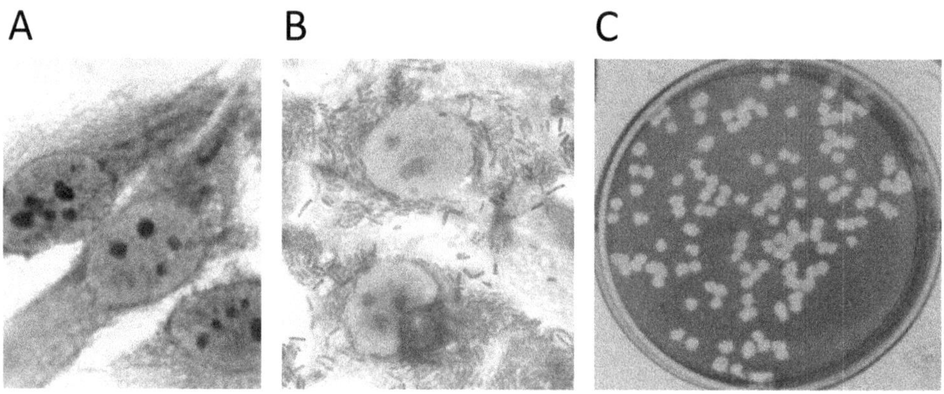

D

Genotype	Iron transport system present	Plaque formation $+O_2$	Plaque formation $-O_2$
wild type	Sit^+, Feo^+, Iuc^+/Iut^+	++	+
feoB⁻, *iucD*⁻	Sit^+	++	+
feoB⁻, *sitA*⁻	Iuc^+/Iut^+	+	–
IucD⁻, *sitA*⁻	Feo^+	–	+
feoB⁻, *iucD*⁻, *sitA*⁻	None	–	–

Figure 2.6. Invasion and plaque assays. In the invasion assay, subconfluent monolayers of cultured epithelial cells (A) are infected with a *Shigella* strain. The bacteria are allowed to penetrate the cells for 30 minutes, and any extracellular bacteria are killed by the addition of gentamicin. Following incubation for 2 hours, the monolayers are stained, and the intracellular bacteria are visualized by microscopy (B). In the plaque assay, invaded epithelial cells are incubated in the presence of gentamicin for an additional 2-4 days, allowing *Shigella* to multiply intracellularly and spread directly to adjacent cells, causing cell death and the formation of plaques in the monolayer, which can be visualized by staining (C). The infection efficiency of various *S. flexneri* iron transport mutants in the presence or absence of oxygen was determined in the plaque assay (D). [++] denotes large plaques, [+] denotes small plaques, and [–] indicates no plaques.

mutant strain largely survived the infection, and were protected against subsequent challenge with the wild type strain. In contrast, mice inoculated with the wild type parental strain all succumbed to the infection (Fisher et al., 2009). Taken together, these data point to a definitive role for the Sit iron uptake system in intracellular growth and virulence of this pathogen.

Aerobactin biosynthesis mutants were tested in the guinea pig Sereny test, which measures inflammation as well as invasion, and in a mouse lung model for *Shigella* virulence. No loss of virulence was noted in either of these models (Fisher et al., 2009; Lawlor et al., 1987), although Nassif et al. noted that the mutant had a reduced ability to provoke inflammatory response in the guinea pig Sereny at lower inocula (Nassif et al., 1987). They also observed less fluid accumulation and less severe mucosal lesions in rabbit ileal loops compared to wild type when smaller inocula were used (Nassif et al., 1987). Thus, although the aerobactin system is not required for intracellular growth, it may provide an advantage for growth in extracellular compartments in the host.

Shigella infection is dependent upon the bacteria being able to successfully compete with the host for iron. The bacteria come armed with siderophores, heme transport systems and transporters for ferrous iron, which allows them to acquire this metal under most environmental conditions. However, it is becoming increasingly apparent that the host innate response can target iron acquisition to help limit the growth of pathogens within the host. Future studies will more fully characterize the components of both pathogen and host that control access to, and acquisition of, this essential element.

References

Abergel, R.J., Clifton, M.C., Pizarro, J.C., Warner, J.A., Shuh, D.K., Strong, R.K., and Raymond, K.N. (2008). The siderocalin/enterobactin interaction: a link between mammalian immunity and bacterial iron transport. J. Am. Chem. Soc. *130*, 11524–11534.

Andrews, S.C. (1998). Iron storage in bacteria. Adv. Microb. Physiol. *40*, 281–351.

Andrews, S.C., Robinson, A.K., and Rodríguez-Quiñones, F. (2003). Bacterial iron homeostasis. FEMS Microbiol. Rev *27*, 215–237.

Bekker, M., Alexeeva, S., Laan, W., Sawers, G., Teixeira de Mattos, J., and Hellingwerf, K. (2010). The ArcBA two-component system of *Escherichia coli* is regulated by the redox state of both the ubiquinone and the menaquinone pool. J. Bacteriol. *192*, 746–754.

Bindereif, A., Braun, V., and Hantke, K. (1982). The cloacin receptor of ColV-bearing *Escherichia coli* is part of the Fe3+-aerobactin transport system. J. Bacteriol. *150*, 1472–1475.

Boulette, M.L., and Payne, S.M. (2007). Anaerobic regulation of *Shigella flexneri* virulence: ArcA regulates Fur and iron acquisition genes. J. Bacteriol *189*, 6957–6967.

Boyd, J., Oza, M.N., and Murphy, J.R. (1990). Molecular cloning and DNA sequence analysis of a diphtheria *tox* iron-dependent regulatory element (*dtxR*) from *Corynebacterium diphtheriae*. Proc. Natl. Acad. Sci. USA *87*, 5968–5972.

Brickman, T.J., and McIntosh, M.A. (1992). Overexpression and purification of ferric enterobactin esterase from *Escherichia coli*. Demonstration of enzymatic hydrolysis of enterobactin and its iron complex. J. Biol. Chem. *267*, 12350–12355.

Broach, W.H., Egan, N., Wing, H.J., Payne, S.M., and Murphy, E.R. (2012). VirF-independent regulation of *Shigella virB* transcription is mediated by the small RNA RyhB. PLoS ONE *7*, e38592.

Burkhard, K.A., and Wilks, A. (2008). Functional characterization of the *Shigella dysenteriae* heme ABC transporter. Biochemistry *47*, 7977–7979.

Calderwood, S.B., and Mekalanos, J.J. (1987). Iron regulation of Shiga-like toxin expression in *Escherichia coli* is mediated by the *fur* locus. J. Bacteriol. *169*, 4759–4764.

Cartron, M.L., Maddocks, S., Gillingham, P., Craven, C.J., and Andrews, S.C. (2006). Feo – Transport of ferrous iron into bacteria. Biometals *19*, 143–157.

Chu, B.C.H., and Vogel, H.J. (2011). A structural and functional analysis of type III periplasmic and substrate binding proteins: their role in bacterial siderophore and heme transport. Biol. Chem. *392*, 39–52.

Davidson, A.L., Dassa, E., Orelle, C., and Chen, J. (2008). Structure, function, and evolution of bacterial ATP-binding cassette systems. Microbiol. Mol. Biol. Rev. *72*, 317–364.

Drakesmith, H., and Prentice, A.M. (2012). Hepcidin and the iron-infection axis. Science *338*, 768–772.

Dubos, R.J., and Geiger, J.W. (1946). Preparation and properties of shiga toxin and toxoid. J. Exp. Med. *84*, 143–156.

Eakanunkul, S., Lukat-Rodgers, G.S., Sumithran, S., Ghosh, A., Rodgers, K.R., Dawson, J.H., and Wilks, A. (2005). Characterization of the periplasmic heme-binding protein shut from the heme uptake system of *Shigella dysenteriae*. Biochemistry *44*, 13179–13191.

Fetherston, J.D., Mier, I., Jr, Truszczynska, H., and Perry, R.D. (2012). The Yfe and Feo transporters are involved in microaerobic growth and virulence of *Yersinia pestis* in bubonic plague. Infect. Immun. *80*, 3880–3891.

Fischbach, M.A., Lin, H., Zhou, L., Yu, Y., Abergel, R.J., Liu, D.R., Raymond, K.N., Wanner, B.L., Strong, R.K., Walsh, C.T., et al. (2006). The pathogen-associated *iroA* gene cluster mediates bacterial evasion of lipocalin 2. Proc. Natl. Acad. Sci. USA *103*, 16502–16507.

Fisher, C.R., Davies, N.M.L.L., Wyckoff, E.E., Feng, Z., Oaks, E.V., and Payne, S.M. (2009). Genetics and virulence association of the *Shigella flexneri sit* iron transport system. Infect. Immun. *77*, 1992–1999.

Flo, T.H., Smith, K.D., Sato, S., Rodriguez, D.J., Holmes, M.A., Strong, R.K., Akira, S., and Aderem, A. (2004). Lipocalin 2 mediates an innate immune response to bacterial infection by sequestrating iron. Nature *432*, 917–921.

Gibson, F., and Magrath, D.I. (1969). The isolation and characterization of a hydroxamic acid (aerobactin) formed by *Aerobacter aerogenes* 62-I. Biochim. Biophys. Acta *192*, 175–184.

Gillespie, J.J., Wattam, A.R., Cammer, S.A., Gabbard, J.L., Shukla, M.P., Dalay, O., Driscoll, T., Hix, D., Mane, S.P., Mao, C., et al. (2011). PATRIC: the comprehensive bacterial bioinformatics resource with a focus on human pathogenic species. Infect. Immun. *79*, 4286–4298.

Goetz, D.H., Holmes, M.A., Borregaard, N., Bluhm, M.E., Raymond, K.N., and Strong, R.K. (2002). The neutrophil lipocalin NGAL is a bacteriostatic agent that interferes with siderophore-mediated iron acquisition. Mol. Cell *10*, 1033–1043.

Gomez, H.F., Ochoa, T.J., Carlin, L.G., and Cleary, T.G. (2003). Human lactoferrin impairs virulence of *Shigella flexneri*. J. Infect. Dis. *187*, 87–95.

Hantke, K. (1983). Identification of an iron uptake system specific for coprogen and rhodotorulic acid in *Escherichia coli* K12. Mol. Gen. Genet. *191*, 301–306.

Hantke, K. (2002). Members of the Fur protein family regulate iron and zinc transport in *E. coli* and characteristics of the Fur-regulated FhuF protein. J. Mol. Microbiol. Biotechnol. *4*, 217–222.

Hantke, K., Nicholson, G., Rabsch, W., and Winkelmann, G. (2003). Salmochelins, siderophores of *Salmonella enterica* and uropathogenic *Escherichia coli* strains, are recognized by the outer membrane receptor IroN. Proc. Natl. Acad. Sci. USA *100*, 3677–3682.
Hung, K.-W., Chang, Y.-W., Eng, E.T., Chen, J.-H., Chen, Y.-C., Sun, Y.-J., Hsiao, C.-D., Dong, G., Spasov, K.A., Unger, V.M., et al. (2010). Structural fold, conservation and Fe(II) binding of the intracellular domain of prokaryote FeoB. J. Struct. Biol. *170*, 501–512.
Hung, K.-W., Tsai, J.-Y., Juan, T.-H., Hsu, Y.-L., Hsiao, C.-D., and Huang, T.-H. (2012). Crystal structure of the Klebsiella pneumoniae NFeoB/FeoC complex and roles of FeoC in regulation of Fe2+ transport by the bacterial Feo system. J. Bacteriol. *194*, 6518–6526.
Iuchi, S. (1993). Phosphorylation/dephosphorylation of the receiver module at the conserved aspartate residue controls transphosphorylation activity of histidine kinase in sensor protein ArcB of *Escherichia coli*. J. Biol. Chem. *268*, 23972–23980.
Kammler, M., Schon, C., and Hantke, K. (1993). Characterization of the ferrous iron uptake system of *Escherichia coli*. J. Bacteriol. *175*, 6212–6219.
Kang, Y., Weber, K.D., Qiu, Y., Kiley, P.J., and Blattner, F.R. (2005). Genome-wide expression analysis indicates that FNR of *Escherichia coli* K-12 regulates a large number of genes of unknown function. J. Bacteriol. *187*, 1135–1160.
Kehres, D.G., Janakiraman, A., Slauch, J.M., and Maguire, M.E. (2002). SitABCD is the alkaline Mn(2+) transporter of *Salmonella enterica* serovar Typhimurium. J. Bacteriol. *184*, 3159–3166.
Kiley, P.J., and Beinert, H. (1998). Oxygen sensing by the global regulator, FNR: the role of the iron-sulfur cluster. FEMS Microbiol. Rev. *22*, 341–352.
Kim, H., Lee, H., and Shin, D. (2012). The FeoA protein is necessary for the FeoB transporter to import ferrous iron. Biochem. Biophys. Res. Commun. *423*, 733–738.
Kim, H., Lee, H., and Shin, D. (2013). The FeoC protein leads to high cellular levels of the Fe(II) transporter FeoB by preventing FtsH protease regulation of FeoB in *Salmonella enterica*. J. Bacteriol. *195*, 3364–3370.
Kos, V., and Ford, R.C. (2009). The ATP-binding cassette family: a structural perspective. Cell. Mol. Life Sci. *66*, 3111–3126.
Kouse, A.B., Righetti, F., Kortmann, J., Narberhaus, F., and Murphy, E.R. (2013). RNA-mediated thermoregulation of iron-acquisition genes in *Shigella dysenteriae* and pathogenic *Escherichia coli*. PLoS ONE *8*, e63781.
Lau, C.K.Y., Ishida, H., Liu, Z., and Vogel, H.J. (2013). Solution structure of *Escherichia coli* FeoA and its potential role in bacterial ferrous iron transport. J. Bacteriol. *195*, 46–55.
Lawlor, K.M., and Payne, S.M. (1984). Aerobactin genes in *Shigella* spp. J. Bacteriol. *160*, 266–272.
Lawlor, K.M., Daskaleros, P.A., Robinson, R.E., and Payne, S.M. (1987). Virulence of iron transport mutants of *Shigella flexneri* and utilization of host iron compounds. Infect Immun *55*, 594–599.
Léveillé, S., Caza, M., Johnson, J.R., Clabots, C., Sabri, M., and Dozois, C.M. (2006). Iha from an *Escherichia coli* urinary tract infection outbreak clonal group A strain is expressed in vivo in the mouse urinary tract and functions as a catecholate siderophore receptor. Infect. Immun. *74*, 3427–3436.
De Lorenzo, V., and Neilands, J.B. (1986). Characterization of *iucA* and *iucC* genes of the aerobactin system of plasmid ColV-K30 in *Escherichia coli*. J. Bacteriol. *167*, 350–355.

De Lorenzo, V., Bindereif, A., Paw, B.H., and Neilands, J.B. (1986). Aerobactin biosynthesis and transport genes of plasmid ColV-K30 in *Escherichia coli* K-12. J. Bacteriol. *165*, 570–578.

Lucchini, S., Liu, H., Jin, Q., Hinton, J.C., and Yu, J. (2005). Transcriptional adaptation of *Shigella flexneri* during infection of macrophages and epithelial cells: insights into the strategies of a cytosolic bacterial pathogen. Infect. Immun. *73*, 88–102.

Luck, S.N., Turner, S.A., Rajakumar, K., Sakellaris, H., and Adler, B. (2001). Ferric dicitrate transport system (Fec) of *Shigella flexneri* 2a YSH6000 is encoded on a novel pathogenicity island carrying multiple antibiotic resistance genes. Infect. Immun. *69*, 6012–6021.

Malpica, R., Franco, B., Rodriguez, C., Kwon, O., and Georgellis, D. (2004). Identification of a quinone-sensitive redox switch in the ArcB sensor kinase. Proc. Natl. Acad. Sci. USA *101*, 13318–13323.

Marlovits, T.C., Haase, W., Herrmann, C., Aller, S.G., and Unger, V.M. (2002). The membrane protein FeoB contains an intramolecular G protein essential for Fe(II) uptake in bacteria. Proc. Natl. Acad. Sci. USA *99*, 16243–16248.

Masse, E., Majdalani, N., and Gottesman, S. (2003a). Regulatory roles for small RNAs in bacteria. Curr Opin Microbiol *6*, 120–124.

Masse, E., Escorcia, F.E., and Gottesman, S. (2003b). Coupled degradation of a small regulatory RNA and its mRNA targets in *Escherichia coli*. Genes Dev *17*, 2374–2383.

Massé, E., and Gottesman, S. (2002). A small RNA regulates the expression of genes involved in iron metabolism in *Escherichia coli*. Proc. Natl. Acad. Sci. USA *99*, 4620–4625.

Matzanke, B.F., Anemüller, S., Schünemann, V., Trautwein, A.X., and Hantke, K. (2004). FhuF, part of a siderophore-reductase system. Biochemistry *43*, 1386–1392.

Miethke, M., Hou, J., and Marahiel, M.A. (2011). The siderophore-interacting protein YqjH acts as a ferric reductase in different iron assimilation pathways of *Escherichia coli*. Biochemistry *50*, 10951–10964.

Mills, M., and Payne, S.M. (1995). Genetics and regulation of heme iron transport in *Shigella dysenteriae* and detection of an analogous system in *Escherichia coli* O157:H7. J. Bacteriol. *177*, 3004–3009.

Mills, M., and Payne, S.M. (1997). Identification of *shuA*, the gene encoding the heme receptor of *Shigella dysenteriae*, and analysis of invasion and intracellular multiplication of a *shuA* mutant. Infect. Immun. *65*, 5358–5363.

Moss, J.E., Cardozo, T.J., Zychlinsky, A., and Groisman, E.A. (1999). The *selC*-associated SHI-2 pathogenicity island of *Shigella flexneri*. Mol. Microbiol. *33*, 74–83.

Murphy, E.R., and Payne, S.M. (2007). RyhB, an iron-responsive small RNA molecule, regulates *Shigella dysenteriae* virulence. Infect Immun *75*, 3470–3477.

Nairz, M., Schroll, A., Sonnweber, T., and Weiss, G. (2010). The struggle for iron - a metal at the host-pathogen interface. Cell. Microbiol. *12*, 1691–1702.

Nassif, X., Mazert, M.C., Mounier, J., and Sansonetti, P.J. (1987). Evaluation with an *iuc*::Tn10 mutant of the role of aerobactin production in the virulence of *Shigella flexneri*. Infect. Immun. *55*, 1963–1969.

Nobles, C.L., and Maresso, A.W. (2011). The theft of host heme by Gram-positive pathogenic bacteria. Metallomics *3*, 788–796.

Noinaj, N., Buchanan, S.K., and Cornelissen, C.N. (2012). The transferrin-iron import system from pathogenic *Neisseria* species. Mol. Microbiol. *86*, 246–257.

O'Brien, I.G., and Gibson, F. (1970). The structure of enterochelin and related 2,3-dihydroxy-N-benzoylserine conjugates from *Escherichia coli*. Biochim. Biophys. Acta *215*, 393–402.

Oglesby, A.G., Murphy, E.R., Iyer, V.R., and Payne, S.M. (2005). Fur regulates acid resistance in *Shigella flexneri* via RyhB and YdeP. Mol. Microbiol. *58*, 1354–1367.
Okeke, I.N., Scaletsky, I.C.A., Soars, E.H., Macfarlane, L.R., and Torres, A.G. (2004). Molecular epidemiology of the iron utilization genes of enteroaggregative *Escherichia coli*. J. Clin. Microbiol. *42*, 36–44.
Payne, S.M., and Mey, A.R. (2004). Pathogenic *Escherichia coli*, *Shigella*, and *Salmonella*. In Iron Transport in Bacteria, (Washington, D.C.: ASM Press),.
Perry, R.D., Mier, I., Jr, and Fetherston, J.D. (2007). Roles of the Yfe and Feo transporters of *Yersinia pestis* in iron uptake and intracellular growth. Biometals *20*, 699–703.
Postle, K., and Larsen, R.A. (2007). TonB-dependent energy transduction between outer and cytoplasmic membranes. Biometals *20*, 453–465.
Purdy, G.E., and Payne, S.M. (2001). The SHI-3 iron transport island of *Shigella boydii* 0-1392 carries the genes for aerobactin synthesis and transport. J. Bacteriol. *183*, 4176–4182.
Reeves, S.A. (2001). Iron Acquisition in the Intracellular Environment of the Host: Multiple Iron Transport Systems in *Shigella dysenteriae* (University of Texas).
Runyen-Janecky, L.J. (2013). Role and regulation of heme iron acquisition in gram-negative pathogens. Front. Cell. Infect. Microbiol. *3*, 55.
Runyen-Janecky, L.J., and Payne, S.M. (2002). Identification of chromosomal *Shigella flexneri* genes induced by the eukaryotic intracellular environment. Infect. Immun. *70*, 4379–4388.
Runyen-Janecky, L., Dazenski, E., Hawkins, S., and Warner, L. (2006). Role and regulation of the *Shigella flexneri* Sit and MntH systems. Infect. Immun. *74*, 4666–4672.
Runyen-Janecky, L.J., Reeves, S.A., Gonzales, E.G., and Payne, S.M. (2003). Contribution of the *Shigella flexneri* Sit, Iuc, and Feo iron acquisition systems to iron acquisition in vitro and in cultured cells. Infect. Immun. *71*, 1919–1928.
Schade, A.L., and Caroline, L. (1944). Raw hen egg white and the role of iron in growth inhibition of *Shigella dysenteriae, Staphylococcus aureus, Escherichia coli* and *Saccharomyces cerevisiae*. Science *100*, 14–15.
Schade, A.L., and Caroline, L. (1946). An iron-binding component in human blood plasma. Science *104*, 340–341.
Schmitt, M.P., and Payne, S.M. (1988). Genetics and regulation of enterobactin genes in *Shigella flexneri*. J. Bacteriol. *170*, 5579–5587.
Schmitt, M.P., and Payne, S.M. (1991). Genetic analysis of the enterobactin gene cluster in *Shigella flexneri*. J. Bacteriol. *173*, 816–825.
Schmitt, M.P., Twiddy, E.M., and Holmes, R.K. (1992). Purification and characterization of the diphtheria toxin repressor. Proc. Natl. Acad. Sci. U.S.A. *89*, 7576–7580.
Smajs, D., and Weinstock, G.M. (2001). The iron- and temperature-regulated *cjrBC* genes of *Shigella* and enteroinvasive *Escherichia coli* strains code for colicin Js uptake. J. Bacteriol. *183*, 3958–3966.
Strahsburger, E., Baeza, M., Monasterio, O., and Lagos, R. (2005). Cooperative uptake of microcin E492 by receptors FepA, Fiu, and Cir and inhibition by the siderophore enterochelin and its dimeric and trimeric hydrolysis products. Antimicrob. Agents Chemother. *49*, 3083–3086.
Struelens, M.J., Mondal, G., Roberts, M., and Williams, P.H. (1990). Role of bacterial and host factors in the pathogenesis of *Shigella* septicemia. Eur. J. Clin. Microbiol. Infect. Dis. *9*, 337–344.
Su, Y.C., Chin, K.H., Hung, H.C., Shen, G.H., Wang, A.H.J., and Chou, S.H. (2010). Structure of Stenotrophomonas maltophilia FeoA complexed with zinc: a unique

prokaryotic SH3-domain protein that possibly acts as a bacterial ferrous iron-transport activating factor. Acta Crystallogr. Sect. F Struct. Biol. Cryst. Commun. *66*, 636–642.

Svinarich, D.M., and Palchaudhuri, S. (1992). Regulation of the SLT-1A toxin operon by a ferric uptake regulatory protein in toxinogenic strains of *Shigella dysenteriae* type 1. J. Diarrhoeal Dis. Res. *10*, 139–145.

Tarr, P.I., Bilge, S.S., Vary, J.C., Jr, Jelacic, S., Habeeb, R.L., Ward, T.R., Baylor, M.R., and Besser, T.E. (2000). Iha: a novel *Escherichia coli* O157:H7 adherence-conferring molecule encoded on a recently acquired chromosomal island of conserved structure. Infect. Immun. *68*, 1400–1407.

Tigyi, Z., Kishore, A.R., Maeland, J.A., Forsgren, A., and Naidu, A.S. (1992). Lactoferrin-binding proteins in *Shigella flexneri*. Infect. Immun. *60*, 2619–2626.

Torres, A.G., and Payne, S.M. (1997). Haem iron-transport system in enterohaemorrhagic *Escherichia coli* O157:H7. Mol. Microbiol. *23*, 825–833.

Vokes, S.A., Reeves, S.A., Torres, A.G., and Payne, S.M. (1999). The aerobactin iron transport system genes in Shigella flexneri are present within a pathogenicity island. Mol Microbiol *33*, 63–73.

Waldminghaus, T., Heidrich, N., Brantl, S., and Narberhaus, F. (2007). FourU: a novel type of RNA thermometer in *Salmonella*. Mol. Microbiol. *65*, 413–424.

Wang, S., Wu, Y., and Outten, F.W. (2011). Fur and the novel regulator YqjI control transcription of the ferric reductase gene *yqjH* in *Escherichia coli*. J. Bacteriol. *193*, 563–574.

Weaver, E.A., Wyckoff, E.E., Mey, A.R., Morrison, R., and Payne, S.M. (2013). FeoA and FeoC are essential components of the *Vibrio cholerae* ferrous iron uptake system, and FeoC interacts with FeoB. J. Bacteriol. *195*, 4826–4835.

Wei, J., Goldberg, M.B., Burland, V., Venkatesan, M.M., Deng, W., Fournier, G., Mayhew, G.F., Plunkett, G., Rose, D.J., Darling, A., et al. (2003). Complete genome sequence and comparative genomics of *Shigella flexneri* serotype 2a strain 2457T. Infect Immun *71*, 2775–2786.

Weinberg, E.D. (1977). Infection and iron metabolism. Am. J. Clin. Nutr. *30*, 1485–1490.

Weinberg, E.D. (2000). Microbial pathogens with impaired ability to acquire host iron. Biometals *13*, 85–89.

Wilks, A. (2001). The ShuS protein of Shigella dysenteriae is a heme-sequestering protein that also binds DNA. Arch. Biochem. Biophys. *387*, 137–142.

Wookey, P., and Rosenberg, H. (1978). Involvement of inner and outer membrane components in the transport of iron and in colicin B action in Escherichia coli. J. Bacteriol. *133*, 661–666.

Wyckoff, E.E., Duncan, D., Torres, A.G., Mills, M., Maase, K., and Payne, S.M. (1998). Structure of the Shigella dysenteriae haem transport locus and its phylogenetic distribution in enteric bacteria. Mol. Microbiol *28*, 1139–1152.

Wyckoff, E.E., Lopreato, G.F., Tipton, K.A., and Payne, S.M. (2005). Shigella dysenteriae ShuS promotes utilization of heme as an iron source and protects against heme toxicity. J. Bacteriol. *187*, 5658–5664.

Xiong, K., Chen, Z., Xiang, G., Wang, J., Rao, X., Hu, F., and Cong, Y. (2012). Deletion of yncD gene in Salmonella enterica ssp. enterica serovar Typhi leads to attenuation in mouse model. FEMS Microbiol. Lett. *328*, 70–77.

Yang, F., Yang, J., Zhang, X., Chen, L., Jiang, Y., Yan, Y., Tang, X., Wang, J., Xiong, Z., Dong, J., et al. (2005). Genome dynamics and diversity of *Shigella species*, the etiologic agents of bacillary dysentery. Nucleic Acids Res *33*, 6445–6458.

Ye, C., and Xu, J. (2001). Prevalence of iron transport gene on pathogenicity-associated island of uropathogenic Escherichia coli in E. coli O157:H7 containing Shiga toxin gene. J. Clin. Microbiol. *39*, 2300–2305.

Zheng, M., Doan, B., Schneider, T.D., and Storz, G. (1999). OxyR and SoxRS regulation of fur. J Bacteriol *181*, 4639–4643.

Zhou, D., Hardt, W.D., and Galán, J.E. (1999). Salmonella typhimurium encodes a putative iron transport system within the centisome 63 pathogenicity island. Infect. Immun. *67*, 1974–1981.

3

Shigella and Antivirulence: the Dark Side of Bacterial Evolution

Kimberly A. Bliven and Anthony T. Maurelli*

Department of Microbiology and Immunology, F. Edward Hébert School of Medicine, Uniformed Services University of the Health Sciences, 4301 Jones Bridge Road, Bethesda, Maryland 20814, USA
*Corresponding author: Anthony T. Maurelli anthony.maurelli@usuhs.edu

Abstract

The acquisition of novel virulence factors and subsequent remodeling of bacteria systems to incorporate these factors are equally important events in pathogen evolution. Genes that are inactivated or lost from an emerging pathogen's genome due to incompatibility with new virulence factors are known as antivirulence genes (AVGs). AVGs were first described in *Shigella*, a human-specific gastrointestinal pathogen that evolved from ancestral *Escherichia coli*. Although AVGs have been described in a variety of bacterial pathogens, including *Salmonella*, *Yersinia*, *Burkholderia*, and *Francisella*, *Shigella* remains perhaps the best model organism for the study of antivirulence due to the vast extent of literature on not only *Shigella*, but also *E. coli*, which retains the functional AVGs that have been inactivated or lost in *Shigella* strains. Currently, there are five documented AVGs in *Shigella*: *cadA*, *nadA/nadB*, *speG*, and *ompT*. In this chapter, we will discuss the discoveries of these AVGs, their effects on virulence, and the events that led to their inactivation or loss in *Shigella*. The development of novel therapeutics and vaccines, an improved understanding of virulence mechanisms, and a greater insight into pathogen evolution are just some of the benefits that may arise from exploration of AVGs in bacterial pathogens.

Introduction

The acquisition of virulence genes via horizontal gene transfer can be crucial to the ability of bacteria to colonize novel niches. Extrachromosomal elements, such as plasmids or bacteriophages, or conjugation between strains to share genes or pathogenicity islands, can all contribute to the incorporation of new DNA into the bacterial genome (Groisman and Ochman, 1996; Lee, 1996). As a result, a pathogen may arise from a formerly harmless commensal, or a pathogen may shift from one specialized niche to another.

Given the complex nature of bacterial gene/protein regulation and function, newly-acquired virulence factors do not always integrate smoothly into established bacterial systems (Sokurenko et al., 1999; Maurelli 2007). Any functional conflict between these new virulence genes and the remainder of the bacterial genome may therefore compromise bacterial fitness. Moreover, colonization of a novel pathogenic niche can add new selective pressures for the bacterium to overcome. If the resulting fitness cost to the bacteria is too extreme, variant strains which avoid or bypass such functional conflict in the new niche will be selected for.

If a bacterium acquires virulence genes that conflict with an existing bacterial network, two outcomes are possible. In the first scenario, a bacterium may simply lose or inactivate the newly-acquired virulence genes. Bacteria that have not spontaneously lost or inactivated these genes could otherwise risk predation or being out-competed by other bacterial strains/species, and will therefore not survive to pass their own genes to the next generation. This concept is known as antagonistic pleiotropy, which states that any gene (or gene system) whose fitness costs outweigh the benefits will be selected against (Williams 1957).

Conversely, if the newly-acquired virulence genes confer significant fitness benefits to the bacterium, they will be more likely to be selected for, and existing genes may be subject to a 'fine-tuning' process to optimally accommodate the new genes. In this second scenario, the bacterium will maintain the newly acquired virulence genes and modify the existing genome through gene loss (gene inactivation/deletion) or positive selection for compensatory mutations in existing genes, a process known as pathoadaptive evolution (Sokurenko et al., 1999).

Optimization of bacterial virulence through pathoadaptive gene loss is a specialized form of pathoadaptive evolution known as antivirulence (Maurelli 1998; Maurelli 2007). An antivirulence gene (AVG) is a gene whose expression in a pathogen is incompatible with the virulence of that pathogen (Bliven and Maurelli, 2012). Consequently, virulence may be enhanced through gene inactivation or partial/full deletions, otherwise known as 'black holes.'

Shigella was the first organism in which AVGs were described. In the past two decades, however, AVGs in other bacterial pathogens have been identified (for a complete review, see Bliven and Maurelli, 2012). To date, there are over 10 known AVGs in at least four other bacterial pathogens, including *Yersinia*, *Salmonella*, *Burkholderia*, and *Francisella* (Erickson et al., 2008; Eroshenko et al., 2010; Eswarappa et al., 2009; Hager et al., 2006; Montminy et al., 2006; Moore et al., 2004; Rebeil et al., 2004; Sun et al., 2008). While this chapter focuses exclusively on AVGs of *Shigella*, studies describing AVGs in other bacterial species are certainly worthy of note and could be a gateway to discoveries of novel AVGs in *Shigella* and other pathogens. These studies also represent the very broad range of molecular functions that must be attenuated in these pathogens to permit virulence, including inactivation or loss of genes involved in biofilm synthesis, metabolism, lipopolysaccharide modification, and host vasoconstriction. This is an excellent reminder that the dynamics of bacterial fitness and pathogenicity are complex, and the breadth of AVGs mirrors that complexity.

Finally, the importance of discovering and describing these AVGs goes beyond merely satisfying academic curiosity and obtaining a better understanding of how these pathogens evolve. There are practical applications. Identification of AVGs may provide a basis for the development of novel therapeutics in the form of inhibitor compounds specific for pathogen virulence factors. With the rise of antimicrobial resistant pathogens, including multi-drug resistant *Shigella* isolates, the development of novel therapeutics is more important than ever (Ahmed et al., 2013; Gu et al., 2013; Yang et al., 2013). New drugs with a high specificity for pathogenic targets are desirable in that they circumvent disruption of commensal flora or the host's own cells. AVG study may also provide beneficial insight into bacterial pathogenicity by helping to uncover novel virulence factors or offer clues to the mechanisms behind already known virulence factors. Lastly, AVGs may aid in vaccine development, which has been experimentally demonstrated with a *Yersinia* AVG, *lxpL* (Montminy et al., 2006).

The evolutionary history of *E. coli* and *Shigella*

Despite boasting a distinct genus classification, *Shigella* is taxonomically grouped as a unique pathovar within *E. coli* (Pupo et al., 2000; Lan et al., 2004). In fact, bioinformatics-based evidence indicates that *Shigella* recently branched from ancestral *E. coli* strains between 35,000 to 270,000 years ago (Pupo et al., 2000). Acquisition of a large virulence plasmid, the defining evolutionary event for *Shigella* evolution, likely occurred up to seven independent times. This suggests that modern *Shigella* strains are descendants of not one but several different ancestral *E. coli* lineages.

In the process of evolution, *Shigella* strains rose to colonize a very unique, specific niche compared to the broader natural habitat that modern non-pathogenic *E. coli* isolates are known to populate, evolving from extracellular ancestors into intracellular pathogens capable of colonizing human intestinal epithelial cells (Figure 3.1; Pupo et al., 2000). This ability was attained primarily via lateral transfer of a large 220 kb virulence plasmid that houses genes which encode for a type three secretion system (T3SS) and various

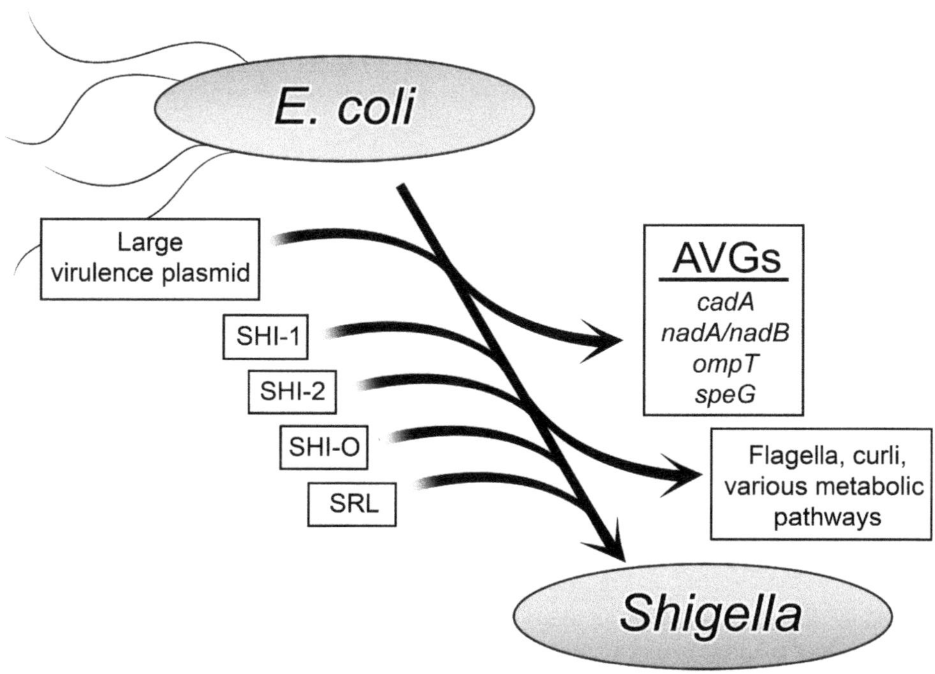

Figure 3.1. Evolution of *Shigella* from *E. coli*. Although *Shigella* acquired several characteristic virulence factors, including a large virulence plasmid and four pathogenicity islands (SHI-1, SHI-2, SHI-O, and SRL), it also lost AVGs (*cadA*, *nadA/nadB*, *ompT*, and *speG*) in addition to flagella, curli, and various metabolic pathways.

secreted effectors (Schroeder and Hilbi, 2006). Furthermore, several distinct *Shigella* pathogenicity islands (SHI-1, SHI-2, SHI-O, and SRL) were transferred to the chromosome, although their presence varies between strains (Peng et al., 2006). These genes encode virulence factors including an enterotoxin, an immunoglobulin A-like cytotoxic protease, a serine protease, a T-cell signaling suppressor, and molecular systems for aerobactin iron acquisition, O antigen modification, and ferric dicitrate uptake, in addition to antimicrobial resistance determinants (Schroeder and Hilbi, 2006). Various smaller plasmids may be present depending on the *Shigella* strain, and have been known to carry additional antibiotic resistance genes which may deliver an edge in bacterial survival within the human gut (Hartman et al., 2003; Chang et al., 2010). Lastly, *S. dysenteriae* serotype 1 strains carry the *stxA/stxB* genes encoding Shiga toxin, an AB_5 toxin capable of cleaving a single adenine residue from the 28S rRNA of the eukaryotic host cell and inhibiting host protein synthesis (Tesh and O'Brien, 1991).

Although modern *E. coli/Shigella* strains share a common core genome, there are obvious genetic and phenotypic differences that distinguish the two organisms. During adaptation to their highly-specific host niche, *Shigella* underwent substantial reductive evolution, and their genomes reveal a high frequency of chromosomal pseudogenization when compared to strains of non-pathogenic *E. coli* (Yang et al., 2005). Shortly following gain of the major virulence plasmid, *Shigella* likely started inactivating AVGs inhibitory to invasion and survival within the bacterium's new intracellular niche (Maurelli 2007). Probably more gradually, *Shigella* strains also began accumulating inactivating mutations in genes unnecessary for survival in the new niche (in other words, neutral genes which confer neither a positive nor negative selective advantage within human intestinal epithelial cells) (Moran, 2004). On average, there are over 200 pseudogenes per strain in the *Shigella* species, including many genes involved in carbon utilization, cell motility, and transporter/membrane proteins (Yang et al., 2005; Feng et al., 2011). In one study, 85 genes were found to be commonly decayed across five sequenced *Shigella* strains, with a rather staggering 1,456 genes inactivated in at least one strain, suggesting that the capacity for genome reduction is high (Feng et al., 2011). Insertion elements are a common form of gene disruption, and *Shigella* strains are rife with them, with anywhere between 300-700 copies of insertion sequence elements per strain (Yang et al., 2005).

It is impossible to discuss *Shigella* in the context of antivirulence without also mentioning enteroinvasive *E. coli* (EIEC), another human-specific pathogen that causes diarrhea in humans. EIEC share many genetic and phenotypic similarities with the shigellae, and are thought to be more recently emerged clones that harbor the same T3SS-encoding large virulence plasmid which allows them to inhabit the same intracellular niche (Lan et al., 2004). Although EIEC are not considered to be ancestral strains of the modern *Shigella* isolates, they may be following a parallel evolutionary pathway (Lan et al., 2004).

Like *Shigella*, EIEC strains are characteristically nonmotile and unable to ferment lactose, although these phenotypes are not as universal across EIEC strains as *Shigella* (Silva et al., 1980). In contrast to *Shigella*, however, EIEC strains share more similarities with commensal *E. coli* isolates, including the ability to ferment xylose and produce gas from glucose (Silva et al., 1980). Although we will not discuss EIEC in this chapter, it is important to note that many of the phenotypes resulting from loss of AVG in *Shigella*, including the inability to decarboxylate lysine, are also characteristic of EIEC strains (Casalino et al., 2003; Silva et al., 1980).

Shigella, the environment, and the human host

As an intracellular, host-specific pathogen that has undergone a substantial amount of reductive evolution, *Shigella* is probably less fit within the environment compared to environmental or commensal *E. coli*. While the majority of *E. coli* isolates are prototrophic and capable of surviving in a wide variety of niches, *Shigella* isolates are universally auxotrophic for nicotinic acid, and certain strains (particularly *S. dysenteriae* isolates) also require supplementation with tryptophan or methionine (Ahmed et al., 1988). Furthermore, *Shigella* strains universally lack a flagellar system and membrane-associated adhesions such as curli (Sakellaris et al., 2000; Tominaga et al., 2005; Mamun et al., 1996). These bacterial appendages may confer environmental survival advantages to *E. coli* strains through enhanced nutrient access or biofilm formation, respectively (Wei et al., 2011; Macarisin et al., 2012). The capacity of *Shigella* to remain virulent following prolonged periods outside its host is thought to be limited, which may be due to inactivation of genes found on the large virulence plasmid, or even loss of portions or all of the virulence plasmid. (Rahman et al., 2011; Faruque et al., 2002). Lastly, maintenance of such a large plasmid can be an energy drain on the bacterium, and plasmid loss may be selected for in *Shigella* isolates outside the host niche, as has been demonstrated during periods of nutrient starvation in *E. coli* (Caldwell et al., 1989).

Conversely, *Shigella* thrives in the replicative niche of the human epithelial cell cytoplasm (Schroeder and Hilbi, 2006). The plasmid-encoded T3SS is an essential requirement for initial entry and subsequent survival of *Shigella* within the human intestinal epithelium. It has diverse functions in epithelial cell invasion, bacterial replication, and cell-to-cell spread, and it also plays a crucial role in regulation of the host inflammatory response necessary for optimal bacterial survival (Schroeder and Hilbi, 2006). The T3SS of *Shigella* likely plays a role in not only virulence, but overall bacterial fitness of the pathogen within the human host.

Although we do not yet know if genes of the large virulence plasmid play any particular role(s) while *Shigella* is in the environment, the necessity of the T3SS within the human host is well-documented, and attenuation of the T3SS in this niche by any AVGs would be a significant impediment to bacterial fitness.

The known AVGs of *Shigella*

cadA

One of the primary factors used to distinguish strains of *Shigella* from *E. coli* in the clinical microbiology laboratory is the inability of the *Shigella* strains to decarboxylate lysine into the polyamine cadaverine (Figure 3.2). In *E. coli* strains, the *cadA* gene encodes for an acid-inducible lysine decarboxylase, which is transcribed in an operon with *cadB*, a gene which encodes for a lysine/cadaverine antiporter. This operon is induced by the transcriptional activator CadC, a member of the ToxR-like family of proteins, which binds the *cadBA* operon in the presence of exogenous lysine and an acidic external pH (Küper and Jung, 2005).

Over 90% of *E. coli* strains encode a functional lysine decarboxylase (Edwards and Erwing, 1972). In distinct contrast, *Shigella* strains have inactivated or even completely lost this function through a staggering variety of different mutations, including partial or full deletions of the *cadBA* operon, nonsense mutations, missense mutations, and the disruption of open reading frames by insertion elements, loci from other chromosomal regions, or even phage genomes (Day et al., 2001). This vast mutational diversity suggests that negative selective pressure against this locus was considerable in the

Shigella ancestor strains, and this locus required inactivation for bacterial fitness in the new niche. Convergent evolution is a common theme among the shigellae with respect to loss of AVGs. Like *Shigella*, EIEC strains are negative for lysine decarboxylase activity and do not produce cadaverine. Unlike *Shigella*, however, some EIEC strains have maintained a functional *cadA* gene (Casalino et al., 2003). The transcriptional activator *cadC* has instead been inactivated in most EIEC strains through open reading frame disruption with different insertion elements. Loss of CadC activity prevents expression of the *cadBA* operon and production of lysine decarboxylase. In the remaining EIEC strains, deletion of the *cadBA* operon has led to the lysine decarboxylase negative phenotype.

When a functional lysine decarboxylase system is restored in a wild type *Shigella* strain in *trans*, several different *Shigella* virulence phenotypes are attenuated (Figure 3.3; Fernandez et al., 2001; Maurelli et al., 1998; McCormick et al., 1999). The ShET1 and ShET2 enterotoxins, encoded on the bacterial chromosome and the large virulence plasmid, respectively, are thought to contribute to the initial phase of watery diarrhea during dysentery (Fasano et al., 1997). In the presence of a functional *cadA* from an *E. coli* K-12 strain, the ability of an *S. flexneri* strain to induce fluid secretion is blocked (Maurelli et al., 1998). Secondly, bacterial phagosome escape, an essential step for subsequent *Shigella* replication in the host cytoplasm, is prevented during *cadA* expression (Fernandez et al., 2001). Finally, polymorphonuclear leukocyte (PMN) influx,

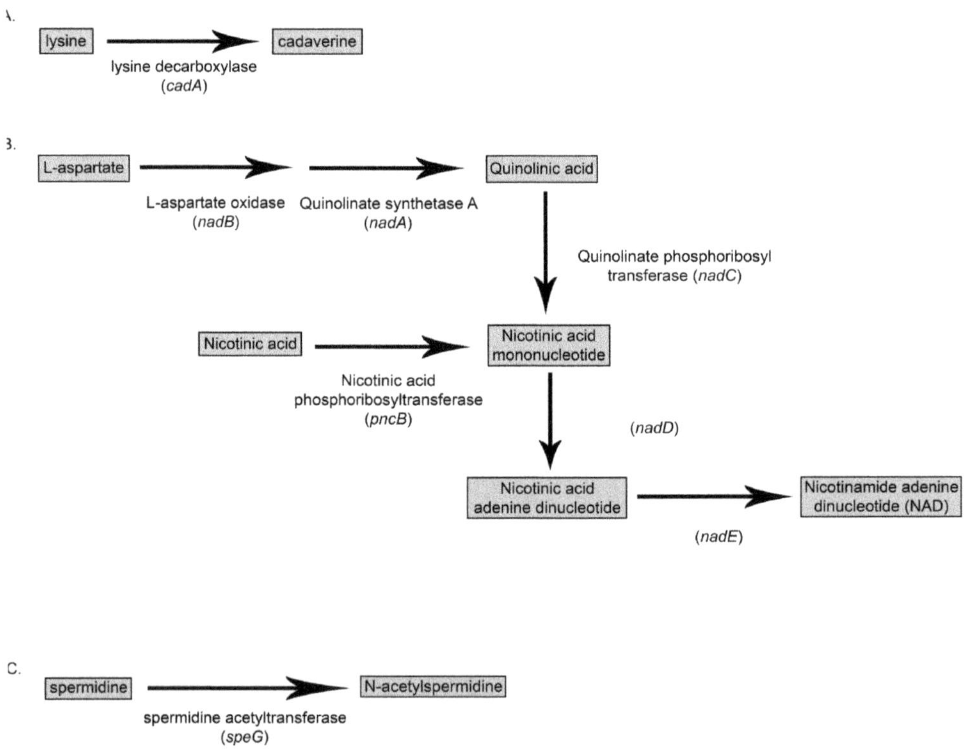

Figure 3.2. Metabolic pathways lost or interrupted in *Shigella*. A. Lysine decarboxylation; B. Biosynthetic and salvage NAD pathways; C. Spermidine metabolism.

an inflammatory event thought to be critical to allow optimal bacterial access to the preferred basolateral side of the intestinal epithelium, is attenuated (McCormick et al., 1999). Inhibition of these same phenotypes is also demonstrated after addition of a physiologically relevant concentration of cadaverine to a wild type *Shigella* strain. This result demonstrates that cadaverine, the end product of the lysine decarboxylase reaction, is an inhibitor of each of these virulence phenotypes.

A mechanism for *cadA*/cadaverine inhibition of *Shigella* pathogenesis has not yet been elucidated. Intriguingly, the *Shigella* virulence phenotypes inhibited are not thought to be linked, which suggests that cadaverine may play multiple roles in the attenuation of *Shigella* pathogenesis, possibly even protecting the host cell prior to *Shigella* exposure. Polyamines such as cadaverine play a wide variety of physiological roles in viral, bacterial, and mammalian cells, primarily due to their cationic nature. The net positive charge of the polyamine is distributed over the length of the molecule rather than at a single point, allowing polyamines to bind to or serve as bridges between negatively-charged molecules such as DNA or RNA (Shah and Swiatlo, 2008). Most research thus far has focused on polyamine activity in plants and mammals, but in bacterial pathogens, polyamines have been demonstrated to influence microbial carcinogenesis, macrophage

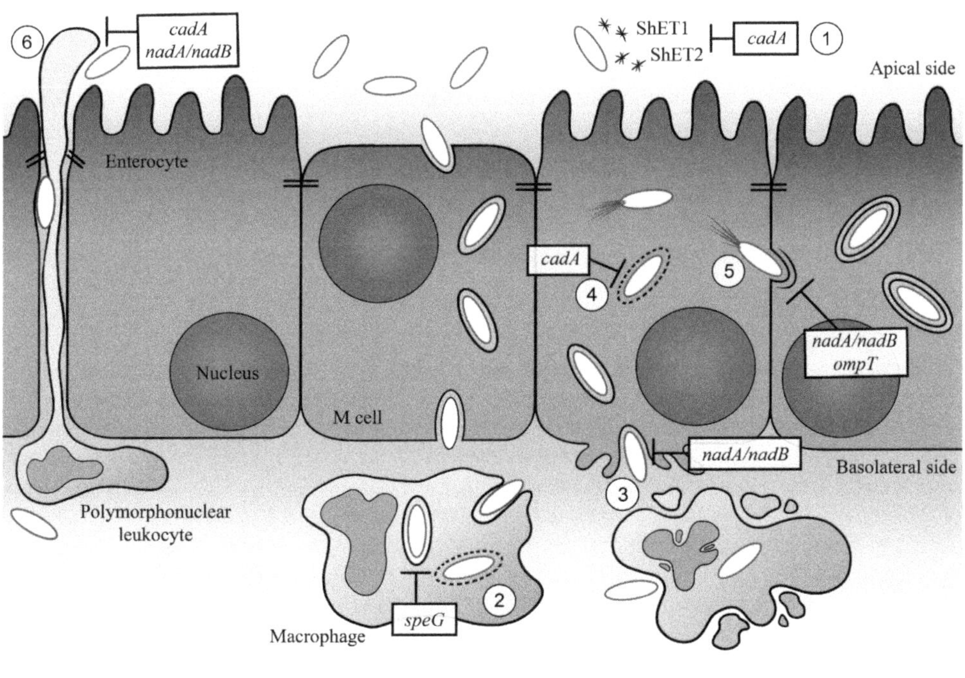

Figure 3.3. Inhibition of pathogenesis in *Shigella* by AVGs. During *Shigella* invasion into the human intestinal epithelium, AVGs inhibit various pathogenesis phenotypes. Cadaverine, the product of the lysine decarboxylase reaction, inhibits ShET1/ShET2 enterotoxin activity (1), phagosome escape (4), and PMN transepithelial migration (6). Quinolinic acid, the product of reactions encoded by the *nadA/nadB* genes, inhibits both *Shigella* invasion (3), intracellular spread (5) and PMN transepithelial migration (6).. Inactivation of the spermidine acetyltransferase *speG* results in the accumulation of spermidine within the phagosome and supports bacterial survival in the macrophage (2). Lastly, the outer membrane protease OmpT cleaves bacterial IcsA, ultimately attenuating actin tail polymerization and inhibiting cell-to-cell spread (5).

apoptosis, biofilm formation, siderophore biosynthesis, and oxidative stress response (Shah and Swiatlo, 2008).

As a curious side note, the Shigella/EIEC strains are not the only members of the *E. coli* pathovar to lose function of *cadA*. Torres and colleagues noted that many enterohemorrhagic *E. coli* (EHEC) and Shiga Toxin-producing *E. coli* (STEC) strains also lack a functional lysine decarboxylase gene (Torres et al., 2005; Vazqeuz-Juarez et al., 2008). When *cadA* is expressed in these strains, bacterial adherence to host cells is significantly decreased. This phenotype is likely due to a corresponding decrease in the expression of outer membrane adhesions such as intimin, although it is not yet known through what mechanism this reduction occurs. In contrast to *Shigella*, the end product cadaverine does not appear to play a role in intimin phenotype. However, as the inactivation or loss of *cadA* in these strains is not universal, *cadA* is not yet considered an AVG in these organisms.

nadA/nadB

Another feature that has long been known to distinguish *Shigella* strains from modern *E. coli* isolates is the almost universal inability of *Shigella* to grow on minimal salts media (Ahmed et al., 1988). 98% of clinical *Shigella* isolates are auxotrophic for nicotinic acid, and strains of *S. dysenteriae* are also auxotrophic for tryptophan and methionine (Ahmed et al., 1988). Although auxotrophy does exist in *E. coli*, many strains are prototrophic and capable of growing without supplementation (Schuetz and Strockbine, 2005). To determine whether or not these auxotrophic phenotypes are indicative of AVG loss, the inability of *Shigella* to synthesize nicotinic acid was studied (Prunier et al., 2007b).

The biosynthetic and salvage pathways for NAD in *E. coli* have been well-described (Figure 3.2; for a review, see Gazzaniga et al., 2009). The enzymes L-aspartate oxidase (encoded by *nadB*) and quinolinate synthetase A (encoded by *nadA*) form a complex that, in *E. coli*, converts L-aspartate to quinolinic acid. Quinolinic acid is then converted to nicotinic acid mononucleotide, one of the precursors of nicotinamide adenine dinucleotide (NAD). In the salvage pathway, bacteria can take up exogenous nicotinic acid, which is then converted to nicotinic acid mononucleotide by nicotinate phosphoribosyltransferase (encoded by *pncB*).

In a study of 12 *Shigella* strains, changes were revealed in the sequences of *nadA*, *nadB*, or both genes within all strains compared to the functional *nadA/nadB* sequences in an *E. coli* K-12 strain (Prunier et al., 2007a). Insertion element interruptions of open reading frames, point mutations, and frameshift mutations all have contributed to the nicotinic acid auxotrophy phenotype in these strains. Two EIEC isolates, known nicotinic acid auxotrophs, also demonstrated insertion element disruptions in the open reading frame of *nadB* and several point mutations within *nadA*. However, the *pncB* gene encodes for a functional transferase in *Shigella*, permitting conversion of exogenous nicotinic acid to nicotinic acid mononucleotide. By utilizing this salvage pathway, the pathogen can completely bypass the *nadA/nadB* biosynthetic pathway to satisfy its requirements for NAD (Gemski et al., 1971).

When a wild type *Shigella* strain is exposed to quinolinic acid, several virulence phenotypes are inhibited, including cell-to-cell spread, host cell invasion, and *Shigella*-induced transepithelial migration of PMNs (Figure 3.3; Prunier et al., 2007b). Similar to cadaverine, there is currently no known mechanism of action for quinolinic acid inhibition of *Shigella* virulence, although investigations remain underway.

Intriguingly, most *Shigella* species appear to retain a functional quinolinate phosphoribosyl transferase (encoded by *nadC*), which can convert quinolinic acid to nicotinic acid mononucleotide. Presumably, in the absence of quinolinic acid biosynthesis, a functional *nadC* may be required to deal with any exogenous quinolinic acid the bacterium encounters, and to neutralize this potentially inhibitory small molecule by metabolizing it into a useful form for the bacterium (Prunier et al., 2007b).

speG

An important role for the *Shigella* T3SS is to promote bacterial survival in and eventual escape from the human macrophage. During infection, *Shigella* is thought to transcytose through M cells into the Peyer's patches of the large intestine, where they are ingested by resident macrophages (Sansonetti et al., 1996). The bacteria quickly escape the phagosome and induce the macrophage to undergo pyroptosis, a specialized form of proinflammatory cell death characterized by caspase-1 activation (Suzuki et al., 2007). This leaves the bacteria free to invade the intestinal epithelium from the basolateral surface which is the more susceptible to invasion than the apical surface (Mounier et al., 1992). To infect its host and effectively replicate, *Shigella* must not only be able to induce pyroptosis in the macrophage, but also survive the hostile conditions of the macrophage phagosome, including reactive oxygen species involved in the oxidative stress response, such as hydrogen peroxide and nitric oxide (Slauch, 2011).

speG encodes for spermidine acetyltransferase, an enzyme which produces N-acetylspermidine through transfer of an acetyl group onto the polyamine spermidine (Figure 3.2). In *Shigella* strains, *speG* has been lost or inactivated via different variations of deletions, non-synonymous missense mutations, and disruptions of the open reading frame by insertion elements (Barbagello et al., 2011). Interestingly, the inactivation or loss of *speG* in *Shigella* is apparently not to prevent production of an inhibitory end product, but to promote buildup of the precursor molecule, spermidine (Barbagallo et al., 2011). In the absence of a functional *speG*, spermidine concentration in *Shigella* strains is up to three-fold higher than in *E. coli* isolates which express *speG*. When a functional *speG* is expressed in *Shigella*, bacteria are less able to withstand the effects of oxidative stress as measured by their survival in the presence of hydrogen peroxide. The expression of a functional *speG* also significantly reduces the ability of *Shigella* to survive within murine macrophages, suggesting that *speG* expression increases *Shigella* sensitivity to the oxidative stress response within the macrophage (Figure 3.3).

Spermidine is one of four major polyamines (putrescine, cadaverine, spermidine, and spermine) found in bacterial cells (Shah and Swiatlo, 2008). Spermidine has been attributed a variety of roles in bacteria, including regulation of gene expression, porin regulation through protein binding, and free radical scavenging (Shah and Swiatlo, 2008). Although the mechanism of action for *speG* antivirulence is not yet fully elucidated, expression of *katG*, a gene encoding hydroperoxidase I, is significantly repressed in the presence of functional *speG*. KatG, part of the bacterial oxidative stress defense system, protects against oxidative stress through conversion of hydrogen peroxide to water and oxygen, and is transcriptionally regulated by the polyamines putrescine and spermidine in *E. coli* (Jung and Kim, 2003). Therefore, loss of the AVG *speG* increases spermidine accumulation and correspondingly induces expression of *katG*, ultimately inhibiting macrophage killing of *Shigella* (Barbagallo et al., 2011).

ompT
One of the hallmarks of *Shigella* pathogenesis is the ability of the bacteria to spread from one epithelial cell to another by hijacking host actin filaments (Egile et al., 1999). IcsA (also known as VirG), induces actin-based motility by binding the host N-WASP protein, which in turn interacts with host F-actin and functions in attachment of the actin tail to the bacterium (Egile et al., 1999).

ompT encodes for an outer membrane protease that cleaves IcsA, effectively attenuating cell-to-cell spread in *Shigella* (Figure 3.3; Nakata et al., 1993). However, not all modern *E. coli* strains carry the *ompT* gene, which makes classification of *ompT* as an AVG difficult. In *E. coli*, *ompT* sits in the midst of a DLP12 cryptic lambdoid-like prophage downstream of the *envY* gene. In a survey of 144 strains in the EcoCyc database (http://ecocyc.org/), 25% of *E. coli* strains do not carry the prophage or *ompT*, and none of the modern *Shigella* isolates studied even carry remnants of the prophage in this position (Keseler et al., 2011; Bliven and Maurelli, 2012). Therefore, we do not know if the *Shigella* ancestral lineages ever carried *ompT*. For that reason, *ompT* is the most ambiguous gene on the current list of *Shigella* AVG, although at the present time, we include *ompT* as a *Shigella* AVG.

Identification of *Shigella* AVG
Identification of novel AVGs remains a slow and rather arduous process. Bioinformatics analyses, while improving, still present significant challenges. Genome annotation for pseudogenes is mediocre at best, and even obvious pseudogenes (such as partial/full gene deletions and premature nonsense mutations) may not be annotated as such. Furthermore, a pseudogene that results from a non-synonymous missense mutation can usually only be documented as non-functional through experimental characterization. An ideal bioinformatics program should be able to predict structural changes resulting from such missense mutations and determine the consequence on protein function. Continuous improvement of bioinformatics platforms is critical to enhance the identification and study of AVGs.

If a putative AVG target is identified, it must then be investigated in the context of the entire network to which it belongs. Many genes are part of larger biosynthetic pathways or bacterial systems, and when one gene is inactivated, the entire pathway or system may be nonfunctional or interrupted. This is especially true for organisms such as *Shigella*, which may still be in the process of genome pseudogenization of unnecessary genes. Furthermore, the same gene within a pathway or system is not always preferentially targeted for inactivation in all strains, suggesting that similar selective pressures may have led to convergent evolution. As an example, although *nadA* and/or *nadB* is always inactivated in *Shigella*, some strains appear to maintain one of the two genes in a functional state (Prunier et al., 2007a). Loss of only one of these two enzymes is sufficient to prevent quinolinic acid from being produced. If only *nadA* had been examined across all *Shigella* strains, this particular AVG locus may have been missed. The decaying flagellar system in *Shigella* is also an excellent example (Tominaga et al., 2005; Mamun et al., 1996). In some isolates, certain flagellar genes, such as *fliC*, are able to fully complement *E. coli* mutants, demonstrating functionality of these genes (Tominaga et al., 1994). However, other genes within the flagellar system (such as the *flhDC* operon) have been inactivated or lost, and *Shigella* strains demonstrate a universally non-motile phenotype (Mamun et al., 1996).

Even when pseudogenes are properly identified, the scientist must then determine whether a gene was inactivated or lost because it was unnecessary to bacterial survival in the new niche, or because it was an AVG that inhibited bacterial virulence. Currently, this distinction is impossible to make without re-introduction of the putative AVG into the bacterium and subsequent characterization of virulence phenotypes in the presence or absence of the AVG.

Concluding remarks

Following lateral acquisition of novel virulence factors, emerging pathogens must optimally incorporate these genes into their existing genome. This is accomplished through loss of genes inhibitory to virulence in the new niche, otherwise known as antivirulence genes (AVGs). *Shigella* is an excellent model organism for the study of AVGs because of the high incidence of pseudogenes present in *Shigella* strains, the wealth of knowledge available for both the pathogenicity and basic biology of *Shigella* and its extant cousin, *E. coli*, and the accessibility of an increasing number of sequenced *Shigella* genomes. The discovery of AVGs in *Shigella* and other bacterial pathogens may help pave the way for the creation of highly pathogen-specific therapeutics, the unearthing of novel virulence targets, and the development of new vaccines.

Acknowledgements

Research on *Shigella* pathogenesis in the Maurelli Lab is supported by grant AI24656 from the National Institute of Allergy and Infectious Diseases. The opinions or assertions contained herein are the private ones of the authors and are not to be construed as official or as reflecting the views of the Department of Defense or the Uniformed Services University.

References

Ahmed, S.F., Klena, J., Husain, T., Monestersky, J., Naquib, A., and Wasfy, M.O. (2013). Genetic characterization of antimicrobial resistance of *Shigella flexneri* 1c isolates from patients in Egypt and Pakistan. Ann. Clin. Microbiol. Antimicrob. *12*, 9.

Ahmed, Z.U., Sarker, M.R., and Sack, D.A. (1988). Nutritional requirements of shigellae for growth in a minimal medium. Infect. Immun. *56*, 1007-1009.

Barbagallo, M., Martino, M.L., Marcocci, L., Pietrangeli, P., Carolis, E.D., Casalino, M., Colonna, B., and Prosseda, G. (2011). A new piece of the *Shigella* pathogenicity puzzle: spermidine accumulation by silencing of the *speG* gene. PLoS One. *6*, e27226.

Caldwell, B.A., Ye, C., Griffiths, R.P., Moyer, C.L., and Morita, R.Y. (1989). Plasmid expression and maintenance during long-term starvation-survival of bacteria in well water. App. Environ. Microbiol. *55*, 1860-1864.

Casalino, M., Latella, M.C., Prosseda, G., and Colonna, B. (2003). CadC is the preferential target of a convergent evolution during enteroinvasive *Escherichia coli* toward a lysine decarboxylase-defective phenotype. Infect. Immun. *71*, 5472-5479.

Chang, C., Lu, P., Lin, C., Lee, T., Tsai, M., and Chang, L. (2010). Integron types, gene cassettes, antimicrobial resistance genes and plasmids of *Shigella sonnei* isolates from outbreaks and sporadic cases in Taiwan. J. Med. Microbiol. *60*, 197-204.

Day, W.A. Jr, Fernández, R.E., and Maurelli, A.T. (2001). Pathoadaptive mutations that enhance virulence: genetic organization of the *cadA* regions of *Shigella* spp. Infect. Immun. *69*, 7471-7480.

Edwards, P.R., and Erwing, W.H. (1972). Identification of Enterobacteriaceae: 3rd Edition. (Minneapolis, MN: Burgess Press).

Egile, C., Loisel, T.P., Laurent, V., Li, R., Pantaloni, D., Sansonetti, P.J., and Carlier, M. (1999). Activation of the Cdc42 effector N-Wasp by the *Shigella flexneri* IcsA protein promotes actin nucleation by Arp2/3 complex and bacterial actin-based motility. J. Cell. Biol. *146*, 1319-1332.

Erickson, D.L., Jarrett, C.O., Callison, J.A., Fischer, E.R., and Hinnebusch, B.J. (2008). Loss of a biofilm-inhibiting glycosyl hydrolase during the emergence of *Yersinia pestis*. J. Bacteriol. *190*, 8163-8170.

Eroshenko, G.A., Vidyaeva, N.A., Kutyrev, V.V. (2010). Comparative analysis of biofilm formation by main and nonmain subspecies of *Yersinia pestis* strains. FEMS Immunol. Med. Microbiol. *59*, 513-520.

Eswarappa, S.M., Karnam, G., Nagarajan, A.G., Chakraborty, S., and Chakrovortty, D. (2009). *lac* repressor is an antivirulence factor of *Salmonella enterica*: its role in the evolution of virulence in *Salmonella*. PLoS One. *4*, e5789.

Faruque, S.M., Khan, R., Kamruzzaman, M., Yamasaki, S., Ahmad, Q.S., Azim, T., Nair, G.B., Takeda, Y., and Sack, D.A. (2002). Isolation of *Shigella dysenteriae* type 1 and *S. flexneri* strains from surface waters in Bangladesh: comparative molecular analysis of environmental *Shigella* isolates versus clinical strains. Appl. Environ. Microbiol. *68*, 3908-13.

Fasano, A., Noriega, F.R., Liao, F.M., Wang, W., and Levine, M.M. (1997). Effect of *Shigella* enterotoxin 1 (ShET1) on rabbit intestine *in vitro* and *in vivo*. Gut. *40*, 505-511.

Feng, Y., Chen, Z., and Liu, S. (2011). Gene decay in *Shigella* as an incipient stage of host-adaptation. PLOS One. *6*, e27754.

Fernandez, I.M., Silva, M., Schuch, R., Walker, W.A., Siber, A.M., Maurelli, A.T., and McCormick, B.A. (2001). Cadaverine prevents the escape of *Shigella flexneri* from the phagolysosome: a connection between bacterial dissemination and neutrophil transepithelial signaling. J. Infect. Dis. *184*, 743-753.

Gazzaniga, F., Stebbins, R., Chang, S.Z., McPeek, M.A., and Brenner, C. (2009). Microbial NAD metabolism: lessons from comparative genomics. Microbiol. Mol. Biol. Rev. *73*, 529-541.

Gemski, P. Jr., Formal, S.B., and Baron, L.S. (1971). Identification of two widely separated loci conferring nicotinic acid dependence on wild-type *Shigella flexneri* 2a.Infect. Immun. *3*, 500-503.

Groisman, E.A., and Ochman, H. (1996). Pathogenicity islands: bacterial evolution in quantum leaps. Cell. *87*, 791-794.

Gu, B., Ke, X., Pan, S., Cao, Y., Zhuang, L., Yu, R., Qian, H., Liu, G., and Tong, M. (2013). Prevalence and trends of aminoglycoside resistance in *Shigella* worldwide, 1999-2013. J. Biomed. Res. *27*, 103-115.

Hager, A.J., Bolton, D.L., Pelletier, M.R., Brittnacher, M.J., Gallagher, L.A., Kaul, R., Skerrett, S.J., Miller, S.I., and Guina, T. (2006). Type IV pili-mediated secretion modulates *Francisella* virulence. Mol. Microbiol. *62*, 227-237.

Hartman, A.B., Essiet, I.I., Isenbarger, D.W., and Lindler, L.E. (2003). Epidemiology of tetracycline resistance determinants in *Shigella* spp. and Enteroinvasive *Escherichia coli*: characterization and dissemination of *tet*(A)-1. J. Clin. Microbiol. *41*, 1023-1032.

Jeong, H.J., Jang, E.S., Han, B.I., Lee, K.H., Ock, M.S., Kong, H.H., Chung, D.I., Seol, S.Y., Cho, D.T., and Yu, H.S. (2007). *Acanthamoeba*: Could it be an environmental host of *Shigella*? Exper. Parasitol. *115*, 181-6.

Jung, I.L., and Kim, I.G. (2003). Transcription of *ahpC*, *katG*, and *katE* genes in *Escherichia coli* is regulated by polyamines: polyamine-deficient mutant sensitive to H_2O_2-induced oxidative damage. Biochem. Biophys. Res. Com. *301*, 915-922.

Keseler, I.M., Collado-Vides, J., Santos-Zavaleta, A., Peralta-Gil, M., Gama-Castro, S., Muniz-Rascado, L., Bonavides-Martinez, C., Paley, S., Krummenacker, M., Altman, T., Kaipa, P., Spaulding, A., Pacheco, J., Latendresse, M., Fulcher, C., Sarker, M., Shearer,

A.G., Mackie, A., Paulsen, I., Gunsalus, R.P., and Karp, P.D. (2011). EcoCyc: a comprehensive database of *Escherichia coli* biology. Nucl. Acids Res. *39*, D583-90.

Küper, C., and Jung, K. (2005). CadC-mediated activation of the *cadBA* promoter in *Escherichia coli*. J. Mol. Microbiol. Biotech. *10*, 26-39.

Lan, R., Alles, M.C., Donohoe, K., Martinez, M.B., and Reeves, P.R. (2004). Molecular evolutionary relationships of enteroinvasive *Escherichia coli* and *Shigella* spp. Infect. Immun. *72*, 5080-5088.

Lee, C.A. (1996). Pathogenicity islands and the evolution of bacterial pathogens. Infect. Agents Dis. *5*, 1-7.

Macarisin, D., Patel, J., Bauchan, G., Giron, J.A., and Sharma, V.K. (2012). Role of curli and cellulose expression in adherence of *Escherichia coli* 0157:H7 to spinach leaves. Foodborne pathogens and disease. *9*, 160-167.

Mamun, A.M., Tominaga, A., and Enomoto, M. (1996). Detection and characterization of the flagellar master operon in the four *Shigella* subgroups. J. Bacteriol. *178*, 3722-3726.

Maurelli, A.T. (2007). Black holes, antivirulence genes, and gene inactivation in the evolution of bacterial pathogens. FEMS Microbiol. Lett. *267*, 1-8.

Maurelli, A.T., Fernández, R.E., Bloch, C.A., Rode, C.K., and Fasano, A. (1998). "Black holes" and bacterial pathogenicity: A large genomic deletion that enhances the virulence of *Shigella* spp. and enteroinvasive *Escherichia coli*. Proc. Natl. Acad. Sci. U.S.A. *95*, 3943-3948.

McCormick, B.A., Fernandez, M.I., Siber, A.M., and Maurelli, A.T. (1999). Inhibition of *Shigella*-induced transepithelial migration of polymorphonuclear leucocytes by cadaverine. Cell Microbiol. *1*, 143-155.

Montminy, S.W., Khan, N., McGrath, S., Walkowicz, M.J., Sharp, F., Conlon, J.E., Fukase, K., Kusumoto, S., Sweet, C., Miyake, K., Akira, S., Cotter, R.J., Goguen, J.D., and Lien, E. (2006). Virulence factors of *Yersinia pestis* are overcome by a strong lipopolysaccharide response. Nat. Immunol. *7*, 1066-1073.

Moore, R.A., Reckseidler-Zenteno, S., Kim, H., Nierman, W., Yu, Y., Tuanyok, A., Warawa, J., DeShazer, D., and Woods, D.E. (2004). Contribution of gene loss to the pathogenic evolution of *Burkholderia pseudomallei* and *Burkholderia mallei*. Infect. Immun. *72*, 4172-4187.

Moran, N.A. (2004). Microbial minimalism: Genome reduction in bacterial pathogens. Cell. *108*, 583-586.

Mounier, J., Vasselon, T., Hellio, R., Lesourd, M., and Sansonetti, P.J. (1992). *Shigella flexneri* enters human colonic Caco-2 epithelial cells through the basolateral pole. Infect. Immun. *60*, 237-248.

Nakata, N., Tobe, T., Fukuda, I., Suzuki, T., Komatsu, K., Yoshikawa, M., and Sasakawa, C. (1993). The absence of a surface protease, OmpT, determines the intercellular spreading ability of *Shigella*: the relationship between the *ompT* and *kcpA* loci. Mol. Microbiol. *9*, 459-468.

Peng, J., Xiaobing, Z., Yang, J., Wang, J., Yang, E., Bin, W., Wei, C., Sun, M., and Jin, Q. (2006). The use of comparative genomic hybridization to characterize genome dynamics and diversity among the serotypes of *Shigella*. BMC Genomics. *7*, 218.

Pupo, G.M., Lan, R., and Reeves, P.R. (2000). Multiple independent origins of *Shigella* clones of *Escherichia coli* and convergent evolution of many of their characteristics. Proc. Natl. Acad. Sci. U.S.A. *97*, 10567-10572.

Prunier, A.L., Schuch, R., Fernández, R.E., and Maurelli, A.T. (2007). Genetic structure of the *nadA* and *nadB* antivirulence loci in *Shigella* spp. J. Bacteriol. *189*, 6482-6486.

Prunier, A.L., Schuch, R., Fernández, R.E., Mumy, K.L., Kohler, H., McCormick, B.A., and Maurelli A.T. (2007). *nadA* and *nadB* of *Shigella flexneri* 5a are antivirulence loci responsible for the synthesis of quinolinate, a small molecule inhibitor of *Shigella* pathogenicity. Microbiol. *153*, 2363-2372.

Rahman, M.Z., Azmuda, N., Hossain, M.J., Sultana, M., Khan, S.I., and Birkeland, N. (2011). Recovery and characterization of environmental variants of *Shigella flexneri* from surface water in Bangladesh. Curr. Microbiol. *63*, 372-376.

Rebeil, R., Ernst, R.K., Gowen, B.B., Miller, S.I., and Hinnebusch, B.J. (2004). Variation in lipid A structure in the pathogenic yersiniae. Mol. Microbiol. *52*, 1363-1373.

Saeed, A., Abd, H., Edvinsson, B., and Sandström, G. (2009). *Acanthamoeba castellanii* an environmental host for *Shigella dysenteriae* and *Shigella sonnei*. Arch. Microbiol. *191*, 83-88.

Sakellaris, H., Hannink, N.K., Rajakumar, K., Bulach, D., Hunt, M., Sasakawa, C., and Adler, B. (2000). Curli loci of *Shigella* spp. Infect. Immun. *68*, 3780-3783.

Sansonetti, P.J., Arondel, J., Cantey, J.R., Prevost, M.C., and Huerre, M. (1996). Infection of rabbit Peyer's patches by *Shigella flexneri*: effect of adhesive of invasive bacterial phenotypes on follicle-associated epithelium. Infect. Immun. *64*, 2752-2764.

Schroeder, G.N., and Hilbi, H. (2008). Molecular pathogenesis of *Shigella* spp.: controlling host cell signaling, invasion, and death by type III secretion. Clin. Microbiol. Rev. *21*, 134-156.

Schuetz, F., and Strockbine, N. (2005). Genus I. *Escherichia*. In Bergey's Manual of Systemic Bacteriology: Second Edition, Volume 2, G.M. Garrity, ed. (New York, NY: Spring Press), pp. 607-624.

Shah, P., and Swiatlo, E. (2008). A multifaceted role for polyamines in bacterial pathogens. Mol. Microbiol. *68*, 4-16.

Silva, R.M., Toledo, M.R, and Trabulsi, L.R. (1980). Biochemical and cultural characteristics of invasive *Escherichia coli*. J. Clin. Microbiol. *11*, 441-444.

Slauch, J.M. (2011). How does the oxidative burst of macrophages kill bacteria? Still an open question. Mol. Microbiol. *80*, 580-583.

Sokurenko, E.V., Hasty, D.L., and Dykhuizen, D.E. (1999). Pathoadaptive mutations: gene loss and variation in bacterial pathogens. Trends Microbiol. *7*, 191-195.

Sun, Y.C., Hinnebusch, B.J., and Darby, C. (2008). Experimental evidence for negative selection in the evolution of a *Yersinia pestis* pseudogene. Proc. Natl. Acad. Sci. U.S.A. *105*, 8097-8101.

Suzuki, T., Franchi, L., Toma, C., Ashida, H., Ogawa, M., Yoshikawa, Y., Mimuro, H., Inohara, N., Sasakawa, C., and Nuñez, G. (2007). Differential regulation of caspase-1 activation, pyroptosis, and autophagy via IpaF and ASC in *Shigella*-infected macrophages. PLoS Path. *3*, e111.

Tesh, V.L., and O'Brien, A.D. (1991). The pathogenic mechanisms of Shiga toxin and the Shiga-like toxins. Mol. Microbiol. *5*, 1817-1822.

Tominaga, A., Lan, R., and Reeses, P.R. (2005). Evolutionary changes of the *flhDC* flagellar master operon in *Shigella* strains. J. Bacteriol. *187*, 4295-4302.

Tominaga. A., Mahmoud, M.A., Mukaihara, T., and Enomoto, M. (1994). Molecular characterization of intact, but cryptic, flagellin genes in the genus *Shigella*. Mol. Microbiol. *12*, 277-285.

Torres, A.G., Vazquez-Juarez, R.C., Tutt, C.B., and Garcia-Gallegos, J.G. (2005). Pathoadaptive mutation that mediates adherence of Shiga Toxin-producing *Escherichia coli* O111. Infect. Immun. *73*, 4766-4776.

Vazqeuz-Juarez, R.C., Kuriakose, J.A., Rasko, D.A., Ritchie, J.M., Kendall, M.M., Slater, T.M., Sinha, M., Luxon, B.A., Popov, V.L., Waldor, M.K., Sperandio, V., and Torres, A.G. (2008). CadA negatively regulates *Escherichia coli* O157:H7 adherence and intestinal colonization. Infect. Immun. *76*, 5072-5081.

Wei, Y., Wang, X., Liu, J., Nememan, I., Singh, A.H., Weiss, H., and Levin, B.R. (2011). The population dynamics of bacteria in physically structured habitats and the adaptive virtue of random motility. Proc. Natl. Acad. Sci. U.S.A. *108*, 4047-4052.

Williams, G.C. (1957). Pleiotropy, natural selection, and the evolution of senescence. Evolution. *11*, 398-411.

Yang. F., Yang, J., Zhang, X., Chen, L., Jiang, Y., Yan, Y., Tang, X., Wang, J., Xiong, Z., Dong, J., Xue, Y., Zhu, Y., Xu, X., Sun, L., Chen, S., Nie, H., Peng, J., Xu, J., Wang, Y., Yuan, Z., Wen, Y., Yao, Z., Shen, Y., Qiang, B., Hou, Y., Yu, J., and Jin, Q. (2005). Genome dynamics and diversity of *Shigella* species, the etiologic agents of bacillary dysentery. Nucl. Acids Res. *33*, 6445-6458.

Yang, H., Chen, G., Zhu, Y., Liu, Y., Cheng, J., Hu, L., Ye, Y., and Li, J. (2013). Surveillance of antimicrobial susceptibility patterns among *Shigella* species isolated in China during the 7-year period of 2005-2011. Ann. Lab. Med. *33*, 111-115.

4

The Genetic Organization and Transcriptional Regulation of *Shigella* Virulence Genes

Natasha Weatherspoon-Griffin, Michael A. Picker and Helen J. Wing

School of Life Sciences, University of Nevada, Las Vegas, Las Vegas, NV 89154-4004. USA, helen.wing@unlv.edu

Abstract

Central to bacterial pathogenicity is the precise and coordinated control of virulence gene expression in response to environmental cues encountered in the human host. This chapter focuses on the transcriptional regulation of *Shigella* virulence genes encoded by the large virulence plasmid, pINV, found in almost all *Shigella* species. We describe the silencing of pINV-encoded virulence genes by the chromosomally encoded nucleoid structuring protein H-NS, which serves as the backdrop for all other regulatory events on this plasmid. We then describe the four-tiered virulence regulatory cascade emphasizing the environmental stimuli, transcriptional regulators and mechanisms that govern virulence gene expression at each tier. Finally, we review other regulatory inputs that are received from the *Shigella* chromosome and their role in modulating virulence gene expression. We begin by describing the genetic organization of *Shigella*, its relationship to *Escherichia coli* and the different forms of pINV carried by *Shigella* species to highlight how the regulatory cascades controlling *Shigella* virulence are, in large part, conserved across species. Overall, this chapter reveals that *Shigella* is a fascinating model for the study of virulence gene regulation, which promotes our understanding of *Shigella* pathogenesis and mechanisms of virulence gene regulation in other bacterial pathogens.

Introduction

Shigella species embark on a remarkable journey to reach the colonic epithelium, their preferred site of invasion in their human and primate hosts. *Shigella* infections are spread by the fecal-oral route and are generally acquired directly by ingestion of contaminated food or water, or indirectly via person-to-person contact. As Shigellae pass through the mouth, stomach and upper and lower intestines, they encounter antibacterial compounds that form part of our immune system and a wide variety of environmental conditions that threaten bacterial survival. Despite these challenges, reports indicate that the infectious dose of *Shigella* is extremely low (ID_{50} of 10-500; (DuPont et al., 1989)) compared to other enteric pathogens (eg. $ID_{50}>10^5$ for *Salmonella* spp.), demonstrating that *Shigella* species are particularly adept at surviving their journey through the gastrointestinal tract and causing an infection. This observation strongly suggests that *Shigella* strains continually monitor their environment and alter their gene expression to allow them to tolerate, resist and/or exploit conditions encountered in the human body, and indeed all evidence suggests that this is the case.

While it is challenging to devise experiments that accurately measure gene expression within the human host (reviewed in Marteyn et al., 2012), careful research has identified some of the most important cues that are likely sensed *in vivo* and how they are responded to by modulation of gene expression. Some of these cues include changes in temperature (Maurelli et al., 1984), osmolarity (Porter and Dorman, 1994) and pH (Nakayama and Watanabe, 1995), the change in freely available iron and oxygen (Marteyn et al., 2010; Murphy and Payne, 2007) and even the activation of the *Shigella* type III secretion system once direct contact with epithelial cells has been made (Demers et al., 1998; Kane et al., 2002; Mavris et al., 2002a). It is reasonable to expect that the bacterial response to these cues directly correlates with the temporo-spatial expression of specific virulence factors required for infection. And, although difficult to assess, it is even conceivable that the order and duration of these cues may help shape the physiology of *Shigella* so that this bacterium is optimally programmed for its life within the human host.
If we are to more thoroughly understand the molecular pathogenesis of *Shigella* spp., a more complete understanding of the environmental cues and genetic networks controlling gene expression in these pathogens is required. In this chapter, our primary focus is the regulatory events that control the transcription of virulence genes carried by the large virulence plasmids found in *Shigella* species. We describe regulatory processes that have allowed the acquisition and maintenance of the virulence plasmid, the regulatory cascade that triggers virulence gene expression on the large plasmid and additional inputs received via some chromosomally encoded regulators. But we first describe the genetic organization of *Shigella* species, so that the reader can better appreciate how our discussion relates to the different strains and species of *Shigella*.

Classification and genetic organization of *Shigella*
In the 1940s, *Shigella* strains were classified as a unique genus because of their pathogenicity and then divided into four species: (1) *S. dysenteriae;* (2) *S. boydii*; (3) *S. sonnei;* and (4) *S. flexneri* (Bensted, 1956; Ewing, 1949, 1986). As early as 1972, however, it was estimated that *Shigella* and *Escherichia coli* were taxonomically indistinguishable at the species level based on DNA hybridization studies (Brenner et al., 1972). More recently, a variety of studies have confirmed that all *Shigella* strains, except possibly *S. boydii* serotype 13, are close phylogenetic relatives of *E. coli* (Ochman et al., 1983; Pupo et al., 1997; Pupo et al., 2000; Rolland et al., 1998). Consequently, the *Shigella* genus is considered a classic example of a "taxon in disguise," meaning that, although originally classified as a separate genus, its evolutionary history reveals that it evolved from another taxonomic unit (Rediers et al., 2004), in this case *E. coli,* which makes *E. coli* paraphyletic to *Shigella* (Lan and Reeves, 2002).

The chromosome of the *Shigella* spp. is approximately the same size (4.3-4.9 Mb) as the *E. coli* K-12 chromosome (4.6 Mb) and is typically more than 97% identical at the DNA sequence level. Genome sequencing, however, has revealed several unique genetic islands on the *Shigella* chromosome, as well as several *E. coli* genes missing from the *Shigella* chromosome (described in depth for *S. flexneri* 2a; (Wei et al., 2003)). Additional analyses reveal that the closest relatives to the four *Shigella* "species" are the enteroinvasive *E. coli* (EIEC) (Lan et al., 2004; Lan and Reeves, 2002; Pupo et al., 1997; Pupo et al., 2000). The primary reason for the clustering of EIEC with the *Shigella* species is the presence of a common 220 kb plasmid (Lan and Reeves, 2002). This large virulence plasmid is essential for the invasive phenotype of these bacteria, and has collectively been given the name pINV, although specific names are given to pINV plasmids residing in different strains (eg. pWR100 of *S. flexneri* serotype 5, pCP301 of *S. flexneri* 2a and pSS120 of *S. sonnei*). Two major forms of the virulence plasmid, referred to as pINV A and pINV B, have been identified based on the DNA sequence of three genes encoded by pINV (*ipgD, mxiC* and *mxiA*) (Lan and Reeves, 2002). Interestingly, the distribution of the two forms correlates well with the variation in chromosomal housekeeping genes (reviewed in Lan et al., 2004; Lan and Reeves, 2002). All *Shigella* spp. contain either form of the plasmid, with the exception of *S. dysenteriae* serotype 1, which contains a recombinant plasmid with a divergent sequence not corresponding to these two forms, and the *S. boydii* serotype 13 strain, which does not contain a plasmid (Table 4.1). All but two EIEC strains analyzed so far carry pINV A (Lan et al., 2004; Pupo et al., 2000).

When the coding regions (those not associated with insertion sequences) of pINV were compared across three *S. flexneri* strains carrying either pINV A or pINV B, numerous polymorphisms were found in many of the virulence genes encoding effector proteins (Lan et al., 2003). These included genes encoding proteins secreted by the *Shigella* type III secretion system (eg. *ipaA, ipaH9.8*, and *ipaD*) and those that enable *Shigella* to move by actin-based motility within the host cell cytoplasm (eg. *icsA* and *virK*) (Lan et al., 2003). The polymorphisms found in these genes frequently led to amino acid substitutions when pINV A was compared to pINV B, but were silent when found on separate examples of the same form of plasmid (Lan et al., 2003). Significantly though, the genes encoding

[1]Table 4.1. Clustering of *Shigella* species and serotypes based on the type of virulence plasmid carried.

Cluster	Strain	Virulence plasmid carried (pINV A or B)
***Shigella* cluster 1**	*S. dysenteriae* serotypes 3-7, 9, 11-13	A
	S. boydii serotypes 1-4, 6, 8, 10, 14, 18	A
	S. flexneri serotype 6 & 6a	A
***Shigella* cluster 2**	*S. boydii* serotypes 5, 7, 9, 15-17	A
	S. boydii serotype 11	B
	S. dysenteriae serotype 2	A
***Shigella* cluster 3**	*S. flexneri* 1-5 and subtypes	B
	S. boydii serotype 12	B
Outliers	S. *sonnei*	Either A or B
	S. *dysenteriae* serotypes 8 & 10	Either A or B
	S. *dysenteriae* serotype 1	Distinct form of pINV
	S. *boydii* serotype 13	No plasmid detected

[1]Adapted from phylogenetic trees that were constructed based on the sequence of three genes in the *ipa-mxi-spa* genetic locus on pINV (*ipgD*, *mxiC* and *mxiA*) (Lan and Reeves, 2002).

transcriptional regulators carried by pINV (*virF, virB, mxiE* and *ipgC*) were found to display far fewer polymorphisms between pINV A and pINV B, and those that were identified were silent mutations (Lan et al., 2003). This strong conservation of the virulence gene regulators strongly suggests that they function in a similar manner in the vast majority of *Shigella* strains regardless of whether they are encoded by pINV A or pINV B. However, it should be remembered that polymorphisms in DNA binding sites recognized by these regulators may lead to different patterns of gene regulation from one strain to another.

In addition to pINV, some *Shigella* strains bear additional plasmids. For instance, during the sequencing of *S. flexneri* serotype 2a genome, strain 2457T, two additional small multi-copy plasmids and a 165 kb plasmid called pSf-R27 were found (Wei et al., 2003). The largest plasmid, which resembled the R27 plasmid typically associated with *S. enterica* serovar Typhi, has not been found in any other *S. flexneri* strains sequenced to date, raising the possibility that this plasmid was either lost from all other *S. flexneri* strains or was acquired shortly after this particular strain was isolated (Wei et al., 2003). Later in this chapter, the impact of a regulator (Sfh) encoded by this plasmid will be discussed in relation to its regulation of virulence genes in 2457T (Beloin et al., 2003).

Despite the presence of pINV, virulence-associated proteins in *Shigella* spp. are also encoded by its chromosome. In general, these genes reside within pathogenicity islands (SHI-1, SHI-2, SHI-3, SHI-O, and SRL) and encode proteins required for adaptation and maintenance within a host (eg. iron acquisition and antibiotic resistance). In contrast, the genes located on pINV typically encode proteins required for invasion of host cells and intercellular spread (reviewed in Schroeder and Hilbi, 2008). Since efficient invasion and infection of *Shigella* spp. requires the timely and coordinated expression of pINV genes, a complex regulatory network involving regulators encoded by both the chromosome and the virulence plasmid has evolved to control the transcription of pINV genes.

The acquisition of any plasmid by a bacterial strain poses a problem of regulation (discussed in Dorman and Porter, 1998) because the newly acquired plasmid genes need to be woven into the existing regulatory networks found within the bacterial genome. In the case of the acquisition of pINV, it appears that the genes encoding regulators specific to pINV were also encoded by the plasmid. But surprisingly, the role of at least some of these regulators appears to have been the relief of transcriptional silencing (repression) mediated by chromosomally encoded nucleoid structuring proteins, which allowed the acquisition and subsequent incorporation of plasmid genes into the already present transcriptional control networks. Of course, with additional time and co-evolution of the plasmid and the chromosome, additional chromosomal inputs are likely to have evolved to help control the expression of the *Shigella* virulence plasmid genes, as is observed today.

Having discussed the genetic composition of *Shigella* strains, our description of the transcriptional regulation of virulence genes can begin. Owing to the complexity of the regulatory system in *Shigella*, this chapter will focus solely on the transcriptional regulation of virulence genes encoded by pINV. We will: (1) describe the silencing of genes on the pINV plasmid by chromosomally encoded and accessory plasmid-encoded nucleoid structuring proteins; (2) provide an overview of the regulatory cascade that triggers virulence gene expression on pINV; (3) describe each of the pINV-encoded regulators that belong to this regulatory cascade, including events that trigger their expression, the proteins themselves and the effect that each has on their immediate downstream targets; and finally (4) discuss additional regulatory inputs made by chromosomally encoded regulators.

Transcriptional silencing of virulence genes encoded by pINV

The virulence genes encoded by pINV remain transcriptionally silent under non-physiological conditions (osmolarities lower than physiological, pH below 7.4 and temperatures below 37°C) (Hromockyj et al., 1992; Maurelli et al., 1984; Maurelli and Sansonetti, 1988; Porter and Dorman, 1994, 1997a). This is advantageous for *Shigella* because inappropriate expression of these genes has been shown to reduce fitness of the organism *ex vivo* (Dorman, 2007; Schuch and Maurelli, 1997). In *Shigella,* like many other enteric bacteria (Lucchini et al., 2006; Navarre et al., 2005), a nucleoid structuring protein called H-NS (histone-like nucleoid structuring protein) prevents the expression of genes on pINV in a process that has been coined xenogeneic silencing (Hromockyj et al., 1992; Maurelli and Sansonetti, 1988 and reviewed in Stoebel et al., 2008). As mentioned before, this silencing is thought to have played an important role in the acquisition and maintenance of pINV (Stoebel et al., 2008), but also provides the context in which all other virulence gene regulation on pINV occurs.

Transcriptional silencing of virulence genes on pINV occurs at multiple levels within the regulatory cascade controlling *Shigella* virulence. Under non-physiological conditions, H-NS directly represses *virF* and *virB,* which encode key transcriptional factors required for the upregulation of pINV genes (see below; Adler et al., 1989; Falconi et al., 1998; Prosseda et al., 2004; Prosseda et al., 1998; Tobe et al., 1991; Tobe et al., 1993), but H-NS also directly represses genes that lie within the VirF and VirB regulons (Beloin and Dorman, 2003; Hromockyj et al., 1992; Porter and Dorman, 1994; Prosseda et al., 2004; Wing et al., 2004). The intimate role that H-NS plays in virulence gene regulation in *Shigella* is further demonstrated by the observation that VirB and, in some cases, VirF function to counteract H-NS-mediated repression rather than functioning as classical activators of transcription (Le Gall et al., 2005; Stoebel et al., 2008; Wing et al., 2004). For this reason, we start our discussion of the transcriptional regulation of virulence genes on pINV by describing the H-NS protein and what is known about its silencing of gene expression. Since *Shigella* strains harbor one (StpA) or sometimes two H-NS paralogues (StpA and Sfh), for the sake of completeness, these proteins are also discussed because under certain circumstances these protein can influence virulence gene regulation in *Shigella* (Beloin et al., 2003; Deighan et al., 2003), and it is possible that these proteins play a direct, yet so far poorly understood, role in the transcriptional silencing of virulence genes residing on pINV.

A. H-NS

H-NS (also referred to as H1 and H1a) was first described in the early 1970s (Cukier-Kahn et al., 1972; Jacquet et al., 1971). Its ability to modulate DNA supercoiling and condense DNA *in vivo* and *in vitro* in a manner similar to eukaryotic histones (Spassky et al., 1984; Spurio et al., 1992) led to its name – the histone-like nucleoid structuring protein, H-NS (Lammi et al., 1984). H-NS, however, does not share sequence homology with histones, so more recently, it has been referred to as a nucleoid associated or nucleoid structuring protein (NAP or NSP) (Atlung and Ingmer, 1997).

H-NS is commonly found throughout the Enterobacteriaceae (Tendeng and Bertin, 2003). While its characterization has primarily been done in *E. coli* K-12 strains, the H-NS protein encoded by *Shigella* spp. is 100% identical at the amino acid level to that produced by *E. coli*. Furthermore, the *hns* genetic locus is also very highly conserved

(99.9%; from the beginning of the *galU* gene to the end of the *tdk* gene, which flank *hns*), suggesting that the expression profiles of *hns* in *E. coli* and *Shigella* are similar. H-NS is a 15 kDa, chromosomally encoded protein that is highly expressed in exponential growth (Deighan et al., 2003; Free and Dorman, 1995), accumulating to approximately 20,000 monomers per cell in stationary phase (Spassky et al., 1984). H-NS can bind to RNA, but it is primarily a DNA binding protein with a preference for curved DNA, which is commonly AT-rich (Brescia et al., 2004; Dame et al., 2001; Prosseda et al., 2004; Tolstorukov et al., 2005; Yamada et al., 1990). In the γ-Proteobacteria, AT-rich DNA sequences are often a hallmark of foreign genes acquired by lateral or horizontal gene transfer (Ochman et al., 2000). Consequently, the newly acquired AT-rich virulence genes located either within chromosomal pathogenicity islands and/or on virulence plasmids of many important pathogens, including *Shigella* (reviewed in Stoebel et al., 2008), are often negatively regulated or silenced by H-NS. While a high affinity DNA binding site for H-NS has been proposed (Bouffartigues et al., 2007; Lang et al., 2007), this site cannot always be found in the vicinity of genes repressed by H-NS (Basta et al., 2013; Gordon et al., 2011). This is most likely because H-NS binds to remotely located, high affinity sites and subsequently spreads along the DNA via interactions with the minor groove to occupy the promoter region (Fang and Rimsky, 2008; Gordon et al., 2011).

The H-NS protein consists of two structured domains separated by a flexible linker (Dorman, 2004; Tendeng and Bertin, 2003). The N-terminal domain contains the major dimerization domain and a region required for oligomerization, while the purified C-terminal domain confers DNA binding (Arold et al., 2010; Dorman et al., 1999; Sette et al., 2009; Williams and Rimsky, 1997; Williams et al., 1996; Wolf et al., 2006). While H-NS is primarily a dimer in solution (Bloch et al., 2003), its ability to form higher order oligomers on DNA (Arold et al., 2010; Badaut et al., 2002; Esposito et al., 2002; Renzoni et al., 2001) has the potential to lead to the formation of a variety of H-NS:DNA complexes (Amit et al., 2003; Dame et al., 2000; Lim et al., 2014; Lim et al., 2012; Maurer et al., 2009).

So far, two DNA binding modes of H-NS have been implicated in the silencing of transcription (Figure 4.1) – a rigid nucleoprotein filament which stiffens the DNA (Amit et al., 2003; Lim et al., 2012) and a DNA-bridging mode (Dame et al., 2005; Dame et al., 2000) where multiple H-NS dimers bind to and crosslink two distal regions of DNA, a process which is proposed to facilitate plectoneme formation (Arold et al., 2010; Rimsky and Travers, 2011). The switch between the nucleoprotein filament and the DNA bridging molecule is poorly understood *in vivo,* but can be controlled by the concentration of divalent cations *in vitro* (Liu et al., 2010). Classically, nucleoprotein filaments have been proposed to prevent transcription by either blocking the access of RNA polymerase to essential promoter elements or functioning as a roadblock in the path of an elongating RNA polymerase ((reviewed in Lim et al., 2012); Figure 4.1a and b). In contrast, DNA bridging by H-NS has classically been proposed to either occlude or trap RNA polymerase at or from target gene promoters ((Dame et al., 2002; Lang et al., 2007); Figure 4.1d and e). Examples of each of these DNA binding modes impacting gene silencing and DNA packaging have been described (Dame et al., 2002; Lang et al., 2007; Lim et al., 2012; Walthers et al., 2011). More recently, however, a more sophisticated view of gene silencing by these two DNA binding modes has been proposed (Lim et al., 2014). In the first, the formation of a rigid H-NS:DNA filament downstream of an active promoter is proposed to increase the level of torque within the DNA helix, causing RNA polymerase to stall during elongation ((Lim et al., 2014); Figure 4.1c). In the second, the diffusion of supercoils generated during transcription elongation (Liu and Wang, 1987) is blocked by plectonemes (Ma et al., 2013) that form as a result of H-NS:DNA bridging, which in turn, leads to the transcriptional silencing of a nearby promoter (Figure 4.1f) because RNA polymerase is inhibited by the newly supercoiled state of the promoter. These scenarios are interesting because if levels of DNA supercoiling can be indirectly modulated or constrained by H-NS, it is tempting to speculate that transcription factors and/or environmental factors that counter these changes may allow the derepression of H-NS-repressed virulence genes.

Although it is clear that H-NS plays a major role in the silencing of *Shigella* virulence genes encoded by pINV, the complete picture of where H-NS binds on pINV, how its binding pattern is remodeled, what triggers this remodeling and how this impacts gene expression remains poorly understood. Furthermore, even though six mechanisms of H-NS-mediated repression have now been proposed ((Dame et al., 2002; Lang et al., 2007; Lim et al., 2012); Figure 4.1a-f), for the most part it remains unclear which mechanisms occur where, or how the resulting repression is relieved to facilitate virulence gene expression in *Shigella*. These topics are the focus of ongoing investigations.

B. Impact of H-NS paralogues StpA and Sfh

Studies of H-NS-mediated repression of genes encoded by pINV is complicated by the fact that all *Shigella* strains carry the chromosomally encoded, paralogous gene *stpA* (Zhang and Belfort, 1992; Zhang et al., 1996) and that *S. flexneri* strain 2457T carries another *hns* paralogue called *sfh* on an R27-like accessory plasmid found exclusively in this strain (Beloin et al., 2003; Deighan et al., 2003). Both StpA and Sfh share significant amino acid homology with H-NS (58% and 59% identity, respectively) (Beloin et al., 2003). The domain structure of these three proteins is very similar (Deighan et al., 2003)

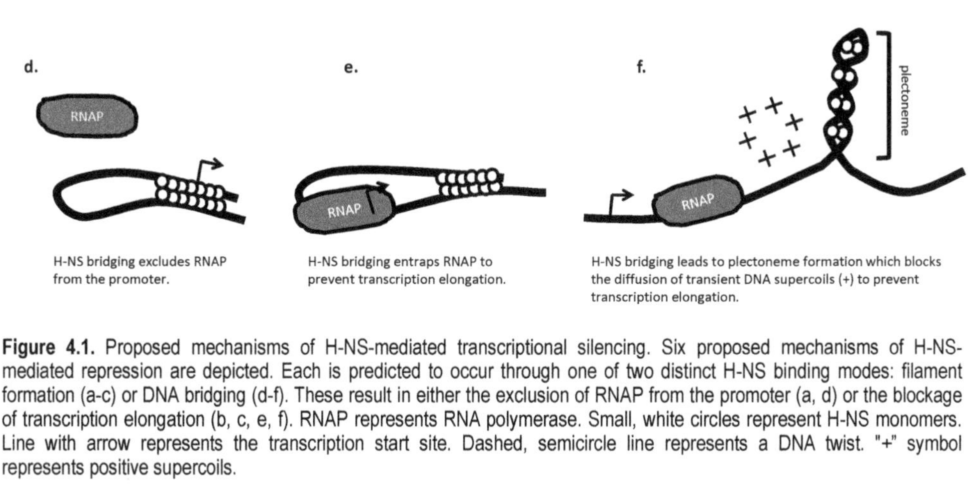

Figure 4.1. Proposed mechanisms of H-NS-mediated transcriptional silencing. Six proposed mechanisms of H-NS-mediated repression are depicted. Each is predicted to occur through one of two distinct H-NS binding modes: filament formation (a-c) or DNA bridging (d-f). These result in either the exclusion of RNAP from the promoter (a, d) or the blockage of transcription elongation (b, c, e, f). RNAP represents RNA polymerase. Small, white circles represent H-NS monomers. Line with arrow represents the transcription start site. Dashed, semicircle line represents a DNA twist. "+" symbol represents positive supercoils.

and to some extent, they appear to share overlapping functional activities with H-NS (discussed more below) (Free et al., 1998; Johansson et al., 1998; Williams et al., 1996; Zhang and Belfort, 1992; Zhang et al., 1996). In addition, these proteins not only auto-repress their own promoters, but also repress the promoters of each other (Deighan et al., 2003). Taken together, these observations have led some to propose that StpA and Sfh function, at least in part, as a back-up system for H-NS (Beloin et al., 2003; Wolf et al., 2006; Zhang et al., 1996).

The negative cross-regulation and the apparent functional redundancy of H-NS and its paralogues have the potential to impact studies of *Shigella* virulence gene expression under certain circumstances. Clean deletions of *hns* can lead to the upregulation of *stpA* and *sfh* (Deighan et al., 2003), which encode proteins that can functionally substitute for H-NS (Beloin et al., 2003; Deighan et al., 2003), thereby masking the effect of the *hns* deletion. To circumvent some of these problems, dominant negative alleles, rather than clean deletions of *hns*, are commonly used by those studying virulence gene expression in *Shigella* (Williams et al., 1996; Wing et al., 2004; Wolf et al., 2006; Yamada et al., 1991). This approach relies on the observation that H-NS, StpA and Sfh not only form homodimers, but can also heterodimerize with each other (Beloin et al., 2003; Deighan et al., 2003). The most commonly used dominant negative forms of H-NS or its paralogues are truncated proteins consisting solely of the N-terminal dimerization domain of the protein (Williams et al., 1996; Wing et al., 2004; Yamada et al., 1991). These proteins retain their ability to heterodimerize with the other H-NS paralogues, thereby significantly hindering the ability of the other protein(s) to bind to DNA and functionally substitute for H-NS. The result is a significant loss of the paralogues' ability to repress transcription. Although it is common for cells bearing dominant negative alleles of *hns* to grow more slowly than wild type cells, their growth is far less impacted than cells that lack all *hns* paralogues (Beloin et al., 2003), which are prohibitively sick. Therefore, even though dominant negative alleles of *hns* can aid studies of H-NS-mediated repression of virulence genes in *Shigella,* it is imperative to interpret data collected carefully and to fully understand the nature of the *hns* mutant in use. Similarly, the tendency for this family of proteins to form heterodimers means that additional care needs to be taken when using inducible systems to express *hns* or its paralogues at super-physiological concentrations, since heterodimer formation may lead to anomalous and/or non-physiological results.

Although functional similarities between the H-NS paralogues exist, these proteins are not completely redundant with one another. For instance, the expression patterns of the three paralogues differ from one another (Deighan et al., 2003; Free and Dorman, 1995, 1997). While H-NS levels remains fairly constant throughout growth in rich broth, StpA levels increase during mid-exponential growth phase and then drop precipitously, whereas Sfh levels increase by 2.5-fold as the cells enter stationary phase (Deighan et al., 2003). The relative abundance of each of these proteins in the cell has the potential to influence the gene regulation mediated by these NSPs, although this possibility has yet to be explored. In addition, at certain genetic loci the regulatory activities of H-NS and StpA differ from one another (Deighan et al., 2000; Suzuki et al., 1996). For example, both StpA and H-NS regulate OmpF porin expression by downregulating the sRNA *micF*, however H-NS negatively regulates the transcription of *micF,* whereas StpA primarily destabilizes *micF* RNA (Deighan et al., 2000). Thus, it is clear that these proteins can serve independent and distinct roles as regulators, which makes sense given their widespread conservation in the genomes of *Shigella* and other enteric species. Finally, since all three NSPs have been shown to form heteromeric complexes with each other (Deighan et al., 2003), it is possible that each of the different heterodimer combinations have different activities, further increasing the potential impact that NSPs play in regulating *Shigella* virulence genes (Cusick and Belfort, 1998; Deighan et al., 2003; Dorman et al., 1999; Free et al., 1998; Williams et al., 1996). Clearly, there is still much to be learned about the impact that H-NS, StpA and Sfh have on the virulence genes residing on pINV. However, we do know that a combination of environmental cues (Falconi et al., 1998; Maurelli et al., 1984; Maurelli and Sansonetti, 1988; Porter and Dorman, 1994) and the regulatory cascade controlling virulence gene expression on pINV (described next; Figure 4.2) is required to overcome transcriptional silencing mediated by H-NS (Beloin and Dorman, 2003; Dorman and Porter, 1998).

Transcriptional control mediated by the pINV-encoded regulators

A. An overview of the pINV-encoded regulatory cascade

The regulatory cascade that controls virulence gene expression on pINV is encoded by pINV and is hierarchical in nature (depicted in Figure 4.2). One of the major roles that this regulatory cascade plays is to counteract transcriptional silencing of pINV genes mediated by H-NS. Each level of the hierarchy (referred to as "tiers" in this discussion) responds to a set of environmental cues encountered in the human host (Dorman and Porter, 1998; Le Gall et al., 2005). It has been proposed that each tier is required to prevent the untimely expression of virulence determinants, which is metabolically expensive and may negatively influence the competitive fitness of the organism (Lucchini et al., 2006; Porter and Dorman, 1997a).

In tier 1, the cascade begins with activation of the *virF* promoter (described in more detail in later sections; (Adler et al., 1989; Falconi et al., 1998; Falconi et al., 2001; Prosseda et al., 2004; Prosseda et al., 1998)). As determined by northern analyses, expression of *virF* can be induced when grown in standard LB medium (Miller, 1972) by the following physiological cues: temperature (37°C, when compared to cells grown at 30°C), pH (7.4, when compared to cells grown at pH 6.0) or osmolarity (physiological i.e. LB, when compared to cells grown in LO, a lower osmolarity growth medium); with maximal expression occurring when each of these environmental cues are present (Porter and Dorman, 1997a).

In tier 2, the VirF protein binds its downstream targets to upregulate the transcription of *virB* (Jost and Adler, 1993; Tobe et al., 1991; Tobe et al., 1993) and *icsA* ((Bernardini et al., 1989; Lett et al., 1989; Sakai et al., 1988; Tran et al., 2011); Figure 4.2). Unlike *virF, virB* expression can be induced by 37°C and physiological osmolarity, but is not influenced by changes in pH, based on northern analyses (Porter and Dorman, 1997a).
In tier 3, the tier 2 gene product VirB binds to multiple downstream targets (tier 3) to counteract H-NS mediated transcriptional repression of various virulence-associated genes ((Adler et al., 1989; Dorman et al., 2001; Tobe et al., 1991; Turner and Dorman, 2007; Uchiya et al., 1995; Wing et al., 2004); Figure 4.2)). Many of the genes regulated by VirB lie within the 30 kb *ipa-mxi-spa* pathogenicity island, while others are found elsewhere on pINV (Castellanos et al., 2009; Santapaola et al., 2002; Uchiya et al., 1995; Wing et al., 2004). Contrary to *virF* and *virB*, tier 3 genes in the *ipa-mxi-spa* locus are only thermally induced (Porter and Dorman, 1997a); they neither respond to changes in pH nor osmolarity, based on northern analyses (Porter and Dorman, 1997a). The VirB protein is also involved in a positive regulatory feedback loop at both the *virF* and *virB* promoters ((Kane and Dorman, 2012); Figure 4.2). This finding demonstrates that VirB is not only capable of directly regulating genes in tier 3 and indirectly regulating genes in tier 4 via its control of *mxiE* expression, but is capable of upregulating the first two tiers of the regulatory cascade too.

In tier 4, MxiE and its co-activator, IpgC, regulate the transcription of genes encoding additional type III effector proteins and other post-invasion functions (Kane et al., 2002; Le Gall et al., 2005; Mavris et al., 2002a; Mavris et al., 2002b). Interestingly, not all tier 4 genes are located on pINV; some are located on the *Shigella* chromosome (Bongrand et al., 2012; Le Gall et al., 2005; Mavris et al., 2002b). Unlike tiers 1-3, which respond to environmental signals, tier 4 genes are expressed once the type III secretion system is activated (Demers et al., 1998; Mavris et al., 2002a; Mavris et al., 2002b). Since *mxiE* mutants are still invasive in tissue culture experiments (Kane et al., 2002), MxiE-regulated genes appear to encode a set of secreted proteins needed for post-invasion events related to virulence, including those that manipulate the host cell to promote bacterial

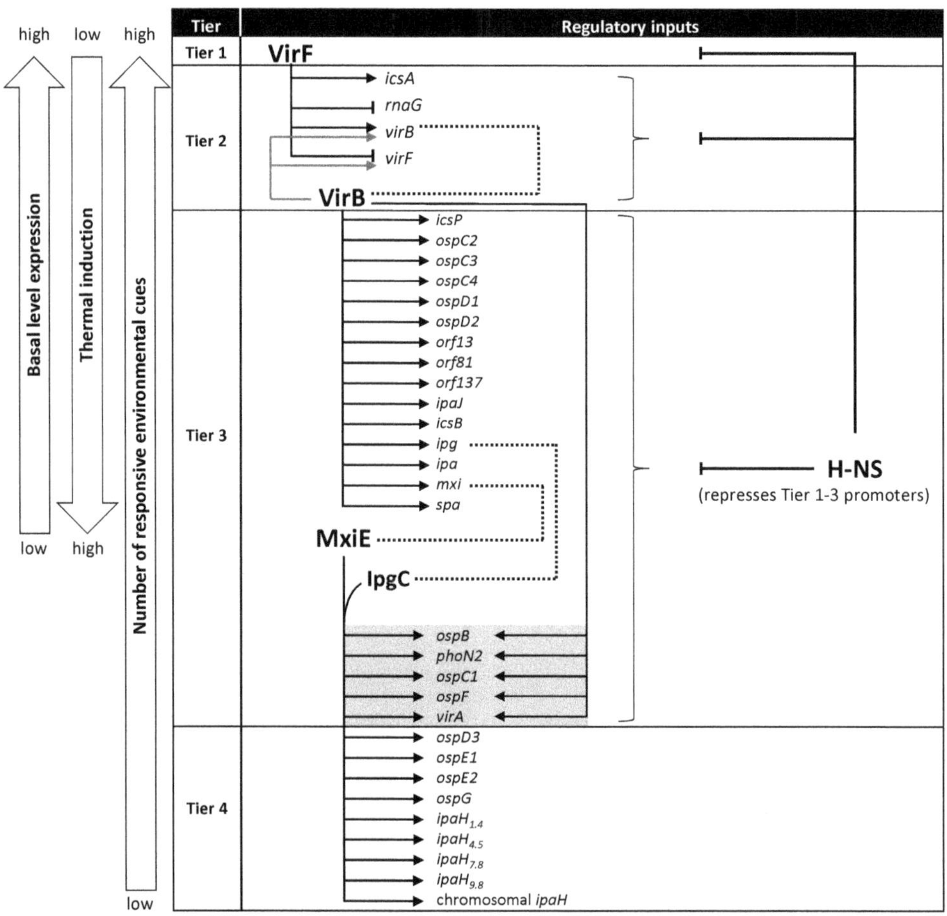

Figure 4.2. Overview of the regulatory cascade controlling *Shigella* virulence. The four tiers of the regulatory cascade that control gene expression on pINV are depicted along with relative levels of gene expression and external regulatory inputs from H-NS. Arrows indicate positive regulation while lines with a terminal, perpendicular line indicate negative regulation. Grey arrows indicate the positive feedback input of VirB. Dotted lines represent protein production by translation. Grey box indicates promoters that are regulated by both VirB and MxiE.

survival within the host cell cytoplasm (Sperandio et al., 2008). Two sub-classes of MxiE-regulated genes have been discovered; those regulated by MxiE alone and those regulated by VirB under conditions of non-secretion and further upregulated by MxiE under conditions of secretion ((Le Gall et al., 2005); Figure 4.2). Thus, based on the hierarchy presented in this chapter, the latter group can be considered tier 3 genes that receive additional regulatory inputs from the tier 4 regulator.

From this overview, it is clear that the four tiers of the regulatory cascade in *Shigella* have distinct levels of stringent and complex regulatory control. In a 1997 study where the first 3 tiers were compared to one another (Porter and Dorman, 1997a), tier 3 promoters were found to be more tightly controlled than tier 2, and tier 2 promoters were more tightly controlled than tier 1. The varying degrees of stringent control were found to manifest in multiple ways: (1) basal level expression, tier 1 > tier 2 > tier 3; (2) level of thermal induction, tier 3 (induced 100-fold) > tier 2 (induced 10-fold) > tier 1 (induced 2-fold); and (3) response to multiple environmental cues (discussed above), tier 1 > tier 2 > tier 3 ((Porter and Dorman, 1997a); Figure 4.2). Tier 4 expression levels were not measured in these studies as MxiE was not discovered or described until later (Kane et al., 2002; Mavris et al., 2002b). Nevertheless, it is clear that MxiE-regulated genes are also prone to stringent control due to their distal position in the regulatory cascade, their induction by a single cue (i.e., type III secretion activation) and the dual regulation of at least a subset of these genes by VirB.

At first glance, the position of VirB in the hierarchy under the control of VirF suggests a less significant role for VirB (Figure 4.2). However, it has been argued that this position allows VirB to serve as a regulatory checkpoint to ensure that appropriate conditions are present prior to committing to full-scale expression of key virulence factors (Dorman, 2009). As mentioned earlier, VirB positively regulates both the *virF* and *virB* promoters (Kane and Dorman, 2012) and co-regulates a subset of MxiE-regulated promoters ((Le Gall et al., 2005); Figure 4.2), which indicates that VirB acts as a regulatory sentinel at every level, integrating additional inputs to determine absolute levels of expression within the regulatory cascade. These findings elevate the importance of VirB in the cascade and somewhat diminish the role of VirF, which was originally coined the "master regulator" (Porter and Dorman, 1997a), even though the *virB* promoter remains completely dependent on VirF for its activity (Tobe et al., 1993).

B. Regulatory control and function of the pINV-encoded regulators

From the overview presented above, it is clear that VirF, VirB and MxiE are key virulence gene regulators in *Shigella* spp. that are required for transcription of genes encoded by pINV. Moreover, in their absence, cells are avirulent in animal models due to the lack of downstream virulence factors (Kane et al., 2002; Sakai et al., 1986; Sasakawa et al., 1988). Thus, systems are in place to ensure the timely and robust expression of these key regulators. We will next discuss these key virulence regulators in the context of how their expression is triggered, how each regulator functions and the direct effect that each regulator has on its regulated genes.

i VirF

Events that trigger the expression of virF

Activation of the *virF* promoter, and hence the regulatory cascade, is primarily triggered by a temperature of 37°C, although moderate osmotic stress (equivalent to physiological saline) and a pH of 7.4 are also needed for maximal activation (Falconi et al., 1998; Porter and Dorman, 1997a). The temperature dependency of the *virF* promoter relies on

i) H-NS and ii) a temperature dependent "hinge" region located within the promoter region itself (Falconi et al., 1998; Prosseda et al., 2004; Prosseda et al., 1998; Ulissi et al., 2014). At temperatures below 32°C, the bending and the rotational orientation of two HNS binding sites centered at -250 and -1 relative to the transcription start site (TSS) of *virF* is optimized for H-NS binding and repression of the promoter (Ulissi et al., 2014). Both sites must be occupied and the template DNA must be supercoiled for full transcriptional repression to occur, although H-NS binding occurs regardless of whether the DNA is supercoiled or not (Falconi et al., 1998). At temperatures higher than 32°C, *in vitro* experiments demonstrate that the angle of the bend in the DNA decreases and its center (found halfway between the two HNS binding sites at 32°C) moves downstream (Prosseda et al., 2004). These events are proposed to favor the removal of H-NS and facilitate the binding of FIS (factor for inversion stimulation) to sites that overlap the H-NS binding sites (Falconi et al., 2001; Prosseda et al., 2004). At 37°C, FIS activates the *virF* promoter (conditions that prevent H-NS-mediated repression), but at the critical temperature of 32°C, FIS is seen to partially counteract H-NS-mediated repression (Falconi et al., 2001).

The transcriptional regulators VirF and VirB can also modulate the activity of the *virF* promoter. VirF autorepresses the *virF* promoter (Porter and Dorman, 1997a), although it is unclear whether this effect is direct or indirect, while VirB boosts the initial activation of *virF* transcription upon thermal upshift to 37°C (Kane and Dorman, 2012). At first glance the regulation of the *virF* promoter by the downstream regulator VirB seems counterintuitive – would VirB even be present if the *virF* promoter hasn't fired? However, low levels of VirB protein have been detected even at non-permissive temperatures in wild type *Shigella* (Beloin et al., 2002; Porter and Dorman, 1997a). So, it is possible that the VirB-dependent effect on the *virF* promoter occurs at very low concentrations of VirB, although this has not been formally tested. It is also interesting to consider how VirB functions to boost *virF* transcription upon thermal upshift. The VirB binding sites that are proposed to be involved in the VirB-dependent upregulation of the *virF* promoter (Kane and Dorman, 2012) are located midway between the two H-NS binding sites and in close proximity to the FIS sites (described above; (Falconi et al., 2001; Prosseda et al., 2004)). Therefore, the role that VirB plays in enhancing *virF* transcription upon thermal induction may be similar to the role that FIS plays at this promoter: one of partially counteracting H-NS-mediated repression (Falconi et al., 2001), perhaps by stabilizing the new, thermally derepressed promoter region.

The *virF* promoter can be further modulated by another chromosomally encoded NSP, integration host factor (IHF) and by other chromosomally encoded transcriptional regulators including CpxR, Fur, Cra, CsrA, and ArcA (see section V). IHF increases *virF* promoter activity by 2-fold in both the logarithmic and the early stationary phases of growth in a temperature-independent manner. Surprisingly, the IHF binding site implicated in mediating this effect is located immediately downstream of the TSS (+45 to +57) (Porter and Dorman, 1997b). In addition, *virF* expression can also be controlled at the post-transcription level. tRNA modification mediated by *tgt* and *miaA* gene products (tRNA guanine transglycosylase and tRNA prenyl transferase, respectively) is required for efficient translation of *virF* mRNA (Durand and Bjork, 2003; Durand et al., 1997; Durand et al., 2000; Durand et al., 1994; Hurt et al., 2007). Thus, it is clear that although temperature plays a major role in the upregulation the *virF* promoter, a multitude of environmental conditions can ultimately influence the amount of *virF* expression and hence VirF protein produced.

The VirF protein

VirF is a 30 kDa protein (Sakai et al., 1988), which is part of the AraC superfamily of transcriptional regulators (Dorman, 1992; Martin and Rosner, 2001; Savelkoul et al., 1990). This group of regulators shares a conserved C-terminal domain containing two HTH motifs (presumed to be required for DNA binding) (Gallegos et al., 1997), and can be divided into three subgroups depending on the nature of the response signal: a chemical, stress or physical signal (Porter and Dorman, 2002). VirF is part of the latter group because like other virulence gene regulators in this subgroup, the gene encoding this regulator is thermally induced (Dorman, 1992; Jordi et al., 1992; Porter et al., 1998). Another member of this subgroup, Rns, shares a 36% amino acid identity with VirF throughout the length of the protein sequence (Munson, 2013; Porter et al., 1998). The genes encoding these regulators are proposed to have been derived from a common ancestor outside of the Enterobacteriaceae family due to their relatively low G+C content (*virF* 30% and *rns* 28%) (Dorman, 1992). Interestingly, the Rns protein present in enterotoxigenic *E. coli* (ETEC) was able to complement a *virF* null mutation in *S. flexneri* to activate transcription of virulence genes (Porter et al., 1998), suggesting that these regulators recognize similar DNA binding sites (Munson and Scott, 1999) and function similarly to activate transcription (Porter et al., 1998). However VirF was unable to complement a deficient *rns* strain, suggesting there are subtle, but significant differences that are required for VirF-dependent activation (Porter et al., 1998). The mechanism in which VirF activates transcription at its target promoters remains elusive, but evidence suggests that VirF monomers can interact (Porter and Dorman, 2002), which may explain the large and continuous VirF-dependent footprints observed at the *virB* promoter region (Tobe et al., 1993). Oligomerization of VirF also lends support for an oligomerization-driven activation hypothesis, where VirF binds to the promoter and then oligomerizes along the DNA in order to activate transcription (Porter and Dorman, 2002).

The regulatory effect of VirF

To date, VirF-mediated regulation has been characterized at the *icsA* (also referred to as *virG*), *rnaG*, *virB* and *virF* promoters (Adler et al., 1989; Jost and Adler, 1993; Sakai et al., 1988; Tobe et al., 1993; Tobe et al., 1995; Tran et al., 2011). IcsA is a 120 kDa outer membrane protein specifically localized to one pole of the bacterium (Goldberg et al., 1993). It mediates the actin tail assembly, which is required for cell-to-cell spread of the bacterium within the human intestinal epithelium, by recruiting host cell actin to the bacterial surface (Bernardini et al., 1989). *In vitro* transcription assays reveal that VirF alleviates H-NS-mediated repression of the *icsA* promoter, but also stimulates transcription of *icsA* directly in the absence of H-NS (Tran et al., 2011). Since *virF* expression increases rapidly above 32°C (described above; (Prosseda et al., 2004)) and the occupancy of the *icsA* promoter by H-NS is lost at 37°C (Tran et al., 2011), it has been proposed that VirF initially functions to alleviate H-NS-mediated repression of the *icsA* promoter, but switches to become an activator of *icsA* transcription once H-NS has been removed from the promoter region (Tran et al., 2011). Regulation of the *icsA* promoter region is further complicated by RnaG, an antisense *cis*-encoded RNA that functions to prematurely terminate *icsA* transcription (Giangrossi et al., 2010; Tran et al., 2011). *In vitro* transcription assays reveal that concentrations of VirF that stimulate the *icsA* promoter also repress the *rnaG* promoter. Therefore, VirF can also indirectly increase *icsA* expression post-transcriptionally by reducing expression of the anti-sense RNA RnaG (Giangrossi et al., 2010; Tran et al., 2011). Importantly, at the *icsA* genetic locus, all sites bound by H-NS and VirF are located downstream or overlapping of the *icsA* TSS. The regulatory architecture of this promoter therefore demonstrates the importance of considering downstream sequences when studying virulence gene regulation in *Shigella* and its close relatives. It is noteworthy that transcriptional regulation by Rns, the VirF homologue in ETEC, has been reported to occur from downstream binding sites (Munson et al., 2001; Munson and Scott, 2000).

At the *virB* promoter, in contrast to the *icsA* promoter, VirF binds to sequences located upstream of the TSS (between -17 and -117) (Jost and Adler, 1993; Tobe et al., 1993). Here, though, VirF functions to activate the *virB* promoter (mechanism described in more detail below).

ii VirB

Events that trigger the expression of virB

A temperature shift to 37°C is the primary trigger for *virB* mRNA expression, while moderate osmotic stress (equivalent to physiological saline), but not neutral pH, is required for maximal expression (Porter and Dorman, 1997a). Several lines of evidence suggest that the *virB* promoter may be repressed by H-NS (Beloin and Dorman, 2003; Hromockyj et al., 1992; Maurelli and Sansonetti, 1988; Porter and Dorman, 1997a; Tobe et al., 1993), but few test this directly. Those that do, use *hns* mutants of *Shigella* in which VirF levels are now known to be elevated due to the dysregulation of the *virF* promoter (described above). Since VirF is a known activator of the *virB* promoter, the *hns* mutant background is likely to cause indirect regulatory effects ((Tobe et al., 1991; Tobe et al., 1993); described more below). The best evidence for direct H-NS-mediated repression of the *virB* promoter comes from *in vitro* DNase I footprinting experiments, where H-NS can be seen to occupy a single region within the *virB* promoter from -20 to +20 (relative to the *virB* TSS) (Tobe et al., 1993). This site of occupancy strongly suggests that when H-NS is bound, RNA polymerase is excluded from the promoter and transcription is repressed.

VirF is absolutely required for induction of the *virB* promoter in wild type and *hns* mutant backgrounds (Tobe et al., 1991; Tobe et al., 1993) and occupies the *virB* promoter *in vitro* from -17 to -105 (relative to the *virB* TSS) (Tobe et al., 1993). VirF does not function as a derepressor that displaces H-NS at the *virB* promoter, but plays a positive role in its regulation, as demonstrated by the requirement for VirF in *in vitro* transcription assays (Tobe et al., 1993). Interestingly though, when VirF levels are artificially elevated to wild type levels and above at the non-permissive temperature of 30°C, *virB* transcription does not reach the level seen in wild type cells grown at 37°C. Three observations strongly suggest that this thermo-regulation is mediated by a temperature-dependent change in the supercoiled state of the *virB* promoter region. Firstly, temperature is known to modulate levels of DNA supercoiling (Dorman et al., 1990; Falconi et al., 1998). Secondly, the *virB* promoter is significantly activated in the presence of VirF *in vitro* when placed within an artificially supercoiled DNA molecule, but not a relaxed or linearized molecule (Tobe et al., 1993). Thirdly, the addition of a gyrase inhibitor novobiocin at 37°C, which relaxes supercoiled DNA *in vivo*, drastically reduces transcription of the *virB* promoter to levels similar to those observed at 30°C (Tobe et al., 1993). Thus, unlike the *virF* promoter where thermal induction triggers a change in DNA topology that removes H-NS and allows FIS to bind (discussed above), at the *virB* promoter thermal induction triggers a change in DNA supercoiling which is predicted to promote the interaction of VirF with RNA polymerase, allowing maximal VirF-dependent activation of the *virB* promoter (Tobe et al., 1993; Tobe et al., 1995).

The *virB* promoter has additional regulatory inputs that work to fine tune its regulatory response. (i) The VirB protein participates in a feedback loop to positively autoregulate its own promoter (Kane and Dorman, 2012). At 30°C, when native *virB* expression level is low, exogenous expression of *virB* can significantly increase the activity of the *virB* promoter to levels similar to those observed in an *hns* null mutant (Kane and Dorman, 2012) by binding to a *cis*-acting binding site within the *virB* promoter (located between -285 and -277 relative to the *virB* TSS) (Kane and Dorman, 2012). During the transition from 30 to 37°C, VirB participates in the full and continuous activation of the *virB* promoter, unlike at the *virF* promoter where VirB only participates in the initial activation (Kane and Dorman, 2012). To date, however, it remains unclear how VirB impacts or is constrained by the regulatory effects mediated by VirF and H-NS at the *virB* promoter. (ii) The integration host factor (IHF, discussed below), is required for the continuous expression of *virB* in stationary phase by binding a region, -171 to -183, located upstream of the essential VirF binding site (Porter and Dorman, 1997b), although the effect is mild (an approximately 2-fold reduction in *virB* expression in the absence of IHF). Notably, exogenous expression of *virB* is able to compensate for the lack of IHF (Porter and Dorman, 1997b). (iii) The iron-responsive *trans*-acting small RNA, RyhB, is able to transcriptionally regulate *virB* independently of VirF when expressed exogenously ((Broach et al., 2012; Murphy and Payne, 2007); see Chapter 5 on Ribo-regulation for further discussion). While RyhB-dependent modulation of *virB* transcription is independent of all factors known to regulate *virB* transcription, DNA sequences in the middle of the *virB* open reading frame appear to be required for this regulation (Broach et al., 2012), which suggests that RyhB-dependent regulation of *virB* transcription is both direct and mechanistically novel.

virB mRNA levels are also post-transcriptionally regulated by the RNA binding protein Hfq (Mitobe et al., 2008, 2009). Under non-physiological temperatures (Mitobe et al., 2008) or osmolarities (Mitobe et al., 2009) *virB* mRNA is relatively unstable when compared to physiological conditions. Deletion of the *hfq* gene increases the stability of *virB* mRNA and consequently increases VirB protein levels and VirB-regulated gene expression under non-physiological conditions (Mitobe et al., 2008, 2009). Therefore, although a description of this post-transcriptional effect may appear misplaced in a chapter that focuses on the transcriptional control of *Shigella* virulence genes, Hfq is clearly involved in the modulation of *virB* mRNA levels in *Shigella.*

The VirB protein

VirB (also called InvE in *S. sonnei*; (Watanabe et al., 1990)) is small (35.4 kDa), basic and encoded by pINV and represents a unique and novel class of transcriptional regulators. It has been extensively characterized as a positive transcriptional regulator of virulence-associated genes (Adler et al., 1989; Le Gall et al., 2005; Watanabe et al., 1990), but does not appear to function as a transcription activator *per se* (Lee et al., 2012). Instead, it functions to counteract H-NS-mediated repression (Stoebel et al., 2008; Wing et al., 2004). VirB shares no homology with currently characterized transcriptional regulators, but shares significant homology with ParB (42.8% amino acid identity over 278 residues) and SopB (30.4% amino acid identity over 168 residues) (Watanabe et al., 1990), which are proteins involved in plasmid partitioning and maintenance of the P1 and F plasmids, respectively (Abeles et al., 1985; Mori et al., 1986). Despite this similarity, which is especially high within the first two thirds of these proteins, VirB does not retain any plasmid partitioning activity (Buchrieser et al., 2000; Radnedge et al., 1997; Taniya et al., 2003).

Two structural domains have been characterized in the VirB protein: the centrally located DNA binding domain, which contains a HTH motif (residues 148-171) and an oligomerization domain, which is predicted to form a trimeric-coiled coil region at the extreme C-terminus (residues 260-309), (Beloin et al., 2002; Gao et al., 2013). Initially, a leucine zipper dimerization domain was also predicted in VirB (residues 193-228) (Beloin et al., 2002), but structural studies did not support this prediction (Gao et al., 2013). Based on high levels of homology with the N-terminus of ParB and SopB, the VirB N-terminal region is likely to be flexible in solution (Vecchiarelli and Funnell, 2013). This probably explains why structural studies of the N-terminus of VirB are lacking, even though a variety of functions have been proposed for it, including transcriptional activation, structural involvement in DNA binding and protein interactions (Beloin et al., 2002).

The DNA binding domain of VirB is 80% identical to the HTH domain of ParB (Beloin et al., 2002; Surtees and Funnell, 2001), which strongly suggests that the DNA binding properties of these two proteins are similar. Not surprisingly, DNA binding studies for VirB have often paralleled those done with ParB (Beloin et al., 2002; Gao et al., 2013; McKenna et al., 2003; Taniya et al., 2003; Turner and Dorman, 2007; Watanabe et al., 1990). To date, the only established VirB binding site (A/G)(A/T)G(G)AAAT (Taniya et al., 2003), is a composite of three sites (located at the *virA*, *spa15*, and *icsB* promoters) that bear similarity to a part of the ParB binding site *parS* on the P1 plasmid (ATTTCAC, also known as A-boxes) (Radnedge et al., 1996). Importantly, this site does not represent a consensus of all VirB binding sites found on pINV, nor is it an *in vitro*-generated protein binding logo. To complicate matters further, the sites required for the VirB-dependent regulation of virulence genes are sometimes organized as inverted repeats (separated by a single nucleotide) (Castellanos et al., 2009; Turner and Dorman, 2007), and sometimes organized as single heptameric sequences (Kane and Dorman, 2012). This raises questions about which DNA sequences are sufficient for VirB binding, and how VirB initially engages these sequences. Although inverted repeats are often bound by dimeric proteins, recent crystallographic studies of the DNA binding domain of VirB bound to two inverted repeats, suggest that VirB does not need to dimerize before engaging its recognition site (Gao et al., 2013). Instead, the DNA binding domain of VirB, which importantly cannot dimerize/oligomerize, interacts with just one half of the inverted repeat, via the half site known as box 2 (Gao et al., 2013). This is in contrast to ParB, which is demonstrated to occupy both half sites within its recognition sequences in similar crystallographic studies (Schumacher and Funnell, 2005). Interestingly, VirB commonly binds to DNA non-specifically *in vitro* (Harrison, 2010; McKenna et al., 2003; Turner and Dorman, 2007), yet is capable of binding to its targets *in vivo* with specificity (Beloin et al., 2002). This strongly suggests that the binding specificity of VirB is determined by a factor or condition present *in vivo* that is absent *in vitro.* So far, however, the identity of this factor or condition remains elusive. Clearly, there is still much that needs to be learned about how VirB recognizes and binds to its DNA targets.

The oligomerization domain of VirB is found in the C-terminus, which is the least similar region of VirB when compared to ParB and SopB. This region enhances DNA binding both *in vivo* and *in vitro* and is necessary for the oligomerization of VirB in solution, because C-terminally truncated proteins have lost this ability (Beloin et al., 2002; Gao et al., 2013). While ParB also oligomerizes in solution and along DNA (Rodionov et al., 1999; Rodionov and Yarmolinsky, 2004; Surtees and Funnell, 1999), this activity is mediated by its extreme N-terminus (Surtees and Funnell, 1999, 2001), demonstrating that different regions within ParB and VirB confer their shared ability to oligomerize. *In vitro* VirB:DNA complexes appear as smears and large molecular weight complexes at the top of the gel when resolved electrophoretically, rather than discrete bands, which supports oligomerization along DNA (Harrison, 2010; McKenna et al., 2003; Turner and

Dorman, 2007). VirB oligomerization in the context of transcriptional derepression appears important, but so far its precise role remains unclear.

Although many questions about VirB-mediated transcriptional derepression remain unanswered, a working model has been proposed. The HTH domain of VirB binds to the *cis*-acting heptameric sequence, which serves as a nucleation point (Beloin et al., 2002; Gao et al., 2013; Kane and Dorman, 2012; Turner and Dorman, 2007). VirB binding introduces a bend in the intrinsically, highly flexible A-tract segment within the binding site, which facilitates the non-specific binding of VirB oligomers along the DNA (Gao et al., 2013). This is a novel feature of VirB because ParB-like proteins are unable to induce bends in the DNA. VirB oligomerization along the DNA strand destabilizes the H-NS-bound DNA complex to allow transcription to proceed (Beloin et al., 2002; Gao et al., 2013; Kane and Dorman, 2012; Turner and Dorman, 2007). Taken together, while it is clear that DNA binding and oligomerization activities of VirB are key for its activity as a transcriptional derepressor of *Shigella* virulence genes, much remains unclear about how this unusual transcriptional regulator counteracts repression mediated by the nucleoid structuring protein H-NS. It is hoped that in depth studies of transcriptional derepression at VirB–regulated promoters will shed light on this non-traditional, yet conserved, regulatory mechanism.

The regulatory effect of VirB

In contrast to VirF, VirB has an extensive regulon. This includes genes of the *ipa-mxi-spa* entry region, genes located outside of this region including *icsP, phoN2, ospB, ospF, ospC1, ospC2/3/4, virA, ospD1, ospD2, orf13, orf81, orf137, ipaJ, virB* and the genes of the VirF and MxiE regulons, which are indirectly regulated by VirB via its direct control of *virF* and *mxiE* (Adler et al., 1989; Berlutti et al., 1998; Bongrand et al., 2012; Le Gall et al., 2005; Santapaola et al., 2002; Taniya et al., 2003; Uchiya et al., 1995; Watanabe et al., 1990; Wing et al., 2004). Of those genes thought to be directly regulated by VirB, binding of VirB to the promoter region has only been demonstrated for a subset of genes, including *icsB*, *icsP, ospZ* and *virB* (Harrison, 2010; Kane and Dorman, 2012; McKenna et al., 2003; Taniya et al., 2003; Turner and Dorman, 2007).

The best characterized VirB-dependent promoter is P*icsB* (Beloin and Dorman, 2003; McKenna et al., 2003; Taniya et al., 2003; Turner and Dorman, 2007; Watanabe et al., 1990). This promoter is found in the entry region and controls the transcriptional expression of the *ipa* operon, of which *icsB* is the first gene. H-NS occupies a large region within the *icsB* promoter from -110 to +25 relative to the *icsB* TSS (Beloin and Dorman, 2003; Porter and Dorman, 1997b; Turner and Dorman, 2007). When bound, H-NS inhibits open complex formation (Turner and Dorman, 2007). Based on its DNase I footprint, however, it seems more likely that H-NS blocks the recruitment RNA polymerase to the *icsB* promoter, rather than blocking open complex formation directly. Located immediately upstream of the H-NS bound region, is the site required for VirB-dependent regulation of P*icsB.* While this site (centered at -114 relative to the TSS; (Porter and Dorman, 1997b; Turner and Dorman, 2007)) is organized as an inverted repeat, mutagenesis studies reveal that only the promoter-distal half of the inverted repeat (Box 2) is required for the VirB-dependent regulation of the *icsB* promoter (Turner and Dorman, 2007). VirB binding to this site changes the DNA topology within the H-NS-occupied region, as judged by DNase I footprinting. It has been proposed that these changes destabilize the H-NS:DNA complex, leading to the displacement of H-NS from the *icsB* promoter (Turner and Dorman, 2007). Since the VirB protein has no effect on RNA polymerase recruitment or formation of the open transcription complex, VirB can be considered a derepressor rather than a positive transcription factor at the *icsB* promoter (Turner and Dorman, 2007).

The *icsP* promoter is located on the opposite side of pINV relative to the entry region. It controls the expression of a monocistronic gene that encodes the outer membrane protease IcsP, which modulates the amount of the actin tail assembly protein IcsA that is associated with the bacterial surface (Egile et al., 1997; Shere et al., 1997; Steinhauer et al., 1999). At P*icsP*, VirB also functions to counteract H-NS-mediated repression of transcription, which remains in effect even after a shift to 37°C (Wing et al., 2004). But here, in contrast to the situation at the *icsB* promoter, both halves of an inverted repeat, made up of two heptamers each closely matching the characterized VirB binding site (Taniya et al., 2003), are required for VirB-dependent regulation (Castellanos et al., 2009). Importantly, these sites are located more than 1 kb upstream of the *icsP* TSS (from position -1144 to -1130, relative to the TSS) (Castellanos et al., 2009), which is striking because traditionally bacterial transcription factors were thought to mediate their effects from sites located within 200 bp of the TSS (Balleza et al., 2009; Collado-Vides et al., 1991). Consequently, this example demonstrates that VirB is capable of mediating long range effects on transcription and raises the possibility that VirB may control the transcription of other pINV-encoded virulence genes from previously overlooked remote sites. More recently, the VirB binding sites found to regulate the *icsP* promoter were also found to regulate transcription from a second *icsP* promoter (Hensley et al., 2011) as well as the divergent *ospZ* promoter (Basta et al., 2013). Interestingly, the VirB binding sites lie closer to the beginning of the *ospZ* gene (425 bp upstream of ATG) than the *icsP* gene (1164 bp upstream of ATG) (Basta et al., 2013). Even so, the regulatory effect mediated by VirB is much stronger on *icsP* than on *ospZ* (Basta et al., 2013; Hensley et al., 2011). Based on 5' truncation analysis of both *icsP* and *ospZ* promoters, the region required for H-NS-mediated repression of the *icsP* and *ospZ* promoters seems to lie immediately downstream of the VirB binding site in the direction of the *icsP* gene located between -893 and -351, relative to the primary *icsP* TSS (Basta et al., 2013; Castellanos et al., 2009; Harrison, 2010). Therefore, unlike the *icsB* promoter, H-NS-mediated repression of *icsP/ospZ* locus require sequences located upstream of the promoter elements. However, similar to the situation at the *icsB* promoter, H-NS-mediated repression of *icsP* and *ospZ* does appear to be mediated by sequences that lie immediately adjacent to the region occupied by VirB.

As described previously, at the *virB* promoter, which is located immediately adjacent and downstream of the *ipa* operon, H-NS binds between -20 and +20 (Tobe et al., 1993) and VirB mediates its effect from a single heptameric sequence rather than an inverted repeat (Kane and Dorman, 2012). This site displays a 7/8 match to the characterized VirB binding site and is centered at position –281.5 relative to the TSS of *virB*. Protection of this site by VirB can be seen by DNase I footprinting. As described above, VirB binding to this site, as cultures are switched from 30 to 37°C, is reported to relieve H-NS-mediated repression of P*virB* more rapidly and in a more sustained manner than is caused by the change in temperature alone (Kane and Dorman, 2012).

Based on these examples, it is clear that there is significant diversity in the architecture of VirB-regulated promoters, not only in the position of the VirB binding site relative to the promoter elements and regions bound by H-NS, but also within the VirB binding sites themselves. Fifty six occurrences of the VirB binding site found at P*icsB* (5'-ATTTCAT-3') have been found on pINV taken from S. *flexneri* serotype 5a (pWR100) and over 1000 sites were found with a single mismatch (Le Gall et al., 2005). Based on these findings, it seems highly unlikely that the characterized VirB binding site is a strong predictor of VirB-regulated targets. Nevertheless, one common theme is that when VirB-mediated regulation is established, the function of the VirB protein appears to be conserved: to counteract H-NS-mediated repression, even though the mechanism driving VirB–dependent derepression remains poorly understood.

As the tier 3 regulator of virulence gene expression on pINV, VirB also regulates the transcription of the genes encoding the tier 4 regulators, MxiE and IpgC, which are encoded by the 30 kb entry region. As described in the next section, MxiE and IpgC regulate the transcription of genes required for post-invasion activities (discussed below); triggering the expression of a unique set of genes (tier 4) as well enhancing the expression of some VirB-regulated genes (tier 3) during conditions of secretion.

iii Regulatory control mediated by the pINV-encoded regulators, MxiE and IpgC

Events that trigger the expression of mxiE, ipgC, ospD1 and spa15

Transcription of tier 4 genes, those expressed in response to activation of the TTSS, relies on a complex circuitry involving tier 3 gene products: the AraC-like transcriptional activator MxiE, its co-activator IpgC, the anti-MxiE activator OspD1, and OspD1 chaperone Spa15 (Parsot et al., 2005). Each of these proteins is encoded by a gene located on pINV. The *mxiE* gene lies within the *mxi* operon of the 30 kb entry region (which is controlled by the P*ipgD*), *ipgC* is located within the divergently transcribed *ipa* operon (which is controlled by P*icsB*), *spa15* is the first gene of the *spa* operon and *ospD1* is encoded by a monocistronic gene located outside of the entry region (residues 20964-21641 on pWR100). As expected by their tier 3 status, *mxiE, ipgC, spa15* and *ospD1* are all maximally expressed at 37°C in a VirB-dependent manner (Le Gall et al., 2005). Importantly though, the amount of MxiE protein made by *Shigella* is determined by regulatory events that occur during transcription elongation of the *mxiE* locus (Penno et al., 2005).

The *mxiE* locus contains two overlapping open reading frames (ORF), *mxiEa* and *mxiEb*, which are out of frame with one another by -1. Transcriptional slippage of RNA polymerase on a run of 9 T residues found in this region can place these two ORFs in the same frame, allowing the full-length *mxiE* gene to be transcribed (Penno et al., 2005). While *mxiEa* (codons 1-43) is often expressed without *mxiEb,* the small MxiEa peptide does not appear to function in the regulatory circuit controlling expression of tier 4 genes (Penno et al., 2005). In contrast, *mxiEb* (codons 44-251) is only ever transcribed in the context of full-length *mxiE,* but is essential because it encodes its DNA binding determinants for MxiE (Penno et al., 2005). Although conditions that influence frameshifting efficiency have the potential to regulate production of MxiE, a small increase in the amount of MxiE is unlikely to significantly impact expression of MxiE-regulated genes. This is because the activity of MxiE is so tightly controlled by the availability of its co-activator IpgC (described below) (Mavris et al., 2002a). If, however, transcriptional slippage was prevented under conditions that trigger the activation of the TTS apparatus, MxiE production would fail and MxiE-regulated genes would not be expressed (Penno et al., 2005). Thus, specific environmental conditions that hinder transcriptional slippage by RNA polymerase during the transcription of *mxiE* have the potential to negatively influence MxiE-regulated gene expression under conditions of secretion. To date, the precise nature of these conditions, if they exist, remains elusive.

The MxiE, IpgC, OspD1 and Spa15 proteins and their interactions

The capacity of MxiE to function as a transcriptional activator is controlled by the secretion status of *Shigella* (Figure 4.3). Key players in the modulation of MxiE-regulatory activity are i) the MxiE co-activator IpgC, which also serves as the chaperone for the TTSS translocators IpaB and IpaC (Menard et al., 1994; Page et al., 2001; Mavris et al., 2002), ii) the MxiE anti-activator OspD1 and iii) the OspD1 chaperone Spa15 (Marteyn et al., 2012; Page et al., 2002). As the TTSS is activated, IpaB and IpaC dissociate from IpgC and OspD1 dissociates from both MxiE and its chaperone Spa15 (Parsot et al., 2005), allowing secretion of IpaB, IpaC and OspD1. As a consequence, levels of liberated MxiE and IpgC rise in the bacterial cytoplasm, where these two proteins are believed to associate with each other to trigger transcriptional activation of MxiE-regulated genes (Parsot et al., 2005). In support of this, MxiE and IpgC have been shown to associate in the cytoplasm of *E. coli* after co-expression of their genes (Pilonieta and Munson, 2008). Furthermore, MxiE and IpgC are sufficient to activate transcription of target promoters in the absence of other virulence plasmid encoded factors (Mavris et al., 2002a; Penno et al., 2005). So far, however, binding of MxiE or even MxiE:IpgC complexes to DNA has not been demonstrated and the molecular details of the interaction between MxiE and its co-activator IpgC remain poorly understood, although it is predicted that the N-terminal part of MxiE, encoded by *mxiEa*, is involved (Penno et al., 2005), based on similarity to the SicA/InvF system in *Salmonella enterica* (Darwin and Miller, 2001). While additional work is required to fully understand the molecular details of the regulatory control mediated by MxiE and IpgC, the sophisticated use of an apparently defective transcriptional activator, a co-activator that doubles as a TTSS chaperone, an anti-activator and its chaperone to control the transcription of MxiE-regulated genes is both evolutionarily impressive and elaborate way to tie the activity of the TTSS to the expression of a second wave of TTSS effector proteins.

The regulatory effect of MxiE and IpgC

Although *mxiE* and *ipgC* are transcribed prior to secretion-specific conditions, most of its regulon members are only transcribed once the TTSS is activated ((Demers et al., 1998; Kane et al., 2002; Mavris et al., 2002b); Figure 4.3). During infection, this activation occurs once host cell contact has been made by *Shigella;* but the addition of Congo red dye to growth media or the use of *ipaB* mutants, in which type III secretion is dysregulated (Kane et al., 2002; Mavris et al., 2002a; Mavris et al., 2002b) can "artificially" trigger the activation the type III secretion system and hence the expression of the MxiE regulon. By combining these approaches, the full extent of the MxiE regulon has been characterized ((Bongrand et al., 2012; Buchrieser et al., 2000; Demers et al., 1998; Kane et al., 2002; Mavris et al., 2002a; Mavris et al., 2002b); Figure 4.2). Thirteen MxiE-regulated genes are encoded by the large virulence plasmid pINV: *ospB, phoN2, ospC1, ospD3, ospE1, ospE2, ospF, ospG, virA, ipaH1.4, ipaH4.5, ipaH7.8 and ipaH9.8* (Bongrand et al., 2012; Le Gall et al., 2005); the remaining MxiE-regulated genes are the chromosomally encoded *ipaH* genes, the number of which varies from strain to strain (Bongrand et al., 2012; Mavris et al., 2002b).

The DNA binding site required for MxiE-mediated activation is 17 bp in length and typically overlaps the -35 region of MxiE-regulated promoters. This site designated the MxiE box (5'-GTATCGTTTTTTTAnAG-3') was originally identified by promoter mapping and deletion analysis of three MxiE-regulated promoters (*ipaH9.8, virA* and *ospC1*) (Mavris et al., 2002b), but similar sites (with up to 3 base pair mismatches) have since been found upstream of all of the designated MxiE-regulated genes (Bongrand et al., 2012). Interestingly, the identification and requirement of MxiE boxes upstream of each one of the chromosomally encoded *ipaH* genes (Bongrand et al., 2012), demonstrates that chromosomal loci can indeed fall under the transcriptional control of a pINV-encoded regulator, raising the possibility that other chromosomal genes may be pINV-regulated too.

By comparing the expression of the MxiE-regulated genes in a variety of *Shigella* mutant backgrounds, two sub-classes of MxiE-regulated genes were discovered (Le Gall et al., 2005): i) those regulated by MxiE alone, and ii) those directly regulated by VirB under conditions of non-secretion and further upregulated by MxiE under conditions of secretion (Figure 4.2). It has been proposed that this hierarchy allows VirB-regulated primary effector proteins to be secreted immediately after the TTSS is activated, and then a subset of these proteins (including *ospB*, *phoN2*, *ospF*, *ospC1* and *virA*) to be further upregulated by MxiE. Thus, MxiE can upregulate expression of effectors initially expressed by VirB, and also regulate a unique set of genes (the true tier 4 genes) in a second wave of effector secretion. The organization of genes into one of these subclasses allows the precise and deliberate secretion of proteins into the host cell cytoplasm, so that the host cell can be manipulated in an exquisite and coordinated manner by these *Shigella* encoded MxiE-regulated effectors.

Transcriptional control mediated by other chromosomally encoded regulators

As we have seen, the expression of virulence genes encoded by pINV relies on a suite of transcriptional regulators that are also encoded by this large virulence plasmid (discussed in section IV). Evolutionarily this makes sense; the horizontally-acquired virulence genes were acquired along with their cognate transcriptional regulators. But, regulatory cross-talk between pINV and the chromosome is evident in *Shigella*. For instance, the chromosomally encoded *ipaH* genes, which encode type III translocators, are directly

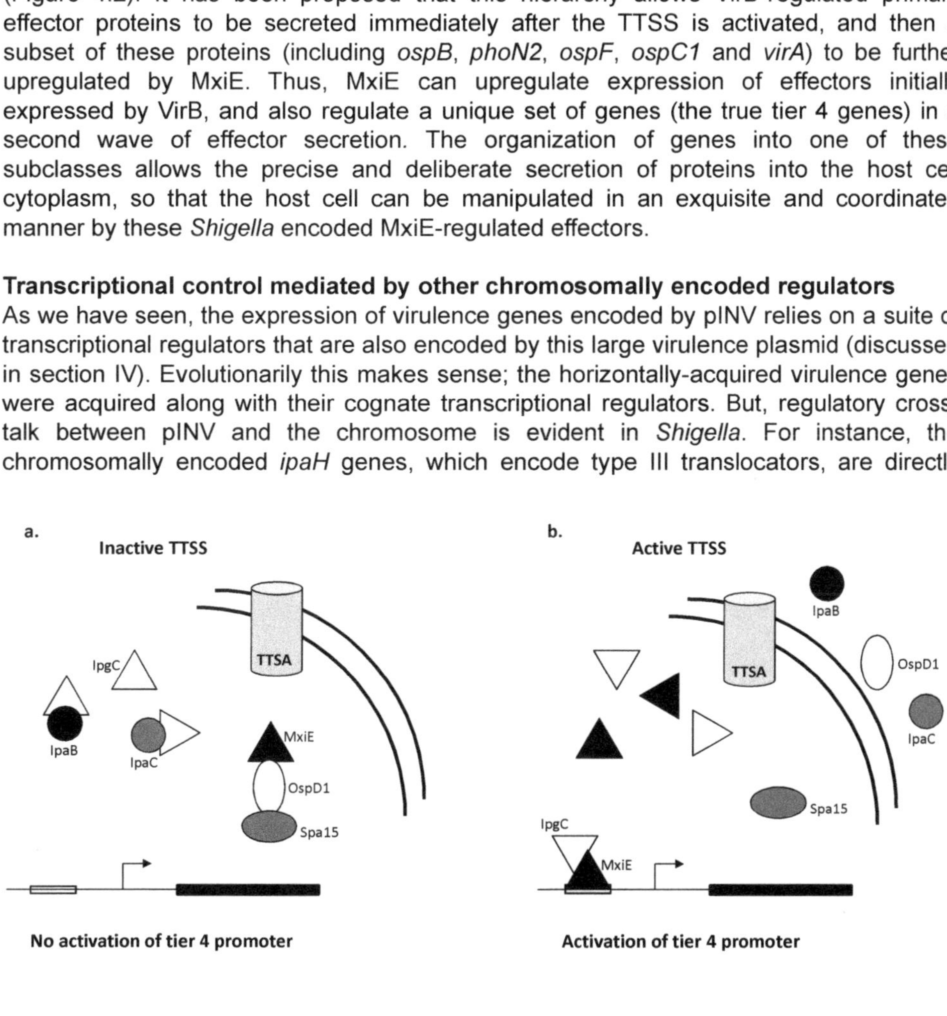

Figure 4.3. Events required for the MxiE-dependent activation of tier 4 promoters. (a) Under conditions in which the type three secretion system (TTSS) is inactive (non-secretion), IpgC serves as a chaperone to associate with the TTSS translocators, IpaB and IpaC, while the MxiE anti-activator, OspD1, is bound to both MxiE and its chaperone, Spa15. In this case, both the activator (MxiE) and co-regulator (IpgC) are sequestered and unable to bind tier 4 promoters to activate transcription. (b) Under conditions in which the TTSS is active (secretion), each protein complex dissociates resulting in secretion of IpaB, IpaC and OspD1. Meanwhile, liberated IpgC and MxiE accumulate within the bacterial cytoplasm, presumably associate and subsequently bind to tier 4 promoters to activate transcription. Open box represents MxiE binding site. Shaded box represents MxiE-dependent gene (tier 4). Line with arrow represents the transcription start site.

regulated by the pINV-encoded transcriptional regulator MxiE (Bongrand et al., 2012). Furthermore, several chromosomally encoded transcriptional regulators have been shown to directly bind to promoters located on the virulence plasmid to regulate their expression (Figure 4.4). In this section, we will focus on the regulatory inputs made by chromosomally encoded transcription regulators, first focusing on those that bind directly to pINV promoters, including FIS, IHF, CpxR, FNR and PhoP (H-NS and its homologs have been previously discussed above) and concluding with those whose influence, thus far, appears to be indirect, such as Fur, CsrA, ArcA and OmpR (Figure 4.4).

A. Chromosomally encoded regulators with a direct role in the transcriptional regulation of virulence genes located on pINV

FIS

FIS, the factor for inversion stimulation, is characterized as a nucleoid structuring protein, like H-NS. In *E. coli,* FIS contributes to the overall cell physiology by participating in many essential processes and by binding to hundreds of DNA targets (discussed in Browning et al., 2010). Consequently, FIS has been given the title "global modulator of metabolism" (Schneider et al., 1999). Although the primary role of FIS is to inhibit transcription of nonessential genes during exponential growth in *E. coli* and *Salmonella enterica,* it does have the capability to activate gene transcription in *E. coli* (Opel et al., 2004; Wilson et al., 2001). In *Shigella*, FIS facilitates the temperature-dependent activation of P*virF* to positively influence *virF* expression (discussed previously; Figure 4.4). After temperature-induced (37°C) changes in the DNA topology has removed H-NS and exposed a FIS binding site, FIS binds to the *virF* promoter to further upregulate transcription (Falconi et al., 2001). Additionally, during the transition between non-permissive (30°C) to permissive (37°C) temperatures, FIS either (1) competes with H-NS for binding sites at P*virF*, since 2 of the 4 identified FIS binding sites overlap with one of the two H-NS binding sites, or (2) forms an intermediate complex with H-NS to promote the transition from a repressed promoter state to an active state (Falconi et al., 2001). Therefore, while FIS directly regulates transcription from the pINV-located *virF* promoter and likely contributes to its robust expression during infection, it is not absolutely essential for *virF* expression. It remains unclear if FIS regulates other genes on pINV at this time.

IHF

The integration host factor, IHF, is another example of a nucleoid structuring protein that influences transcription of pINV genes. The alpha and beta subunits of this heterodimeric protein are encoded by the *ihfA* and *ihfB* genes, respectively. In *E. coli*, the primary role of IHF is to sharply bend DNA (more than 160°) (Rice et al., 1996) to cooperatively work with other NSPs or transcription factors to facilitate gene transcription, DNA recombination or DNA replication. Moreover, IHF can contribute to promoter activation by directly interacting with the alpha subunit of RNA polymerase (Goosen and van de Putte, 1995). Evidence suggests IHF binds directly to both tier 1 (P*virF*) and tier 2 promoters (P*virB* and P*icsA*) on pINV (Figure 4.4), based on the formation of IHF-dependent protein DNA complexes in *in vitro* binding assays from crude extracts, and the presence of putative IHF binding sites in the DNA surrounding these promoters (*virF* +45 to +57; *icsA* +109 to +122; *virB* -171 to -183, relative to the respective TSS (Porter and Dorman, 1997b)). The role of IHF at these promoters appears to be one of modulation rather than essentiality, but its regulatory function appears to vary; being an activator at P*virF* and P*icsA* during exponential growth phase and an alleviator of H-NS-mediated repression at P*virB* during stationary phase (Porter and Dorman, 1997b).

CpxR

The CpxR/CpxA two component system is a signal transducing pathway that is initiated in response to extracytoplasmic stress, including, but not limited to, cell envelope perturbations, misfolded proteins and changes in pH (recently reviewed in Raivio, 2014). As a two component system, the presence of a specific signal is recognized by the sensor kinase CpxA, which phosphorylates its cognate response regulator CpxR, which in turn regulates the expression of several genes needed to overcome the stress. In *Shigella,* the CpxR/CpxA system was first implicated in the modulation of virulence gene expression when an *E. coli cpxA* mutant led to the pH-dependent dysregulation of a *virF'-'lacZ* translational fusion, at both pH 7.4 and 6.0, but particularly at pH 6.0 (Nakayama and Watanabe, 1995). A subsequent study, established CpxR as an essential activator of *virF* expression and demonstrated that CpxR binds to sequences located between -103 to -37 relative to the TSS of the *virF* gene (Nakayama and Watanabe, 1998). In this study, however, disparate phenotypes were associated with the *cpxR* and *cpxA* mutants; the *cpxR* mutant abolished *virF* expression, while the *cpxA* mutant retained comparatively high levels of *virF* expression, regardless of pH. Two explanations were offered (Nakayama and Watanabe, 1998). First, in the absence of *cpxA,* CpxR may have become phosphorylated by another phosphate donor. Second, phosphorylated CpxR may not be the only factor that contributes to *virF* expression *in vivo,* which we now know to be true (Falconi et al., 2001; Kane and Dorman, 2012). A later study, characterized another *cpxA* mutant, which had a transposon insertion closer to the 5' end of the *cpxA* gene (Mitobe et al., 2005), than the originally described mutant (Nakayama and Watanabe, 1995). Strikingly, this *cpxA* mutant did not show a difference in *virF* expression profiles compared to wild type at neutral pH, despite this mutation causing a down-regulation of the TTSS in *Shigella* (Mitobe et al., 2005). Instead, this *cpxA* mutant exhibited a decrease in the posttranscriptional processing of *virB* mRNA (InvE) (Mitobe et

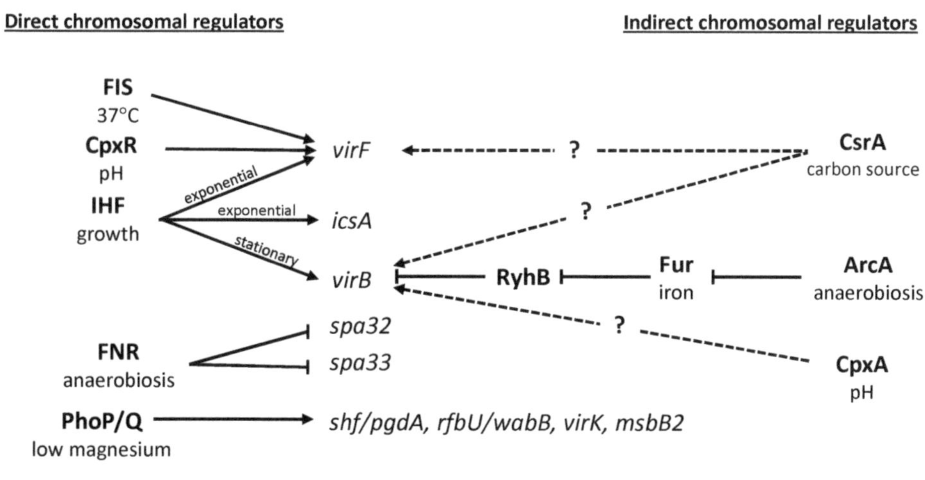

Figure 4.4. Chromosomally encoded regulators that influence expression of *Shigella* virulence genes on pINV. Direct and indirect regulatory inputs are illustrated. The environmental cue triggering the activity of each regulator is given directly below each regulator and the specific gene(s) whose expression is modulated is depicted. Straight lines indicate direct regulation while dashed lines indicate indirect regulation. Arrows indicate positive regulation while lines with a terminal, perpendicular line indicate negative regulation. Question marks indicate unidentified components that may regulate the gene of interest. OmpR/EnvZ has been excluded since it is unclear where the virulence regulatory input is received.

al., 2005), which suggest that CpxA mediates post-transcriptional regulation of *virB* via an unidentified factor. Based on these three studies (Mitobe et al., 2005; Nakayama and Watanabe, 1995; Nakayama and Watanabe, 1998), it is clear that the chromosomally encoded CpxR/CpxA system plays an important role in the activation of the *Shigella* virulence regulatory cascade (Figure 4.4): CpxR functions as a direct and essential activator of *virF* gene expression (Nakayama and Watanabe, 1998), while CpxA is involved in the pH-dependent modulation of *virF* expression and the post-transcriptional modulation of *virB* mRNA levels (Mitobe et al., 2005; Nakayama and Watanabe, 1995).

FNR

The anaerobic transcriptional regulator, FNR (fumarate and nitrate reduction), is required for the invasive phenotype of *Shigella* in anaerobic environments, such as those encountered in the host gastrointestinal (GI) tract lumen (Marteyn et al., 2010). In fact, cells grown anaerobically exhibit an increase in cell entry into epithelial cells when compared to cells grown aerobically (Marteyn et al., 2010). This is because *Shigella* becomes "primed" during anoxic conditions via FNR. FNR directly represses the promoters of the tier 3 genes *spa32* and *spa33* ((Marteyn et al., 2010); Figure 4.4). Spa32 controls the production of proteins that mediate TTSS needle length and the selection of substrates for secretion (Magdalena et al., 2002; Tamano et al., 2002) while dysregulation of Spa33 blocks Ipa secretion (Schuch and Maurelli, 2001). Therefore, FNR-mediated repression of *spa32* and *spa33* causes TTSS needle length to increase, which is predicted to facilitate contact with host cells during infection, and prevents type III secretion, thus allowing effectors to accumulate in the bacterial cytoplasm (Marteyn et al., 2010). It is proposed that as the "primed" *Shigella* approaches the epithelial layer of the lower GI tract, there is sufficient oxygen to inactivate FNR, allowing type III secretion to occur from the longer and more numerous needles in contact with host cells, ultimately leading to enhanced host cell entry. These proposed events are congruent with multiple lines of investigation, including microelectrode monitoring of the surface of epithelial cells that have recently been moved to an anaerobic cabinet, oxygen reporters used within the rabbit ileal loop model of infection and SEM imaging of *Shigella* taken from aerobic or anaerobic cultures (Marteyn et al., 2010). In summary, the chromosomally encoded transcription factor FNR transcriptionally regulates the pINV-located genes, *spa32* and *spa33,* to facilitate the priming of bacterial cells in anoxic conditions, like those found in the lower intestine, so that *Shigella* is primed and ready for invasion of the colonic epithelium.

PhoP

In many Gram-negative pathogens, the PhoP/PhoQ two component regulatory system modulates the expression of genes essential for virulence. This system is thought to be activated intracellularly when PhoQ senses the low magnesium environment of the phagosome. In *Shigella, phoP* mutants are more susceptible to killing by polymorphonuclear leukocytes (PMNs) and cationic antimicrobial peptides, which play important roles in late stage infections (Moss et al., 2000), but the role of PhoP in the invasion and intercellular spread of *Shigella* remains somewhat ambiguous (Cai et al., 2011; Moss et al., 2000; Sidik et al., 2014). Although the genes responsible for the phenotypes observed in the original Moss et al. study were not identified, at least one operon on pINV, which is comprised of four genes (*shf/pgdA, rfbU/wabB, virK, msbB2*), has subsequently been shown to be regulated by the PhoP/Q system ((Goldman et al., 2008); Figure 4.4). Upstream of this operon and overlapping the -35 promoter element lies a putative PhoP binding site, mutations in which lead to diminished PhoP-dependent promoter activity (Goldman et al., 2008). Furthermore, mutations in the last two genes of the operon, *msbB2* and *virK,* generate phenotypes that are consistent with other studies of *phoP* mutants (Cai et al., 2011; Goldman et al., 2008; Moss et al., 2000; Sidik et al., 2014). Briefly, the *msbB2* gene, encodes a myristoyl transferase enzyme that catalyzes the last step in the synthesis of the lipid A moiety of LPS. Expression of *msbB2* is detected in low magnesium ion conditions, but not in the presence of high magnesium ion concentration (Goldman et al., 2008). Under low magnesium conditions, MsbB2 functions to increase the overall hexa-acylation of lipid A (Goldman et al., 2008), which is proposed to confer increased resistance to antimicrobial peptides: a phenotype observed by Moss et al. (Moss et al., 2000). On the other hand, the upstream gene, *virK,* was initially implicated in both intracellular and intercellular spread because a *virK*::Tn10 mutation led to a decrease in *Shigella*-associated IcsA (Nakata et al., 1992), the protein required for actin-based motility. Subsequent analyses revealed that this mutation caused an increase in the amount of IcsP, an outer membrane protease that functions to proteolytically cleave the IcsA determinant for actin-based motility from the bacterial surface (Wing et al., 2005). More recently, *virK* has also been implicated in relieving bacterial envelope stress that occurs during growth at 37°C (Sidik et al., 2014). Although the precise activity of VirK remains unclear at this time, it is likely that all the phenotypes associated with *virK* mutants will be explained once it is discovered (Nakata et al., 1992; Sidik et al., 2014; Wing et al., 2005). In summary, the chromosomally encoded PhoP/Q system is essential for *Shigella* virulence because it plays an important role in the regulation of pINV genes encoding traits such as actin-based intracellular and intercellular spread, resistance to antimicrobial peptides and PMN-mediated killing. Whether PhoP mediates control over these virulence traits solely via its direct regulation of the *shf/pgdA, rfbU/wabB, virK, msbB2* operon, or whether other PhoP-regulated virulence loci on pINV are directly or indirectly involved, still needs to be established.

B. Chromosomally encoded regulators with an indirect role in the expression of genes located on pINV

Fur

In the presence of iron, the global transcriptional repressor of bacterial iron acquisition, Fur (Ferric uptake regulator), complexes with iron and binds target DNA sites to repress transcription of iron-responsive genes. Meanwhile, in iron-limiting conditions, Fur-mediated repression is alleviated, which permits the transcription of iron-responsive genes. In *Shigella* and *E. coli*, Fur represses expression of the sRNA RyhB in the presence of iron (Massé and Gottesman, 2002; Oglesby et al., 2005). Strikingly, in *Shigella,* production of RyhB not only decreases mRNA levels of targets known to be regulated by RyhB in *E. coli,* but it also decreases *virB* mRNA levels, presumably by RyhB binding to a complementary sequence within the *virB* coding region (discussed above; (Broach et al., 2012; Murphy and Payne, 2007)). Consequently, RyhB production leads to the downregulation of tier 3 VirB-dependent loci, including *ipaABCDE, ipgABCE, mxiACDEJL, virA* and *icsP* (Africa et al., 2011; Murphy and Payne, 2007). Thus, under iron-limiting conditions, which are likely encountered within the intracellular compartment of host cells, the expression of *icsP* (Africa et al., 2011), and presumably other VirB-regulated genes, is downregulated (Figure 4.4). Therefore, transcriptional repression of P*ryhB* by Fur, the chromosomally encoded iron responsive transcription factor, indirectly leads to the upregulation of the pINV-encoded virulence regulator gene *virB* and its downstream targets on pINV, demonstrating yet another layer of a stringent control of tier 3 promoters.

CsrA

The carbon storage regulator CsrA has been shown to positively influence the expression of *virF* and *virB* ((Gore and Payne, 2010); Figure 4.4). Although this glycolysis activator/ gluconeogenesis repressor has been demonstrated as a RNA binding protein in *E. coli* (Baker et al., 2002; Dubey et al., 2003; Liu et al., 1995; Romeo, 1998; Sabnis et al., 1995), the mechanism in which it modulates *virF* and *virB* expression is unclear. What is known is that a *csrA* null mutation results in an inability of cells to attach and invade Henle cells, likely due to the reduction of *virF* and *virB* expression, which directly correlates to a reduction of invasion proteins (Gore and Payne, 2010). Meanwhile, the opposing carbon metabolism regulator Cra, which serves as a glycolysis repressor/ gluconeogenesis activator (Saier and Ramseier, 1996), has an opposite effect regarding cell invasion. *cra* mutants are able to attach and invade more Henle cells when compared to wild type (Gore and Payne, 2010). These observations suggest that *Shigella* invasion and virulence is intimately connected to glycolysis and carbon metabolism via the regulators controlling these metabolic pathways (also discussed in Eisenreich et al., 2010). Indeed, these conclusions were supported by a recent proteomic survey in which high levels of proteins involved in glycolysis were found in bacteria localized within Henle cells (Pieper et al., 2013). Therefore, although the role of carbon metabolism regulators on pINV-encoded virulence genes needs further investigation, there is undoubtedly a connection between carbon metabolism and virulence gene expression.

ArcA

The presence of an additional anaerobic transcriptional regulator ArcA has the potential to add another layer of complexity onto the existing virulence regulatory circuit. As previously discussed, cells are primed during anoxic growth via FNR-dependent repression of *spa32* and *spa33,* which prepares bacterial cells for invasion by inhibiting the secretion of effector proteins and by increasing the length and number of the type III secretion system needles (Marteyn et al., 2010). Furthermore, ArcA can bind to the *fur* promoter and repress its transcription during anaerobiosis (Boulette and Payne, 2007) which has the potential to increase production of RyhB (described above; (Massé and Gottesman, 2002; Oglesby et al., 2005)) and decrease *virB* transcription (Broach et al., 2012; Murphy and Payne, 2007). Therefore, while ArcA may indirectly influence expression of pINV-encoded virulence genes (Figure 4.4), it is apparent that it plays a role in fine-tuning and shaping the overall physiology of the bacterium to optimize infection and invasion strategies.

OmpR

The OmpR/EnvZ two component system, encoded by the *ompB* locus of *Shigella,* has been demonstrated to facilitate expression of tier 3 virulence genes of the entry region in response to changes in osmolarity and is required for *Shigella* virulence (Bernardini et al., 1990). Subsequent studies, however, demonstrated that constitutive expression of the OmpR-dependent *ompC* gene, which encodes an outer membrane porin, bypassed the need for OmpR/EnvZ and restored virulence phenotypes to *Shigella* (Bernardini et al., 1993). As such, the role of OmpR/EnvZ in the regulation of genes of the entry region appears to be indirect, but exactly how the modulation of OmpC by OmpR/EnvZ impacts virulence gene expression has yet to be characterized.

Conclusions

The regulation of *Shigella* virulence genes is a broad topic, which can be viewed succinctly as the mechanism by which *Shigella* adapts to environmental conditions encountered 'en route' through its human host. The continual sensing of the surrounding environment by *Shigella* that are in-transit triggers a whole range of bacterial responses that occur at every level of gene expression (transcriptional, post-transcriptional, translational and post-translational) to ensure that *Shigella* is both competitive and successful within its human host. In this chapter, the primary focus has been the transcriptional regulation of virulence genes carried by the large virulence plasmid pINV, which is found in almost all pathogenic *Shigella* strains (with the exception of *S. dysenteriae* serotype 1 and *S. boydii* serotype 13, see section II). This emphasis has allowed us to highlight the environmental cues, regulatory hierarchies, transcriptional regulators and the regulatory mechanisms that modulate the first step of virulence gene expression in *Shigella.* Moreover, it has demonstrated that transcription of the virulence genes carried by pINV is orchestrated by so much more than simple transcription factor:DNA interactions. To emphasize some of this complexity, we conclude by reviewing the central themes of virulence gene expression on pINV, namely transcriptional silencing and anti-silencing, which commonly controls virulence gene expression in other bacterial pathogens. We also highlight some of the regulatory mechanisms that allow virulence gene expression in *Shigella* to be tied to cues encountered within the host environment, and briefly discuss some of the more unusual and intriguing features of virulence gene expression on pINV, which merit further investigation. Finally, we describe the challenge of identifying where and when specific genes are expressed in the human host, as well as some recent advances that may significantly improve our understanding of this topic. This section will cumulatively reveal what an excellent model system *Shigella* is for the study of transcriptional regulation of virulence genes, but also demonstrate, more generally, what a fascinating organism it is for the study of regulatory mechanisms controlling bacterial transcription.

As we have seen, the virulence genes carried by pINV are transcriptionally silenced by the nucleoid structuring protein, H-NS. The H-NS-mediated transcriptional silencing of horizontally acquired genes on large plasmids, like pINV, is relatively common among the Enterobacteriaceae. Indeed, it has been proposed that the silencing of these foreign genes (xenogeneic silencing) was central in the evolution of *Shigella* species and other enteric pathogens because it allowed the acquisition and maintenance of large plasmids like pINV in many bacterial species (Stoebel et al., 2008). Transcriptional silencing occurs at ambient temperatures (below 37°C), when H-NS nucleoprotein complexes form to organize large DNA molecules like pINV. This renders the DNA inaccessible to sequence-specific DNA binding proteins, like RNA polymerase, that are needed for transcription to proceed. Although many questions remain about the extent and mechanism of H-NS-mediated transcriptional repression on pINV, it is clear that the transcriptional silencing of pINV is central to the genetic circuitry and cellular physiology of *Shigella.* Transcriptional silencing by H-NS also provides the context in which almost all other virulence gene regulation occurs on pINV. Consequently, *Shigella* is an excellent model system for studying mechanisms of transcriptional repression by H-NS, as well as mechanisms of transcriptional derepression. In *Shigella*, the major protein involved in transcriptional derepression of virulence genes is the anomalous transcriptional regulator VirB. This protein appears to have been co-opted from its former role as a plasmid partitioning protein to its new role as a transcriptional regulator. Although the array of transcriptional derepressors employed by bacterial pathogens is impressive (Stoebel et al., 2008), evidence suggests that the various mechanisms of transcriptional derepression may display some common features (see text below). Hence, studies of VirB-mediated transcriptional derepression in *Shigella* are likely to promote our understanding of virulence gene expression in *Shigella* and provide valuable insight into mechanisms of transcriptional derepression mediated by other DNA binding proteins in other bacteria.

The *Shigella* system exemplifies how the transcriptional regulation of virulence gene expression can be exquisitely tied to environmental conditions and/or cues encountered in the human host. Examples of this can be seen throughout *Shigella's* journey through its host. While the initial expression of virulence genes carried by pINV can be enhanced by physiological pH and osmolarities found in the human body (Porter and Dorman, 1994, 1997a), undoubtedly the primary cue that triggers the first three tiers of the regulatory cascade controlling virulence is an upshift to 37°C (Hromockyj et al., 1992; Maurelli et al., 1984; Maurelli and Sansonetti, 1988; Porter and Dorman, 1994, 1997a). This increase in temperature triggers a variety of regulatory events and mechanisms at each of the first 3 levels of the cascade that vary in their complexity. First, as seen at the *virF* promoter, a shift to 37°C triggers a change in DNA curvature that is sufficient to destabilize H-NS:DNA complexes that mediate H-NS-dependent repression of the *virF* gene ((Falconi et al., 1998; Falconi et al., 2001; Prosseda et al., 2004); Figure 4.5). This simple and yet elegant way of tying transcriptional derepression to a thermal upshift relies on nothing more than the DNA molecule itself, and yet it triggers the production of VirF, the first regulator of the transcriptional cascade. Second, at the *virB* promoter, even though VirF is essential for the activation of the *virB* promoter, the genome-wide increase in DNA supercoiling that occurs in response to a temperature of 37°C, maximizes transcription of the *virB* gene ((Tobe et al., 1991; Tobe et al., 1993; Tobe et al., 1995); Figure 4.5). Hence, a thermally induced change in DNA topology again leads to the upregulation of a gene within the regulatory cascade, but here the change in DNA topology is modulated indirectly via the essential enzyme DNA gyrase, which ultimately facilitates activation of the *virB* promoter by VirF. Third, the expression of the *virB* gene in response to a thermal upshift ultimately allows VirB to bind to its DNA recognition site at tier 3 promoters, where it relieves H-NS-mediated repression ((Castellanos et al., 2009; Stoebel et al., 2008; Turner and Dorman, 2007; Wing et al., 2004); Figure 4.5). Although the mechanistic details of VirB-mediated transcriptional derepression remain elusive, transcription derepression also appears to be accompanied by a change in DNA topology, at least at the *icsB* promoter (Turner and Dorman, 2007). These three examples demonstrate that *Shigella* has evolved a variety of mechanisms to tie virulence gene expression to human body temperature. Regardless, all three examples involve the modulation of DNA topology, although the importance of this in the context of transcriptional derepression by VirB remains unclear at this time.

Another example of environmental conditions impacting the transcription of virulence genes occurs in the anaerobic environment of the lower intestine and allows the "anoxic priming" of *Shigella* for cell entry (Marteyn et al., 2010). In this case, FNR-mediated repression of two genes found in the entry region of pINV increases the number and length of type III secretion needles, which facilitates contact with epithelial cells, and promotes the build-up of type III effector proteins in the bacterial cytosol, so they are ready for immediate release once host cell contact is made. Perhaps the best example of a cue that triggers the transcription of virulence genes in the host environment, however, occurs during the upregulation of tier 4 genes (Kane et al., 2002; Le Gall et al., 2005; Mavris et al., 2002a), where contact with host epithelial cells activates the *Shigella* type III secretion system. This allows the effector proteins IpaB and IpaC to be secreted from the cell and liberates their chaperone, IpgC, so that it can now function as a co-regulator with the otherwise incapacitated MxiE transcription factor. This elegant system allows the precise and coordinated expression of tier 4 genes to occur only after the type III secretion system is active and the primary wave of effectors have been secreted into the host. Clearly then, *Shigella* has evolved a variety of mechanisms to tie virulence gene expression to environmental cues encountered within its host. This ultimately allows virulence genes to be expressed in a deliberate and controlled manner, so that virulence gene products are synthesized and ready for use when needed by the bacterial pathogen.

The *Shigella* system also provides plenty of scope for those interested in the more unusual regulatory scenarios and/or processes. **H-NS and its paralogues.** While much is still to be learned about H-NS mediated transcriptional repression, the presence of one, or sometimes two, H-NS paralogues (StpA and Sfh) in different *Shigella* strains, adds additional complexity to the *Shigella* system, making it worthy of investigation (Beloin et al., 2003; Deighan et al., 2003). While these paralogues appear to act as functional backups for H-NS, at least in the context of virulence gene expression, it remains possible that they serve additional roles either by functioning on their own or by forming complexes with the other H-NS paralogues (Beloin et al., 2003; Deighan et al., 2003). **sRNA-mediated regulation.** While it is firmly established that sRNA-mediated regulation plays a significant role in bacterial gene expression (Waters and Storz, 2009), the extent by which it controls *Shigella* virulence remains unclear. The presence of novel

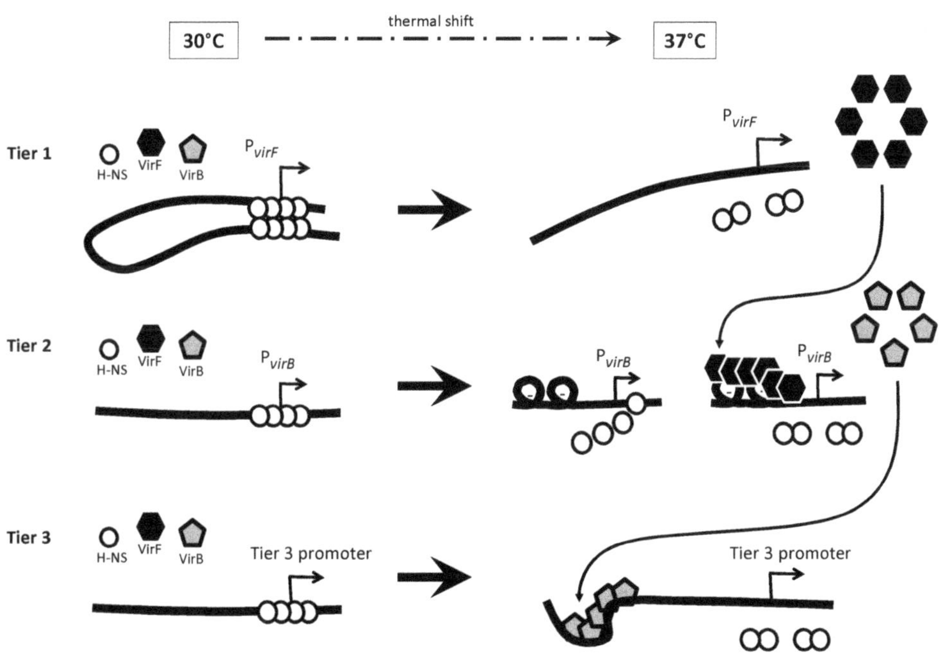

Figure 4.5. Temperature-dependent events that upregulate the first three tiers of the pINV regulatory cascade. A temperature switch to 37°C upregulates tier 1-3 promoters. At the *virF* promoter (tier 1), a thermally induced change in local DNA structure alleviates H-NS-mediated repression. At the *virB* promoter (tier 2), DNA gyrase increases negative supercoiling of the promoter region, which promotes maximal VirF-dependent transcription activation. At tier 3 promoters, VirB, which is maximally expressed at 37°C, alleviates H-NS-mediated repression through a mechanisms predicted to involve DNA wrapping and VirB oligomerization (Gao et al., 2013). Note: the positive regulatory feedback loops mediated by VirB at 37°C (Kane and Dorman, 2012) are not shown for simplicity sake. Additional regulatory inputs occur at each tier depicted (see text), but these have been omitted because they are not directly modulated by temperature. This is indicated in the figure. Line with arrow represents the transcription start site. "-" symbol represents negative supercoils.

mechanisms of sRNA-mediated regulation in *Shigella* is also possible, based on recent findings regarding RyhB-dependent regulation of *virB* transcription (Africa et al., 2011; Broach et al., 2012; Murphy and Payne, 2007), although further studies are required to support this unusual regulation. ***mxiE* expression**. Although MxiE plays an important role in tying the expression of tier 4 genes to the activation of the type III secretion, the transcriptional slippage that must occur in order for the full-length *mxiE* gene to be transcribed is another fascinating example of transcriptional control in *Shigella* (Penno et al., 2005). While the purpose of this regulation remains unclear, it certainly is both novel and interesting from a gene expression standpoint. **pINV entry region**. Finally, the entry region of pINV, which contains the *ipa-mxi-spa* operons (which appear to range from approximately 7-11 kb in length), raises questions about the transcripts generated from this locus; what is their length, and how is the length of each transcript regulated during the transcription of these extremely long operons?

Throughout this chapter we have presented our current understanding of the transcriptional control of virulence genes on pINV, and highlighted the extent and/or limit of our knowledge to facilitate future research questions. However, the topic of where and when genes of pINV are transcribed in the human host has only been briefly discussed. This is because studies that specifically address this question are relatively rare. Of course, the ideal solution to this problem is to study transcriptional regulation of virulence genes *in situ* within human tissues. Although a lofty goal, recent advances have perhaps made this quest more attainable (Campbell-Valois and Sansonetti, 2014; Campbell-Valois et al., 2014). The use of a fast maturing green fluorescent protein as a transcriptional reporter of the MxiE-regulated *ipaH7.8* gene has revealed the dynamic and temporal regulation of this gene during the course of infection, being initially upregulated upon cell entry and subsequently downregulated when bacteria gain access to the host cell cytoplasm (Campbell-Valois et al., 2014). It is proposed that this pattern of transcription allows the replenishment of bacterial stores of this type III effector so that the cell is prepared for invasion of the neighboring cell (Campbell-Valois et al., 2014). While these studies were primarily done in the colonic epithelial cell line TC7 cells, it seems reasonable to predict that with the right animal model, suitable reporter protein and high resolution imaging system, it would be possible to observe the expression of various virulence genes in real time within animal tissue. This future direction is an exciting prospect for those interested in *Shigella* pathogenicity because it would not only allow us to identify where and when a particular gene is expressed, but may also allow us to determine the sequential expression of *Shigella* virulence genes in the human host. This information has the potential to be extremely powerful for those interested in novel drug design, because if we know that a specific gene product is only expressed in the host cell cytoplasm, the importance of delivering a drug that inhibits the action of this gene product to the host cell cytoplasm becomes abundantly clear.

In summary then, *Shigella* provides an excellent model system in which to study transcriptional silencing and anti-silencing, regulatory processes that not only control virulence gene expression in *Shigella* but also in many other bacterial pathogens. The *Shigella* system also yields an exciting array of transcriptional mechanisms that tie virulence gene expression to environmental conditions and cues encountered in the human host, making it a fascinating organism for the study of regulatory mechanisms controlling bacterial transcription. While the transcription regulation of virulence genes may seem disjointed from bacterial pathogenesis at first glance, in this chapter we have highlighted how the precise and coordinated transcriptional regulation of *Shigella* virulence is essential for the success and pathogenicity of *Shigella* species. Therefore, studies that improve understanding of transcriptional regulation of virulence genes will also undoubtedly advance our knowledge of *Shigella* pathogenesis and its interplay with its human host.

Footnote: For those interested in a broader view of virulence gene regulation, including a description of regulatory events that occur at and beyond the transcriptional level, a recent review entitled "*Shigella*: a model of virulence regulation *in vivo*" describes these events from the perspective of regulation of the T3SS; an essential virulence structure and secretory system in the pathogenicity of *Shigella* (Marteyn et al., 2012). Those specifically interested in the regulatory pathways that control the expression of virulence gene encoded by the *Shigella* chromosome are referred to recent article from Dr. Shelley Payne's laboratory that presents a helpful summary (Pieper et al., 2013).

Acknowledgements
We wish to thank Drs. Claude Parsot and Sylvie Rimsky who discussed various aspects of this chapter with us during its preparation. We also thank the editor for the opportunity to present our view of this field. Our research has been and is supported by the NIH grant AI 090573.

References

Abeles, A.L., Friedman, S.A., and Austin, S.J. (1985). Partition of unit-copy miniplasmids to daughter cells. III. The DNA sequence and functional organization of the P1 partition region. Journal of molecular biology *185*, 261-272.

Adler, B., Sasakawa, C., Tobe, T., Makino, S., Komatsu, K., and Yoshikawa, M. (1989). A dual transcriptional activation system for the 230 kb plasmid genes coding for virulence-associated antigens of *Shigella flexneri*. Molecular Microbiology *3*, 627-635.

Africa, L.A., Murphy, E.R., Egan, N.R., Wigley, A.F., and Wing, H.J. (2011). The iron-responsive Fur/RyhB regulatory cascade modulates the *Shigella* outer membrane protease IcsP. Infection and immunity *79*, 4543-4549.

Amit, R., Oppenheim, A.B., and Stavans, J. (2003). Increased bending rigidity of single DNA molecules by H-NS, a temperature and osmolarity sensor. Biophysical journal *84*, 2467-2473.

Arold, S.T., Leonard, P.G., Parkinson, G.N., and Ladbury, J.E. (2010). H-NS forms a superhelical protein scaffold for DNA condensation. Proceedings of the National Academy of Sciences of the United States of America *107*, 15728-15732.

Atlung, T., and Ingmer, H. (1997). H-NS: a modulator of environmentally regulated gene expression. Mol Microbiol *24*, 7-17.

Badaut, C., Williams, R., Arluison, V., Bouffartigues, E., Robert, B., Buc, H., and Rimsky, S. (2002). The degree of oligomerization of the H-NS nucleoid structuring protein is related to specific binding to DNA. J Biol Chem *277*, 41657-41666.

Baker, C.S., Morozov, I., Suzuki, K., Romeo, T., and Babitzke, P. (2002). CsrA regulates glycogen biosynthesis by preventing translation of glgC in *Escherichia coli*. Mol Microbiol *44*, 1599-1610.

Balleza, E., Lopez-Bojorquez, L.N., Martinez-Antonio, A., Resendis-Antonio, O., Lozada-Chavez, I., Balderas-Martinez, Y.I., Encarnacion, S., and Collado-Vides, J. (2009). Regulation by transcription factors in bacteria: beyond description. FEMS Microbiol Rev *33*, 133-151.

Basta, D.W., Pew, K.L., Immak, J.A., Park, H.S., Picker, M.A., Wigley, A.F., Hensley, C.T., Pearson, J.S., Hartland, E.L., and Wing, H.J. (2013). Characterization of the *ospZ* promoter in *Shigella flexneri* and its regulation by VirB and H-NS. Journal of bacteriology *195*, 2562-2572.

Beloin, C., Deighan, P., Doyle, M., and Dorman, C.J. (2003). *Shigella flexneri* 2a strain 2457T expresses three members of the H-NS-like protein family: characterization of the Sfh protein. Molecular genetics and genomics : MGG *270*, 66-77.

Beloin, C., and Dorman, C.J. (2003). An extended role for the nucleoid structuring protein H-NS in the virulence gene regulatory cascade of *Shigella flexneri*. Mol Microbiol *47*, 825-838.

Beloin, C., McKenna, S., and Dorman, C.J. (2002). Molecular Dissection of VirB, a Key Regulator of the Virulence Cascade of *Shigella flexneri*. Journal of Biological Chemistry *277*, 15333-15344.

Bensted, H.J. (1956). Dysentery bacilli-*Shigella*; a brief historical review. Canadian journal of microbiology *2*, 163-174.

Berlutti, F., Casalino, M., Zagaglia, C., Fradiani, P.A., Visca, P., and Nicoletti, M. (1998). Expression of the virulence plasmid-carried apyrase gene (*apy*) of enteroinvasive *Escherichia coli* and *Shigella flexneri* is under the control of H-NS and the VirF and VirB regulatory cascade. Infection and immunity *66*, 4957-4964.

Bernardini, M.L., Fontaine, A., and Sansonetti, P.J. (1990). The two-component regulatory system *ompR-envZ* controls the virulence of *Shigella flexneri*. Journal of bacteriology *172*, 6274-6281.

Bernardini, M.L., Mounier, J., d'Hauteville, H., Coquis-Rondon, M., and Sansonetti, P.J. (1989). Identification of *icsA*, a Plasmid Locus of *Shigella flexneri* that Governs Bacterial Intra- and Intercellular Spread through Interaction with F-Actin. Proceedings of the National Academy of Sciences of the United States of America *86*, 3867-3871.

Bernardini, M.L., Sanna, M.G., Fontaine, A., and Sansonetti, P.J. (1993). OmpC is involved in invasion of epithelial cells by Shigella flexneri. Infection and immunity *61*, 3625-3635.

Bongrand, C., Sansonetti, P.J., and Parsot, C. (2012). Characterization of the promoter, MxiE box and 5' UTR of genes controlled by the activity of the type III secretion apparatus in *Shigella flexneri*. PloS one *7*, e32862.

Bouffartigues, E., Buckle, M., Badaut, C., Travers, A., and Rimsky, S. (2007). H-NS cooperative binding to high-affinity sites in a regulatory element results in transcriptional silencing. Nature structural and molecular biology *14*, 441-448.

Boulette, M.L., and Payne, S.M. (2007). Anaerobic regulation of *Shigella flexneri* virulence: ArcA regulates Fur and iron acquisition genes. Journal of bacteriology *189*, 6957-6967.

Brenner, D.J., Fanning, G.R., Steigerwalt, A.G., Orskov, I., and Orskov, F. (1972). Polynucleotide sequence relatedness among three groups of pathogenic *Escherichia coli* strains. Infection and immunity *6*, 308-315.

Brescia, C.C., Kaw, M.K., and Sledjeski, D.D. (2004). The DNA binding protein H-NS binds to and alters the stability of RNA *in vitro* and *in vivo*. Journal of molecular biology *339*, 505-514.

Broach, W.H., Egan, N., Wing, H.J., Payne, S.M., and Murphy, E.R. (2012). VirF-independent regulation of *Shigella virB* transcription is mediated by the small RNA RyhB. PloS one *7*, e38592.

Browning, D.F., Grainger, D.C., and Busby, S.J. (2010). Effects of nucleoid-associated proteins on bacterial chromosome structure and gene expression. Current opinion in microbiology *13*, 773-780.

Buchrieser, C., Glaser, P., Rusniok, C., Nedjari, H., D'Hauteville, H., Kunst, F., Sansonetti, P., and Parsot, C. (2000). The virulence plasmid pWR100 and the repertoire of proteins secreted by the type III secretion apparatus of *Shigella flexneri*. Molecular Microbiology *38*, 760-771.

Cai, X., Zhang, J., Chen, M., Wu, Y., Wang, X., Chen, J., Zhang, J., Shen, X., Qu, D., and Jiang, H. (2011). The effect of the potential PhoQ histidine kinase inhibitors on *Shigella flexneri* virulence. PloS one *6*, e23100.

Campbell-Valois, F.X., and Sansonetti, P.J. (2014). Tracking bacterial pathogens with genetically-encoded reporters. FEBS letters *588*, 2428-2436.

Campbell-Valois, F.X., Schnupf, P., Nigro, G., Sachse, M., Sansonetti, P.J., and Parsot, C. (2014). A fluorescent reporter reveals on/off regulation of the *Shigella* type III secretion apparatus during entry and cell-to-cell spread. Cell host and microbe *15*, 177-189.

Castellanos, M.I., Harrison, D.J., Smith, J.M., Labahn, S.K., Levy, K.M., and Wing, H.J. (2009). VirB alleviates H-NS repression of the *icsP* promoter in *Shigella flexneri* from sites more than one kilobase upstream of the transcription start site. Journal of bacteriology *191*, 4047-4050.

Collado-Vides, J., Magasanik, B., and Gralla, J.D. (1991). Control site location and transcriptional regulation in *Escherichia coli*. Microbiological reviews *55*, 371-394.

Cukier-Kahn, R., Jacquet, M., and Gros, F. (1972). Two heat-resistant, low molecular weight proteins from *Escherichia coli* that stimulate DNA-directed RNA synthesis. Proceedings of the National Academy of Sciences of the United States of America *69*, 3643-3647.

Cusick, M.E., and Belfort, M. (1998). Domain structure and RNA annealing activity of the *Escherichia coli* regulatory protein StpA. Mol Microbiol *28*, 847-857.

Dame, R.T., Luijsterburg, M.S., Krin, E., Bertin, P.N., Wagner, R., and Wuite, G.J. (2005). DNA bridging: a property shared among H-NS-like proteins. Journal of bacteriology *187*, 1845-1848.

Dame, R.T., Wyman, C., and Goosen, N. (2000). H-NS mediated compaction of DNA visualised by atomic force microscopy. Nucleic Acids Res *28*, 3504-3510.

Dame, R.T., Wyman, C., and Goosen, N. (2001). Structural basis for preferential binding of H-NS to curved DNA. Biochimie *83*, 231-234.

Dame, R.T., Wyman, C., Wurm, R., Wagner, R., and Goosen, N. (2002). Structural basis for H-NS-mediated trapping of RNA polymerase in the open initiation complex at the *rrnB* P1. J Biol Chem *277*, 2146-2150.

Darwin, K.H., and Miller, V.L. (2001). Type III secretion chaperone-dependent regulation: activation of virulence genes by SicA and InvF in *Salmonella typhimurium*. Embo J *20*, 1850-1862.

Deighan, P., Beloin, C., and Dorman, C.J. (2003). Three-way interactions among the Sfh, StpA and H-NS nucleoid-structuring proteins of *Shigella flexneri* 2a strain 2457T. Mol Microbiol *48*, 1401-1416.

Deighan, P., Free, A., and Dorman, C.J. (2000). A role for the *Escherichia coli* H-NS-like protein StpA in OmpF porin expression through modulation of *micF* RNA stability. Mol Microbiol *38*, 126-139.

Demers, B., Sansonetti, P.J., and Parsot, C. (1998). Induction of type III secretion in *Shigella flexneri* is associated with differential control of transcription of genes encoding secreted proteins. EMBO J *17*, 2894-2903.

Dorman, C.J. (1992). The VirF protein from *Shigella flexneri* is a member of the AraC transcription factor superfamily and is highly homologous to Rns, a positive regulator of virulence genes in enterotoxigenic *Escherichia coli*. Mol Microbiol *6*, 1575.

Dorman, C.J. (2004). H-NS: a universal regulator for a dynamic genome. Nat Rev Microbiol *2*, 391-400.

Dorman, C.J. (2007). H-NS, the genome sentinel. Nat Rev Microbiol *5*, 157-161.

Dorman, C.J. (2009). The virulence plasmids of *Shigella flexneri*. In Microbial megaplasmids (Springer), pp. 151-170.

Dorman, C.J., Bhriain, N.N., and Higgins, C.F. (1990). DNA supercoiling and environmental regulation of virulence gene expression in *Shigella flexneri*. Nature *344*, 789-792.

Dorman, C.J., Hinton, J.C., and Free, A. (1999). Domain organization and oligomerization among H-NS-like nucleoid-associated proteins in bacteria. Trends Microbiol *7*, 124-128.

Dorman, C.J., McKenna, S., and Beloin, C. (2001). Regulation of virulence gene expression in *Shigella flexneri*, a facultative intracellular pathogen. Int J Med Microbiol *291*, 89-96.

Dorman, C.J., and Porter, M.E. (1998). The *Shigella* virulence gene regulatory cascade: a paradigm of bacterial gene control mechanisms. Molecular Microbiology *29*, 677-684.

Dubey, A.K., Baker, C.S., Suzuki, K., Jones, A.D., Pandit, P., Romeo, T., and Babitzke, P. (2003). CsrA regulates translation of the *Escherichia coli* carbon starvation gene, *cstA*, by blocking ribosome access to the *cstA* transcript. Journal of bacteriology *185*, 4450-4460.

DuPont, H.L., Levine, M.M., Hornick, R.B., and Formal, S.B. (1989). Inoculum size in shigellosis and implications for expected mode of transmission. The Journal of infectious diseases *159*, 1126-1128.

Durand, J.M., and Bjork, G.R. (2003). Putrescine or a combination of methionine and arginine restores virulence gene expression in a tRNA modification-deficient mutant of *Shigella flexneri*: a possible role in adaptation of virulence. Mol Microbiol *47*, 519-527.

Durand, J.M., Bjork, G.R., Kuwae, A., Yoshikawa, M., and Sasakawa, C. (1997). The modified nucleoside 2-methylthio-N6-isopentenyladenosine in tRNA of *Shigella flexneri* is required for expression of virulence genes. Journal of bacteriology *179*, 5777-5782.

Durand, J.M., Dagberg, B., Uhlin, B.E., and Bjork, G.R. (2000). Transfer RNA modification, temperature and DNA superhelicity have a common target in the regulatory network of the virulence of *Shigella flexneri*: the expression of the *virF* gene. Mol Microbiol *35*, 924-935.

Durand, J.M., Okada, N., Tobe, T., Watarai, M., Fukuda, I., Suzuki, T., Nakata, N., Komatsu, K., Yoshikawa, M., and Sasakawa, C. (1994). *vacC*, a virulence-associated chromosomal locus of *Shigella flexneri,* is homologous to *tgt*, a gene encoding tRNA-guanine transglycosylase (Tgt) of *Escherichia coli* K-12. Journal of bacteriology *176*, 4627-4634.

Egile, C., d'Hauteville, H., Parsot, C., and Sansonetti, P.J. (1997). SopA, the outer membrane protease responsible for polar localization of IcsA in *Shigella flexneri*. Mol Microbiol *23*, 1063-1073.

Eisenreich, W., Dandekar, T., Heesemann, J., and Goebel, W. (2010). Carbon metabolism of intracellular bacterial pathogens and possible links to virulence. Nat Rev Microbiol *8*, 401-412.

Esposito, D., Petrovic, A., Harris, R., Ono, S., Eccleston, J.F., Mbabaali, A., Haq, I., Higgins, C.F., Hinton, J.C., Driscoll, P.C.*, et al.* (2002). H-NS oligomerization domain structure reveals the mechanism for high order self-association of the intact protein. Journal of molecular biology *324*, 841-850.

Ewing, W.H. (1949). *Shigella* Nomenclature. Journal of bacteriology *57*, 633-638.

Ewing, W.H. (1986). Edward and Ewing's Identification of the Enterobacteriaceae (Elsevier Science Publishers, Amsterdam, The Netherlands).

Falconi, M., Colonna, B., Prosseda, G., Micheli, G., and Gualerzi, C.O. (1998). Thermoregulation of *Shigella* and *Escherichia coli* EIEC pathogenicity. A temperature-dependent structural transition of DNA modulates accessibility of *virF* promoter to transcriptional repressor H-NS. EMBO J *17*, 7033-7043.

Falconi, M., Prosseda, G., Giangrossi, M., Beghetto, E., and Colonna, B. (2001). Involvement of FIS in the H-NS-mediated regulation of *virF* gene of *Shigella* and enteroinvasive *Escherichia coli*. Mol Microbiol *42*, 439-452.
Fang, F.C., and Rimsky, S. (2008). New insights into transcriptional regulation by H-NS. Current opinion in microbiology *11*, 113-120.
Free, A., and Dorman, C.J. (1995). Coupling of *Escherichia coli hns* mRNA levels to DNA synthesis by autoregulation: implications for growth phase control. Mol Microbiol *18*, 101-113.
Free, A., and Dorman, C.J. (1997). The *Escherichia coli stpA* gene is transiently expressed during growth in rich medium and is induced in minimal medium and by stress conditions. Journal of bacteriology *179*, 909-918.
Free, A., Williams, R.M., and Dorman, C.J. (1998). The StpA protein functions as a molecular adapter to mediate repression of the *bgl* operon by truncated H-NS in *Escherichia coli*. Journal of bacteriology *180*, 994-997.
Gallegos, M.T., Schleif, R., Bairoch, A., Hofmann, K., and Ramos, J.L. (1997). Arac/XylS family of transcriptional regulators. Microbiology and molecular biology reviews : MMBR *61*, 393-410.
Gao, X., Zou, T., Mu, Z., Qin, B., Yang, J., Waltersperger, S., Wang, M., Cui, S., and Jin, Q. (2013). Structural insights into VirB-DNA complexes reveal mechanism of transcriptional activation of virulence genes. Nucleic Acids Res.
Giangrossi, M., Prosseda, G., Tran, C.N., Brandi, A., Colonna, B., and Falconi, M. (2010). A novel antisense RNA regulates at transcriptional level the virulence gene *icsA* of *Shigella flexneri*. Nucleic Acids Res *38*, 3362-3375.
Goldberg, M.B., Barzu, O., Parsot, C., and Sansonetti, P.J. (1993). Unipolar localization and ATPase activity of IcsA, a *Shigella flexneri* protein involved in intracellular movement. Infectious agents and disease *2*, 210-211.
Goldman, S.R., Tu, Y., and Goldberg, M.B. (2008). Differential regulation by magnesium of the two MsbB paralogs of *Shigella flexneri*. Journal of bacteriology *190*, 3526-3537.
Goosen, N., and van de Putte, P. (1995). The regulation of transcription initiation by integration host factor. Molecular Microbiology *16*, 1-7.
Gordon, B.R., Li, Y., Cote, A., Weirauch, M.T., Ding, P., Hughes, T.R., Navarre, W.W., Xia, B., and Liu, J. (2011). Structural basis for recognition of AT-rich DNA by unrelated xenogeneic silencing proteins. Proceedings of the National Academy of Sciences of the United States of America *108*, 10690-10695.
Gore, A.L., and Payne, S.M. (2010). CsrA and Cra Influence *Shigella flexneri* Pathogenesis. Infection and immunity *78*, 4674-4682.
Harrison, D.J. (2010). Transcriptional regulation of the *Shigella flexneri icsP* promoter: Silencing and anti-silencing by H-NS and VirB. In School of Life Sciences (Las Vegas, Nevada, USA.: University of Nevada, Las Vegas), pp. 145.
Hensley, C.T., Kamneva, O.K., Levy, K.M., Labahn, S.K., Africa, L.A., and Wing, H.J. (2011). Two promoters and two translation start sites control the expression of the *Shigella flexneri* outer membrane protease IcsP. Arch Microbiol *193*, 263-274.
Hromockyj, A.E., Tucker, S.C., and Maurelli, A.T. (1992). Temperature regulation of *Shigella* virulence: identification of the repressor gene *virR*, an analogue of *hns*, and partial complementation by tyrosyl transfer RNA (tRNA1(Tyr)). Mol Microbiol *6*, 2113-2124.
Hurt, J.K., Olgen, S., and Garcia, G.A. (2007). Site-specific modification of *Shigella flexneri virF* mRNA by tRNA-guanine transglycosylase *in vitro*. Nucleic Acids Res *35*, 4905-4913.

Jacquet, M., Cukier-Kahn, R., Pla, J., and Gros, F. (1971). A thermostable protein factor acting on *in vitro* DNA transcription. Biochemical and biophysical research communications *45*, 1597-1607.

Johansson, J., Dagberg, B., Richet, E., and Uhlin, B.E. (1998). H-NS and StpA proteins stimulate expression of the maltose regulon in *Escherichia coli*. Journal of bacteriology *180*, 6117-6125.

Jordi, B.J., Dagberg, B., de Haan, L.A., Hamers, A.M., van der Zeijst, B.A., Gaastra, W., and Uhlin, B.E. (1992). The positive regulator CfaD overcomes the repression mediated by histone-like protein H-NS (H1) in the CFA/I fimbrial operon of *Escherichia coli*. EMBO J *11*, 2627-2632.

Jost, B.H., and Adler, B. (1993). Site of transcriptional activation of *virB* on the large plasmid of *Shigella flexneri* 2a by VirF, a member of the AraC family of transcriptional activators. Microb Pathog *14*, 481-488.

Kane, C.D., Schuch, R., Day, W.A., Jr., and Maurelli, A.T. (2002). MxiE regulates intracellular expression of factors secreted by the *Shigella flexneri* 2a type III secretion system. Journal of bacteriology *184*, 4409-4419.

Kane, K.A., and Dorman, C.J. (2012). VirB-Mediated Positive Feedback Control of the Virulence Gene Regulatory Cascade of *Shigella flexneri*. Journal of bacteriology *194*, 5264-5273.

Lammi, M., Paci, M., Pon, C.L., Losso, M.A., Miano, A., Pawlik, R.T., Gianfranceschi, G.L., and Gualerzi, C.O. (1984). Proteins from the prokaryotic nucleoid: biochemical and 1H NMR studies on three bacterial histone-like proteins. Advances in experimental medicine and biology *179*, 467-477.

Lan, R., Alles, M.C., Donohoe, K., Martinez, M.B., and Reeves, P.R. (2004). Molecular evolutionary relationships of enteroinvasive *Escherichia coli* and *Shigella* spp. Infection and immunity *72*, 5080-5088.

Lan, R., and Reeves, P.R. (2002). *Escherichia coli* in disguise: molecular origins of *Shigella*. Microbes and infection / Institut Pasteur *4*, 1125-1132.

Lan, R., Stevenson, G., and Reeves, P.R. (2003). Comparison of two major forms of the *Shigella* virulence plasmid pINV: positive selection is a major force driving the divergence. Infection and immunity *71*, 6298-6306.

Lang, B., Blot, N., Bouffartigues, E., Buckle, M., Geertz, M., Gualerzi, C.O., Mavathur, R., Muskhelishvili, G., Pon, C.L., Rimsky, S., *et al.* (2007). High-affinity DNA binding sites for H-NS provide a molecular basis for selective silencing within proteobacterial genomes. Nucleic Acids Res *35*, 6330-6337.

Le Gall, T., Mavris, M., Martino, M.C., Bernardini, M.L., Denamur, E., and Parsot, C. (2005). Analysis of virulence plasmid gene expression defines three classes of effectors in the type III secretion system of *Shigella flexneri*. Microbiology *151*, 951-962.

Lee, D.J., Minchin, S.D., and Busby, S.J. (2012). Activating transcription in bacteria. Annual review of microbiology *66*, 125-152.

Lett, M.C., Sasakawa, C., Okada, N., Sakai, T., Makino, S., Yamada, M., Komatsu, K., and Yoshikawa, M. (1989). *virG*, a plasmid-coded virulence gene of *Shigella flexneri*: identification of the *virG* protein and determination of the complete coding sequence. Journal of bacteriology *171*, 353-359.

Lim, C.J., Kenney, L.J., and Yan, J. (2014). Single-molecule studies on the mechanical interplay between DNA supercoiling and H-NS DNA architectural properties. Nucleic Acids Res *42*, 8369-8378.

Lim, C.J., Lee, S.Y., Kenney, L.J., and Yan, J. (2012). Nucleoprotein filament formation is the structural basis for bacterial protein H-NS gene silencing. Scientific reports *2*, 509.

Liu, L.F., and Wang, J.C. (1987). Supercoiling of the DNA template during transcription. Proceedings of the National Academy of Sciences of the United States of America *84*, 7024-7027.

Liu, M.Y., Yang, H., and Romeo, T. (1995). The product of the pleiotropic *Escherichia coli gene csrA* modulates glycogen biosynthesis via effects on mRNA stability. Journal of bacteriology *177*, 2663-2672.

Liu, Y., Chen, H., Kenney, L.J., and Yan, J. (2010). A divalent switch drives H-NS/DNA-binding conformations between stiffening and bridging modes. Genes and development *24*, 339-344.

Lucchini, S., Rowley, G., Goldberg, M.D., Hurd, D., Harrison, M., and Hinton, J.C.D. (2006). H-NS Mediates the Silencing of Laterally Acquired Genes in Bacteria. PLoS Pathogens *2*, 0746-0752.

Magdalena, J., Hachani, A., Chamekh, M., Jouihri, N., Gounon, P., Blocker, A., and Allaoui, A. (2002). Spa32 regulates a switch in substrate specificity of the type III secreton of *Shigella flexneri* from needle components to Ipa proteins. Journal of bacteriology *184*, 3433-3441.

Marteyn, B., Gazi, A., and Sansonetti, P. (2012). *Shigella*: a model of virulence regulation *in vivo*. Gut microbes *3*, 104-120.

Marteyn, B., West, N.P., Browning, D.F., Cole, J.A., Shaw, J.G., Palm, F., Mounier, J., Prevost, M.C., Sansonetti, P., and Tang, C.M. (2010). Modulation of *Shigella* virulence in response to available oxygen *in vivo*. Nature *465*, 355-358.

Martin, R.G., and Rosner, J.L. (2001). The AraC transcriptional activators. Current opinion in microbiology *4*, 132-137.

Massé, E., and Gottesman, S. (2002). A small RNA regulates the expression of genes involved in iron metabolism in *Escherichia coli*. Proceedings of the National Academy of Sciences *99*, 4620-4625.

Maurelli, A.T., Blackmon, B., and Curtiss, R., 3rd (1984). Temperature-dependent expression of virulence genes in *Shigella* species. Infection and immunity *43*, 195-201.

Maurelli, A.T., and Sansonetti, P.J. (1988). Identification of a Chromosomal Gene Controlling Temperature-Regulated Expression of *Shigella* Virulence. Proceedings of the National Academy of Sciences of the United States of America *85*, 2820-2824.

Maurer, S., Fritz, J., and Muskhelishvili, G. (2009). A systematic *in vitro* study of nucleoprotein complexes formed by bacterial nucleoid-associated proteins revealing novel types of DNA organization. Journal of molecular biology *387*, 1261-1276.

Mavris, M., Page, A.L., Tournebize, R., Demers, B., Sansonetti, P., and Parsot, C. (2002a). Regulation of transcription by the activity of the *Shigella flexneri* type III secretion apparatus. Mol Microbiol *43*, 1543-1553.

Mavris, M., Sansonetti, P.J., and Parsot, C. (2002b). Identification of the *cis*-acting site involved in activation of promoters regulated by activity of the type III secretion apparatus in *Shigella flexneri*. Journal of bacteriology *184*, 6751-6759.

McKenna, S., Beloin, C., and Dorman, C.J. (2003). *In vitro* DNA-binding properties of VirB, the *Shigella flexneri* virulence regulatory protein. FEBS letters *545*, 183-187.

Menard, R., Sansonetti, P., Parsot, C., and Vasselon, T. (1994). Extracellular association and cytoplasmic partitioning of the IpaB and IpaC invasins of *S. flexneri*. Cell *79*, 515-525.

Miller, J. (1972). Experiments in Molecular Genetics. (Cold Spring Harbor, NY: Cold Spring Harbor Laboratory Press).

Mitobe, J., Arakawa, E., and Watanabe, H. (2005). A sensor of the two-component system CpxA affects expression of the type III secretion system through posttranscriptional processing of InvE. Journal of bacteriology *187*, 107-113.

Mitobe, J., Morita-Ishihara, T., Ishihama, A., and Watanabe, H. (2008). Involvement of RNA-binding protein Hfq in the post-transcriptional regulation of *invE* gene expression in *Shigella sonnei*. J Biol Chem *283*, 5738-5747.

Mitobe, J., Morita-Ishihara, T., Ishihama, A., and Watanabe, H. (2009). Involvement of RNA-binding protein Hfq in the osmotic-response regulation of *invE* gene expression in *Shigella sonnei*. BMC microbiology *9*, 110.

Mori, H., Kondo, A., Ohshima, A., Ogura, T., and Hiraga, S. (1986). Structure and function of the F plasmid genes essential for partitioning. Journal of molecular biology *192*, 1-15.

Moss, J.E., Fisher, P.E., Vick, B., Groisman, E.A., and Zychlinsky, A. (2000). The regulatory protein PhoP controls susceptibility to the host inflammatory response in *Shigella flexneri*. Cellular microbiology 2, 443-452.

Munson, G.P. (2013). Virulence regulons of enterotoxigenic *Escherichia coli*. Immunologic research *57*, 229-236.

Munson, G.P., Holcomb, L.G., and Scott, J.R. (2001). Novel group of virulence activators within the AraC family that are not restricted to upstream binding sites. Infection and immunity *69*, 186-193.

Munson, G.P., and Scott, J.R. (1999). Binding site recognition by Rns, a virulence regulator in the AraC family. Journal of bacteriology *181*, 2110-2117.

Munson, G.P., and Scott, J.R. (2000). Rns, a virulence regulator within the AraC family, requires binding sites upstream and downstream of its own promoter to function as an activator. Mol Microbiol *36*, 1391-1402.

Murphy, E.R., and Payne, S.M. (2007). RyhB, an iron-responsive small RNA molecule, regulates *Shigella dysenteriae* virulence. Infection and immunity *75*, 3470-3477.

Nakata, N., Sasakawa, C., Okada, N., Tobe, T., Fukuda, I., Suzuki, T., Komatsu, K., and Yoshikawa, M. (1992). Identification and characterization of *virK*, a virulence-associated large plasmid gene essential for intercellular spreading of *Shigella flexneri*. Mol Microbiol *6*, 2387-2395.

Nakayama, S., and Watanabe, H. (1995). Involvement of *cpxA*, a sensor of a two-component regulatory system, in the pH-dependent regulation of expression of *Shigella sonnei virF* gene. Journal of bacteriology *177*, 5062-5069.

Nakayama, S., and Watanabe, H. (1998). Identification of *cpxR* as a positive regulator essential for expression of the *Shigella sonnei virF* gene. Journal of bacteriology *180*, 3522-3528.

Navarre, W.W., Halsey, T.A., Walthers, D., Frye, J., McClelland, M., Potter, J.L., Kenney, L.J., Gunn, J.S., Fang, F.C., and Libby, S.J. (2005). Co-regulation of *Salmonella enterica* genes required for virulence and resistance to antimicrobial peptides by SlyA and PhoP/PhoQ. Mol Microbiol *56*, 492-508.

Ochman, H., Lawrence, J.G., and Groisman, E.A. (2000). Lateral gene transfer and the nature of bacterial innovation. Nature *405*, 299-304.

Ochman, H., Whittam, T.S., Caugant, D.A., and Selander, R.K. (1983). Enzyme polymorphism and genetic population structure in *Escherichia coli* and *Shigella*. Journal of general microbiology *129*, 2715-2726.

Oglesby, A.G., Murphy, E.R., Iyer, V.R., and Payne, S.M. (2005). Fur regulates acid resistance in *Shigella flexneri* via RyhB and ydeP. Mol Microbiol *58*, 1354-1367.

Opel, M.L., Aeling, K.A., Holmes, W.M., Johnson, R.C., Benham, C.J., and Hatfield, G.W. (2004). Activation of transcription initiation from a stable RNA promoter by a Fis protein-mediated DNA structural transmission mechanism. Mol Microbiol *53*, 665-674.

Page, A.L., Fromont-Racine, M., Sansonetti, P., Legrain, P., and Parsot, C. (2001). Characterization of the interaction partners of secreted proteins and chaperones of *Shigella flexneri*. Mol Microbiol *42*, 1133-1145.

Page, A.L., Sansonetti, P., and Parsot, C. (2002). Spa15 of *Shigella flexneri*, a third type of chaperone in the type III secretion pathway. Mol Microbiol *43*, 1533-1542.

Parsot, C., Ageron, E., Penno, C., Mavris, M., Jamoussi, K., d'Hauteville, H., Sansonetti, P., and Demers, B. (2005). A secreted anti-activator, OspD1, and its chaperone, Spa15, are involved in the control of transcription by the type III secretion apparatus activity in *Shigella flexneri*. Mol Microbiol *56*, 1627-1635.

Penno, C., Sansonetti, P., and Parsot, C. (2005). Frameshifting by transcriptional slippage is involved in production of MxiE, the transcription activator regulated by the activity of the type III secretion apparatus in *Shigella flexneri*. Mol Microbiol *56*, 204-214.

Pieper, R., Fisher, C.R., Suh, M.J., Huang, S.T., Parmar, P., and Payne, S.M. (2013). Analysis of the proteome of intracellular *Shigella flexneri* reveals pathways important for intracellular growth. Infection and immunity *81*, 4635-4648.

Pilonieta, M.C., and Munson, G.P. (2008). The chaperone IpgC copurifies with the virulence regulator MxiE. Journal of bacteriology *190*, 2249-2251.

Porter, M.E., and Dorman, C.J. (1994). A role for H-NS in the thermo-osmotic regulation of virulence gene expression in *Shigella flexneri*. Journal of bacteriology *176*, 4187-4191.

Porter, M.E., and Dorman, C.J. (1997a). Differential regulation of the plasmid-encoded genes in the *Shigella flexneri* virulence regulon. Molec Gen Genet *256*, 93-103.

Porter, M.E., and Dorman, C.J. (1997b). Positive regulation of *Shigella flexneri* virulence genes by integration host factor. Journal of bacteriology *179*, 6537-6550.

Porter, M.E., and Dorman, C.J. (2002). *In Vivo* DNA-Binding and Oligomerization Properties of the *Shigella flexneri* AraC-Like Transcriptional Regulator VirF as Identified by Random and Site-Specific Mutagenesis. Journal of bacteriology *184*, 531-539.

Porter, M.E., Smith, S.G., and Dorman, C.J. (1998). Two highly related regulatory proteins, *Shigella flexneri* VirF and enterotoxigenic *Escherichia coli* Rns, have common and distinct regulatory properties. FEMS microbiology letters *162*, 303-309.

Prosseda, G., Falconi, M., Giangrossi, M., Gualerzi, C.O., Micheli, G., and Colonna, B. (2004). The *virF* promoter in *Shigella*: more than just a curved DNA stretch. Mol Microbiol *51*, 523-537.

Prosseda, G., Fradiani, P.A., Di Lorenzo, M., Falconi, M., Micheli, G., Casalino, M., Nicoletti, M., and Colonna, B. (1998). A role for H-NS in the regulation of the *virF* gene of *Shigella* and enteroinvasive *Escherichia coli*. Res Microbiol *149*, 15-25.

Pupo, G.M., Karaolis, D.K., Lan, R., and Reeves, P.R. (1997). Evolutionary relationships among pathogenic and nonpathogenic *Escherichia coli* strains inferred from multilocus enzyme electrophoresis and *mdh* sequence studies. Infection and immunity *65*, 2685-2692.

Pupo, G.M., Lan, R., and Reeves, P.R. (2000). Multiple independent origins of *Shigella* clones of *Escherichia coli* and convergent evolution of many of their characteristics. Proceedings of the National Academy of Sciences of the United States of America *97*, 10567-10572.

Radnedge, L., Davis, M.A., and Austin, S.J. (1996). P1 and P7 plasmid partition: ParB protein bound to its partition site makes a separate discriminator contact with the DNA that determines species specificity. Embo J *15*, 1155-1162.

Radnedge, L., Davis, M.A., Youngren, B., and Austin, S.J. (1997). Plasmid maintenance functions of the large virulence plasmid of *Shigella flexneri*. Journal of bacteriology *179*, 3670-3675.

Raivio, T.L. (2014). Everything old is new again: an update on current research on the Cpx envelope stress response. Biochimica et biophysica acta *1843*, 1529-1541.

Rediers, H., Vanderleyden, J., and De Mot, R. (2004). *Azotobacter vinelandii*: a Pseudomonas in disguise? Microbiology *150*, 1117-1119.

Renzoni, D., Esposito, D., Pfuhl, M., Hinton, J.C., Higgins, C.F., Driscoll, P.C., and Ladbury, J.E. (2001). Structural characterization of the N-terminal oligomerization domain of the bacterial chromatin-structuring protein, H-NS. Journal of molecular biology *306*, 1127-1137.

Rice, P.A., Yang, S., Mizuuchi, K., and Nash, H.A. (1996). Crystal structure of an IHF-DNA complex: a protein-induced DNA U-turn. Cell *87*, 1295-1306.

Rimsky, S., and Travers, A. (2011). Pervasive regulation of nucleoid structure and function by nucleoid-associated proteins. Current opinion in microbiology *14*, 136-141.

Rodionov, O., Lobocka, M., and Yarmolinsky, M. (1999). Silencing of genes flanking the P1 plasmid centromere. Science *283*, 546-549.

Rodionov, O., and Yarmolinsky, M. (2004). Plasmid partitioning and the spreading of P1 partition protein ParB. Mol Microbiol *52*, 1215-1223.

Rolland, K., Lambert-Zechovsky, N., Picard, B., and Denamur, E. (1998). *Shigella* and enteroinvasive *Escherichia coli* strains are derived from distinct ancestral strains of *E. coli*. Microbiology *144 (Pt 9)*, 2667-2672.

Romeo, T. (1998). Global regulation by the small RNA-binding protein CsrA and the non-coding RNA molecule CsrB. Mol Microbiol *29*, 1321-1330.

Sabnis, N.A., Yang, H., and Romeo, T. (1995). Pleiotropic regulation of central carbohydrate metabolism in *Escherichia coli* via the gene *csrA*. J Biol Chem *270*, 29096-29104.

Saier, M.H., Jr., and Ramseier, T.M. (1996). The catabolite repressor/activator (Cra) protein of enteric bacteria. Journal of bacteriology *178*, 3411-3417.

Sakai, T., Sasakawa, C., Makino, S., Kamata, K., and Yoshikawa, M. (1986). Molecular cloning of a genetic determinant for Congo red binding ability which is essential for the virulence of *Shigella flexneri*. Infection and immunity *51*, 476-482.

Sakai, T., Sasakawa, C., and Yoshikawa, M. (1988). Expression of four virulence antigens of *Shigella flexneri* is positively regulated at the transcriptional level by the 30 kilo Dalton *virF* protein. Molecular Microbiology *2*, 589-597.

Santapaola, D., Casalino, M., Petrucca, A., Presutti, C., Zagaglia, C., Berlutti, F., Colonna, B., and Nicoletti, M. (2002). Enteroinvasive *Escherichia coli* virulence-plasmid-carried apyrase (*apy*) and *ospB* genes are organized as a bicistronic operon and are subject to differential expression. Microbiology *148*, 2519-2529.

Sasakawa, C., Kamata, K., Sakai, T., Makino, S., Yamada, M., Okada, N., and Yoshikawa, M. (1988). Virulence-associated genetic regions comprising 31 kilobases of the 230-kilobase plasmid in *Shigella flexneri* 2a. Journal of bacteriology *170*, 2480-2484.

Savelkoul, P.H., Willshaw, G.A., McConnell, M.M., Smith, H.R., Hamers, A.M., van der Zeijst, B.A., and Gaastra, W. (1990). Expression of CFA/I fimbriae is positively regulated. Microb Pathog *8*, 91-99.

Schneider, R., Travers, A., Kutateladze, T., and Muskhelishvili, G. (1999). A DNA architectural protein couples cellular physiology and DNA topology in *Escherichia coli*. Mol Microbiol *34*, 953-964.

Schroeder, G.N., and Hilbi, H. (2008). Molecular Pathogenesis of *Shigella* spp.: Controlling Host Cell Signaling, Invasion, and Death by Type III Secretion. Clinical Microbiology Reviews *21*, 134-156.

Schuch, R., and Maurelli, A.T. (1997). Virulence plasmid instability in *Shigella flexneri* 2a is induced by virulence gene expression. Infection and immunity *65*, 3686-3692.

Schuch, R., and Maurelli, A.T. (2001). Spa33, a cell surface-associated subunit of the Mxi-Spa type III secretory pathway of *Shigella flexneri*, regulates Ipa protein traffic. Infection and immunity *69*, 2180-2189.

Schumacher, M.A., and Funnell, B.E. (2005). Structures of ParB bound to DNA reveal mechanism of partition complex formation. Nature *438*, 516-519.

Sette, M., Spurio, R., Trotta, E., Brandizi, C., Brandi, A., Pon, C.L., Barbato, G., Boelens, R., and Gualerzi, C.O. (2009). Sequence-specific recognition of DNA by the C-terminal domain of nucleoid-associated protein H-NS. J Biol Chem *284*, 30453-30462.

Shere, K.D., Sallustio, S., Manessis, A., D'Aversa, T.G., and Goldberg, M.B. (1997). Disruption of IcsP, the major *Shigella* protease that cleaves IcsA, accelerates actin-based motility. Mol Microbiol *25*, 451-462.

Sidik, S., Kottwitz, H., Benjamin, J., Ryu, J., Jarrar, A., Garduno, R., and Rohde, J.R. (2014). A *Shigella flexneri* Virulence Plasmid Encoded Factor Controls Production of Outer Membrane Vesicles. G3 *4*, 2493-2503.

Spassky, A., Rimsky, S., Garreau, H., and Buc, H. (1984). H1a, an *E. coli* DNA-binding protein which accumulates in stationary phase, strongly compacts DNA *in vitro*. Nucleic Acids Res *12*, 5321-5340.

Sperandio, B., Regnault, B., Guo, J., Zhang, Z., Stanley, S.L., Jr., Sansonetti, P.J., and Pedron, T. (2008). Virulent *Shigella flexneri* subverts the host innate immune response through manipulation of antimicrobial peptide gene expression. The Journal of experimental medicine *205*, 1121-1132.

Spurio, R., Durrenberger, M., Falconi, M., La Teana, A., Pon, C.L., and Gualerzi, C.O. (1992). Lethal overproduction of the *Escherichia coli* nucleoid protein H-NS: ultramicroscopic and molecular autopsy. Molecular and general genetics : MGG *231*, 201-211.

Steinhauer, J., Agha, R., Pham, T., Varga, A.W., and Goldberg, M.B. (1999). The unipolar *Shigella* surface protein IcsA is targeted directly to the bacterial old pole: IcsP cleavage of IcsA occurs over the entire bacterial surface. Mol Microbiol *32*, 367-377.

Stoebel, D.M., Free, A., and Dorman, C.J. (2008). Anti-silencing: overcoming H-NS-mediated repression of transcription in Gram-negative enteric bacteria. Microbiology *154*, 2533-2545.

Surtees, J.A., and Funnell, B.E. (1999). P1 ParB domain structure includes two independent multimerization domains. Journal of bacteriology *181*, 5898-5908.

Surtees, J.A., and Funnell, B.E. (2001). The DNA binding domains of P1 ParB and the architecture of the P1 plasmid partition complex. J Biol Chem *276*, 12385-12394.

Suzuki, T., Ueguchi, C., and Mizuno, T. (1996). H-NS regulates OmpF expression through *micF* antisense RNA in *Escherichia coli*. Journal of bacteriology *178*, 3650-3653.

Tamano, K., Katayama, E., Toyotome, T., and Sasakawa, C. (2002). *Shigella* Spa32 is an essential secretory protein for functional type III secretion machinery and uniformity of its needle length. Journal of bacteriology *184*, 1244-1252.

Taniya, T., Mitobe, J., Nakayama, S., Mingshan, Q., Okuda, K., and Watanabe, H. (2003). Determination of the InvE binding site required for expression of IpaB of the *Shigella sonnei* virulence plasmid: involvement of a ParB boxA-like sequence. Journal of bacteriology *185*, 5158-5165.

Tendeng, C., and Bertin, P.N. (2003). H-NS in Gram-negative bacteria: a family of multifaceted proteins. Trends Microbiol *11*, 511-518.

Tobe, T., Nagai, S., Okada, N., Adler, B., Yoshikawa, M., and Sasakawa, C. (1991). Temperature-regulated expression of invasion genes in *Shigella flexneri* is controlled through the transcriptional activation of the *virB* gene on the large plasmid. Mol Microbiol *5*, 887-893.

Tobe, T., Yoshikawa, M., Mizuno, T., and Sasakawa, C. (1993). Transcriptional control of the invasion regulatory gene *virB* of *Shigella flexneri*: activation by *virF* and repression by H-NS. Journal of bacteriology *175*, 6142-6149.

Tobe, T., Yoshikawa, M., and Sasakawa, C. (1995). Thermoregulation of *virB* transcription in *Shigella flexneri* by sensing of changes in local DNA superhelicity. Journal of bacteriology *177*, 1094-1097.

Tolstorukov, M.Y., Virnik, K.M., Adhya, S., and Zhurkin, V.B. (2005). A-tract clusters may facilitate DNA packaging in bacterial nucleoid. Nucleic Acids Res *33*, 3907-3918.

Tran, C.N., Giangrossi, M., Prosseda, G., Brandi, A., Di Martino, M.L., Colonna, B., and Falconi, M. (2011). A multifactor regulatory circuit involving H-NS, VirF and an antisense RNA modulates transcription of the virulence gene *icsA* of *Shigella flexneri*. Nucleic Acids Res *39*, 8122-8134.

Turner, E.C., and Dorman, C.J. (2007). H-NS antagonism in *Shigella flexneri* by VirB, a virulence gene transcription regulator that is closely related to plasmid partition factors. Journal of bacteriology *189*, 3403-3413.

Uchiya, K., Tobe, T., Komatsu, K., Suzuki, T., Watarai, M., Fukuda, I., Yoshikawa, M., and Sasakawa, C. (1995). Identification of a novel virulence gene, *virA*, on the large plasmid of *Shigella*, involved in invasion and intercellular spreading. Mol Microbiol *17*, 241-250.

Ulissi, U., Fabbretti, A., Sette, M., Giuliodori, A.M., and Spurio, R. (2014). Time-resolved assembly of a nucleoprotein complex between *Shigella flexneri virF* promoter and its transcriptional repressor H-NS. Nucleic Acids Res *42*, 13039-13050.

Vecchiarelli, A.G., and Funnell, B.E. (2013). Probing the N-terminus of ParB using cysteine-scanning mutagenesis and thiol modification. Plasmid *70*, 86-93.

Walthers, D., Li, Y., Liu, Y., Anand, G., Yan, J., and Kenney, L.J. (2011). *Salmonella enterica* response regulator SsrB relieves H-NS silencing by displacing H-NS bound in polymerization mode and directly activates transcription. J Biol Chem *286*, 1895-1902.

Watanabe, H., Arakawa, E., Ito, K., Kato, J., and Nakamura, A. (1990). Genetic analysis of an invasion region by use of a Tn3-lac transposon and identification of a second positive regulator gene, *invE*, for cell invasion of *Shigella sonnei*: significant homology of *invE* with ParB of plasmid P1. Journal of bacteriology *172*, 619-629.

Waters, L.S., and Storz, G. (2009). Regulatory RNAs in bacteria. Cell *136*, 615-628.

Wei, J., Goldberg, M.B., Burland, V., Venkatesan, M.M., Deng, W., Fournier, G., Mayhew, G.F., Plunkett, G., 3rd, Rose, D.J., Darling, A., *et al.* (2003). Complete genome sequence and comparative genomics of *Shigella flexneri* serotype 2a strain 2457T. Infection and immunity *71*, 2775-2786.

Williams, R.M., and Rimsky, S. (1997). Molecular aspects of the *E. coli* nucleoid protein, H-NS: a central controller of gene regulatory networks. FEMS microbiology letters *156*, 175-185.

Williams, R.M., Rimsky, S., and Buc, H. (1996). Probing the structure, function, and interactions of the *Escherichia coli* H-NS and StpA proteins by using dominant negative derivatives. Journal of bacteriology *178*, 4335-4343.

Wilson, R.L., Libby, S.J., Freet, A.M., Boddicker, J.D., Fahlen, T.F., and Jones, B.D. (2001). Fis, a DNA nucleoid-associated protein, is involved in *Salmonella typhimurium* SPI-1 invasion gene expression. Mol Microbiol *39*, 79-88.

Wing, H.J., Goldman, S.R., Ally, S., and Goldberg, M.B. (2005). Modulation of an outer membrane protease contributes to the virulence defect of *Shigella flexneri* strains carrying a mutation in the *virK* locus. Infection and immunity *73*, 1217-1220.

Wing, H.J., Yan, A.W., Goldman, S.R., and Goldberg, M.B. (2004). Regulation of IcsP, the outer membrane protease of the *Shigella* actin tail assembly protein IcsA, by virulence plasmid regulators VirF and VirB. Journal of bacteriology *186*, 699-705.

Wolf, T., Janzen, W., Blum, C., and Schnetz, K. (2006). Differential dependence of StpA on H-NS in autoregulation of *stpA* and in regulation of *bgl*. Journal of bacteriology *188*, 6728-6738.

Yamada, H., Muramatsu, S., and Mizuno, T. (1990). An *Escherichia coli* protein that preferentially binds to sharply curved DNA. Journal of biochemistry *108*, 420-425.

Yamada, H., Yoshida, T., Tanaka, K., Sasakawa, C., and Mizuno, T. (1991). Molecular analysis of the *Escherichia coli hns* gene encoding a DNA-binding protein, which preferentially recognizes curved DNA sequences. Molecular and general genetics : MGG *230*, 332-336.

Zhang, A., and Belfort, M. (1992). Nucleotide sequence of a newly-identified *Escherichia coli* gene, *stpA*, encoding an H-NS-like protein. Nucleic Acids Res *20*, 6735.

Zhang, A., Rimsky, S., Reaban, M.E., Buc, H., and Belfort, M. (1996). *Escherichia coli* protein analogs StpA and H-NS: regulatory loops, similar and disparate effects on nucleic acid dynamics. EMBO J *15*, 1340-1349.

5

Ribo-regulation in *Shigella*

Erin R. Murphy[1*], William H. Broach[2] and Andrew B. Kouse[3]

[1*]Ohio University Heritage College of Osteopathic Medicine, Department of Biomedical Sciences, Ohio, USA , murphye@ohio.edu
[2]Department of Cell Biology, Microbiology and Molecular Biology, University of South Florida,USA
[3]Cell Biology and Metabolism Program, Eunice Kennedy Shriver, National Institutes of Health, USA

Abstract
Bacterial ribo-regulators are a diverse collection of molecules defined here as *any RNA molecule that functions to specifically control the expression of a target gene(s), or the activity of a target protein.* While the role of regulatory proteins in controlling virulence gene expression in *Shigella* species has long been appreciated, the role of RNA-mediated regulation on this process is just now coming into focus. Recent advances have revealed that ribo-regulators play a central role in modulating *Shigella* gene expression in response to changes in specific environmental conditions, a process that is essential to the survival and virulence of these important pathogens. This chapter will present a general overview of bacterial ribo-regulators, followed by a detailed discussion of each *Shigella* ribo-regulator characterized to date. Special emphasis will be placed on the role that each known *Shigella* ribo-regulator plays in facilitating the survival and/or virulence of these significant pathogens.

Introduction

Bacteria of the genus *Shigella* are Gram-negative, facultative intracellular pathogens, and the causative agents of shigellosis, a severe diarrheal disease in humans (Jennison and Verma, 2004; Niyogi, 2005). Shigellosis is a highly infectious disease transmitted via the fecal oral route. The process of transmission is initiated when a species of *Shigella* is shed in the stool of an infected host. Once shed, the bacterium must survive within the non-host environment until it gains entry into a new host. Upon ingestion by a new host, typically by the consumption of contaminated food or water, *Shigella* transit the length of the gastro-intestinal tract to reach the colon. Only after *Shigella* species breach the colonic epithelium and evade phagocytic killing do they initiate true infection by active invasion of cells within the colonic epithelium (Niyogi, 2005; Parsot, 2009; Philpott et al., 2000). Once within the intracellular environment, *Shigella* species replicate and spread into neighboring cells via actin based motility (Goldberg, 2001). Physical damage caused by the intracellular spread of *Shigella* species between cells of the colonic epithelium, along with the resulting inflammatory response, disrupts proper functioning of the organ and results in the clinical symptoms of shigellosis, shedding of the pathogen, and thus, re-initiation of the infection cycle (Niyogi, 2005; Philpott et al., 2000).

The ability of *Shigella* species to cause disease is strictly dependent upon the ability of these pathogens to sense and adapt to the varied and often extreme environmental conditions encountered during the complex infection cycle. While the dramatic changes in environmental conditions encountered during infection and transmission can present a challenge to bacterial survival, these changes also act as important signals that influence the expression of virulence-associated genes. It is well established that the expression of *Shigella* virulence genes, many of which are encoded on the large virulence plasmid, is highly regulated in response to specific environmental conditions including pH, temperature and osmolarity, as well as the availability of oxygen, iron and other nutrients (Buchrieser et al., 2000; Dorman and Porter, 1998; Jennison and Verma, 2004; Payne et al., 2006; Schroeder and Hilbi, 2008; Suzuki and Sasakawa, 2001). The cumulative effect of this extensive regulation is that *Shigella* species produce the factors that are required to initiate and promote infection under conditions that mimic those encountered within the human host, the precise environment where these factors are most advantageous to the invading pathogen.

The role of regulatory proteins in controlling virulence-associated gene expression in *Shigella* species is well established (Jennison and Verma, 2004). Recent advances have revealed that RNA-mediated regulation of *Shigella* gene expression also plays a significant role in facilitating the survival and virulence of these important pathogens (Africa et al., 2011; Broach et al., 2012; Giangrossi et al., 2010; Gore and Payne, 2010; Kouse et al., 2013; Murphy and Payne, 2007; Oglesby et al., 2005; Tran et al., 2011). Such RNA mediated regulation is carried out by a diverse collection of molecules collectively termed ribo-regulators. This chapter will present a general overview of bacterial ribo-regulators, followed by a detailed discussion of each *Shigella* ribo-regulator characterized to date. Special emphasis will be placed on the role that each *Shigella* ribo-regulator plays in facilitating the survival and/or pathogenesis of these important pathogens.

Bacterial Regulatory RNA molecules

Bacterial ribo-regulators are a diverse collection of molecules broadly defined here as *any RNA molecule that functions to specifically control the expression of a target gene(s), or the activity of a target protein* (Lioliou et al., 2010; Liu and Camilli, 2010; Mandal and Breaker, 2004; Narberhaus et al., 2006; Smith et al., 2010; Waters and Storz, 2009). The importance of bacterial ribo-regulators is highlighted by the fact that these molecules have been implicated in controlling fundamental processes such as stress response, nutrient acquisition, carbon metabolism and virulence, in a wide range of bacterial species (Geissmann et al., 2006; Gottesman et al., 2006; Masse et al., 2007; Oglesby-Sherrouse and Murphy, 2013; Romby and Charpentier, 2010; Romeo et al., 2013). The molecular mechanism by which bacterial ribo-regulators mediate their effects varies. The majority of bacterial ribo-regulators characterized to date control target gene expression directly by altering transcription rate, transcript stability and/or translational efficiency, while others mediate an indirect effect on gene expression by modifying the activity of a target regulatory protein (Frohlich and Vogel, 2009; Georg and Hess, 2011; Romeo et al., 2013; Storz et al., 2011; Timmermans and Van Melderen, 2010). While the underlying regulatory mechanisms vary, important features are conserved amongst the vast majority of bacterial ribo-regulators characterized to date. Specifically, most ribo-regulators are known to mediate precise, fine-tuned control over target gene expression in a manner more closely resembling that of a rheostat as opposed to that of an on/off switch more typical for protein-based regulation. Additionally, the production and/or activity of many bacterial ribo-regulators is modulated in response to one or more specific environmental condition (Wassarman, 2002). Together, these conserved features underlie the ability of bacterial ribo-regulators to precisely control target gene expression in response to specific environmental signals.

Bacterial ribo-regulators can be divided into two broad categories based strictly on where the molecule is encoded relative to its direct regulatory target or targets (Figure 5.1). *Trans*-encoded ribo-regulators are encoded distal to their target(s), while *cis*-encoded regulatory RNAs are encoded within, and thus as part of, the regulated transcript or as a separate transcript encoded anti-sense to the target. Ribo-regulators within the broad categories of *trans*- and *cis*-encoded can be further sub-divided by type (small RNA, small anti-sense RNA, riboswitch or RNA thermometer), and finally by the step or steps of gene expression that they regulate (transcription, transcript stability, translation or protein activity) (Figure 5.1). It is important to note that a single ribo-regulator may function to regulate more than one step of target gene expression. For example, a small RNA that binds to a target mRNA molecule may influence both stability of, and translation efficiency from, that transcript. While much knowledge has been gained regarding the molecular mechanisms underlying the activity of both *trans*- and *cis*-encoded bacterial ribo-regulators, there is little doubt that additional mechanisms will be revealed as advances are made in this dynamic field of study.

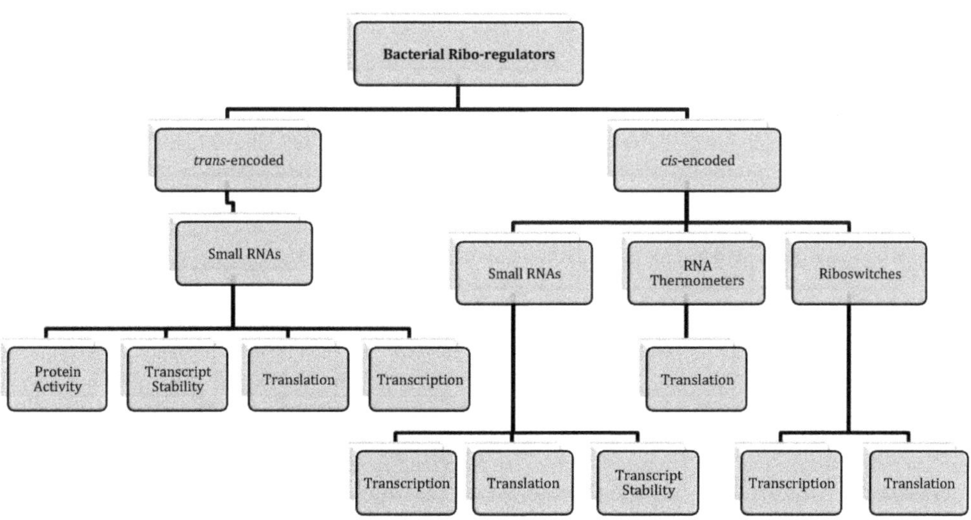

Figure 5.1. Organization schema of bacterial ribo-regulators: All bacterial ribo-regulators can be broadly divided based on where they are encoded relative to their direct regulatory target(s). Each of these broad categories can be further divided by regulator type and finally by the bacterial process that they function to regulate.

Trans-encoded bacterial regulatory RNAs

The category of *trans*-encoded bacterial regulatory RNAs is dominated by small RNAs, molecules that are encoded distally to their targets, are generally less than 500 nt in length and that function to control the expression of target genes or the activity of target proteins (Liu and Camilli, 2010). Bacterial small RNAs are themselves a collection of diverse regulators that differ in structure, expression pattern and regulatory mechanism. The various mechanisms by which *trans*-encoded bacterial small RNA molecules have been shown to regulate target gene expression are depicted in Figure 5.2. The majority of *trans*-encoded bacterial small RNA molecules characterized to date mediate their regulatory effect by interacting directly with target transcripts via sequence specific binding that is facilitated by the RNA binding protein Hfq (Aiba, 2007; Beisel and Storz, 2010; Caron et al., 2010; Frohlich and Vogel, 2009). Binding between a bacterial small RNA molecule and a specific transcript can alter expression from that transcript by influencing the stability of, and/or translation from, the target transcript (Figure 5.2 A-C). While the precise amount and nature of the nucleic acid complementarity required for a functional interaction between a bacterial small RNA and a target transcript remains an area of active investigation, it is clear that neither perfect nor extensive complementarity is required. The apparent flexibility in the nature and amount of complementarity required between a bacterial small RNA and its target is though to facilitate the ability of many of these ribo-regulators to directly control the expression of multiple genes (Papenfort and Vogel, 2009).

Regardless of the extent or nature of the nucleic acid complementarity between a bacterial small RNA and its target transcript, a productive interaction between the two RNA molecules is often dependent upon the activity of the RNA binding protein Hfq (Brennan and Link, 2007; De Lay et al., 2013; Hajnsdorf and Boni, 2012). The exact mechanism by which Hfq facilitates binding between a small RNA molecule and its mRNA target remains an area of dispute; at least two non-mutually exclusive hypotheses exist. One leading hypothesis suggests that by binding to both the small RNA and the target transcript, Hfq functions to increase the local concentration of each molecule, and by doing so, increases the probability of a sequence-dependent interaction (De Lay et al., 2013). Another hypothesis states that Hfq restructures both the small RNA and the target transcript such that each is in a conformation that is more favorable for interaction (Brennan and Link, 2007). Regardless of the precise mechanism of action, the importance of Hfq to the regulation of bacterial gene expression is highlighted by the fact that over 50% of all sequenced bacterial genomes encode a homolog of this protein (Brennan and Link, 2007).

Not all bacterial small RNAs mediate a direct effect on gene expression via an interaction with a target transcript or transcripts. Some small RNA molecules mediate an indirect effect on gene expression by directly altering the function of a target regulatory protein (Figure 5.2D) (Romeo et al., 2013; Storz et al., 2011). In each characterized example, binding of a small RNA to a regulatory protein functionally inactivates the protein by physically sequestering it away from its regulatory targets. Small RNA-dependent sequestration of a regulatory protein results in the coordinated regulation of each gene within the regulon of the target protein (Romeo et al., 2013). For this reason, small RNA molecules that function to modify target protein activity tend to influence the expression of a relatively large number of genes.

Cis-encoded bacterial regulatory RNAs

As the name suggests, *cis*-encoded ribo-regulators are encoded within, or anti-sense to, their regulatory target. Specifically, *cis*-encoded ribo-regulators can take the form of structural elements within a regulated transcript, as is the case with RNA thermometers

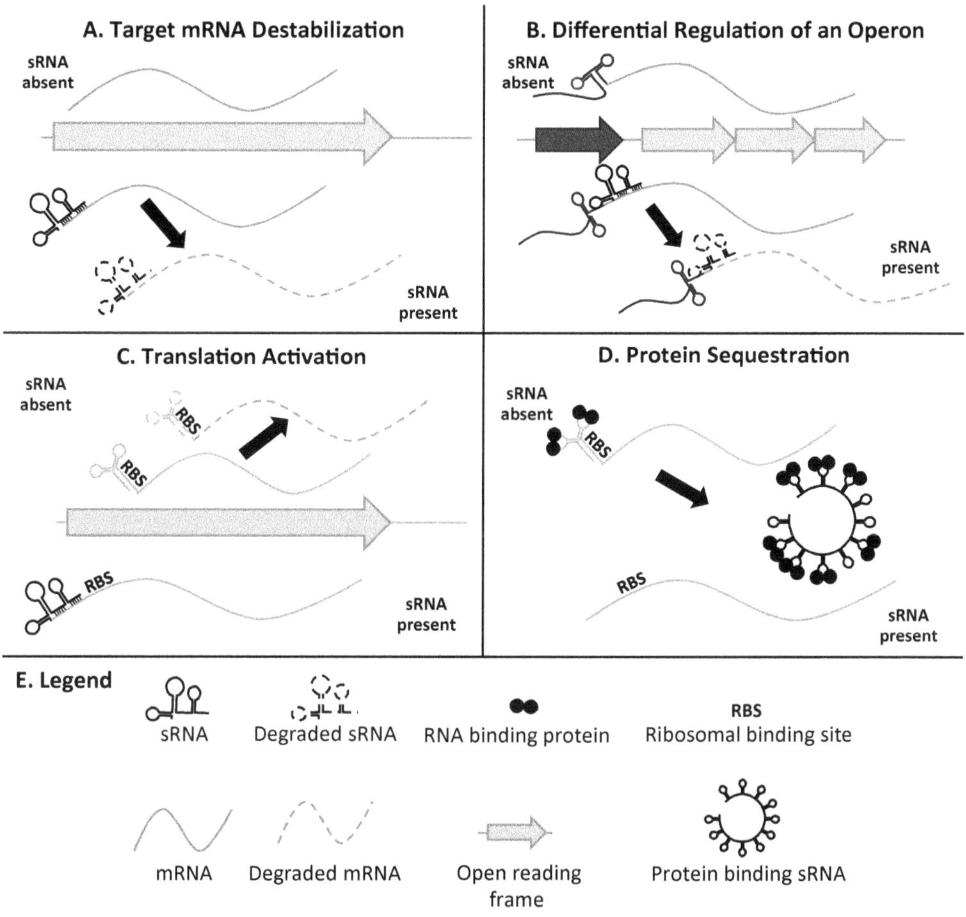

Figure 5.2. General regulatory mechanisms of *trans*-encoded small RNA molecules. Each panel represents a mechanism by which small RNA molecules (sRNA) have been demonstrated to control target gene expression. Target gene expression is depicted in both the absence (top) and presence (bottom) of the small RNA molecule. A) Target mRNA Destabilization. The binding of a small RNA molecule to a mRNA can destabilize the target transcript by initiating a process leading to the degradation of both RNA molecules. Not pictured, a small RNA can bind to its mRNA target and stabilize the transcript, thus promoting expression by increasing the half-life of the target. B) Differential Regulation of an Operon. Binding of a small RNA molecule within an untranslated region within a polycistronic message can induce a change in the native folding of the target and induce degradation of a portion of the polycistronic transcript. C) Translation Activation. A small RNA molecule can bind to a specific mRNA molecule and induce changes in the folding of the target transcript that expose the ribosomal binding site and allow translation to occur. Not pictured, a small RNA molecule can bind to its mRNA target and induce changes in the structure of the transcript such that the ribosomal binding site is occluded and translation is inhibited. D) Protein Sequestration. Binding of a regulatory protein to specific binding sites within a small RNA molecule can functionally sequester the target protein and prevent it from interacting with its regulatory targets. E) Figure Key. Each of the symbols used in Figure 5.3 is shown in this panel.

and riboswitches, or can be encoded on the opposite DNA strand to the regulated target, as is the case for small anti-sense RNA molecules (Georg and Hess, 2011; Narberhaus et al., 2006; Smith et al., 2010). Anti-sense RNA molecules resemble *trans*-encoded small RNAs in several important ways, including their diversity of function, regulatory mechanism and expression profile (Georg and Hess, 2011). Despite these similarities, small anti-sense RNA molecules differ from their *trans*-encoded counterparts in specific important aspects. In addition to being able to influence target transcript stability and/or translation efficiency, *cis*-encoded anti-sense RNA molecules have been shown to control target gene expression via at least two molecular mechanisms unique to this class of ribo-regulators (Figure 5.3). Specifically, small anti-sense RNAs can directly modulate target gene transcription following an interaction with a growing transcript (Figure 5.3A), a function not yet assigned to any *trans*-encoded small RNA molecule (Giangrossi et al., 2010). Additionally, the unique arrangement of the genes encoding the small anti-sense RNA and its regulated target has, in at least one example, been shown to facilitate regulation of target gene expression via the process of promoter interference (Figure 5.3B) (Giangrossi et al., 2010). Additionally, while *trans*-encoded small RNA molecules often control the expression of several gene targets, *cis*-encoded small anti-sense RNA molecules exclusively regulate the expression of the gene to which they are encoded opposite to, and to which they share extensive and perfect nucleic acid complementarity.

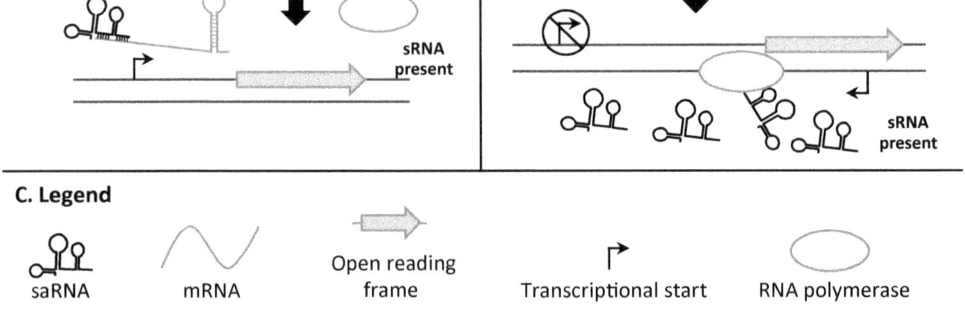

Figure 5.3. Regulatory mechanisms unique to small anti-sense RNA molecules. Each panel represents a regulatory mechanism that is unique to small anti-sense RNA molecules (saRNA). Target gene expression is depicted in both the absence (top) and presence (bottom) of the small anti-sense RNA molecule. A) Transcription Attenuation. A small anti-sense RNA molecule can bind to the nascent mRNA molecule as it exits the RNA polymerase and alter the structure of that growing transcript such that transcription is halted by the formation of a terminator structure within the target molecule. B) Promoter Interference. When a small anti-sense RNA molecule and its target transcript are encoded on opposite strands of DNA with overlapping 5' regions, the activity of the target gene promoter activity can be influenced by the activity of the promoter controlling production of the small anti-sense RNA. C) Figure Key. Each of the symbols used in Figure 5.3 is shown in this panel.

The majority of *cis*-encoded ribo-regulators characterized to date are not small anti-sense RNA molecules, but instead are regulatory elements embedded within a target transcript itself. The simplest of these *cis*-encoded regulatory elements is the RNA thermometer. RNA thermometers are located within the 5' untranslated region of the regulated transcript and modulate translation initiation via temperature-induced changes in structure (Figure 5.4) (Kortmann and Narberhaus, 2012; Narberhaus et al., 2006). Specifically, at non-permissive temperatures a hairpin is formed by a sequence-dependent interaction between the ribosomal binding site and an upstream region of the regulated transcript. Formation of the inhibitory hairpin physically occludes ribosomal binding and thus prevents translation initiation from the regulated transcript. At permissive temperatures, the hydrogen bonds stabilizing the inhibitory hairpin break, the structure opens to expose the ribosomal binding site, and translation proceeds unobstructed (Kortmann and Narberhaus, 2012). Multiple classes of RNA thermometers have been identified in bacteria, each marked by characteristic primary sequences and/or secondary structures (Kortmann and Narberhaus, 2012; Waldminghaus et al., 2005; Waldminghaus et al., 2007). While the sequence and structure of different classes of RNA thermometers varies, the same exquisitely simple regulatory mechanism underlies the ability of each to modulate target gene expression in response to temperature (Kortmann and Narberhaus, 2012).

Riboswitches are superficially similar to RNA thermometers in that they are regulatory RNA elements harbored within the 5' end of a target transcript that control expression via structural changes in the molecule (Smith et al., 2010). Unlike RNA thermometers that only influence translation efficiency, ribo-switches can influence different steps of target

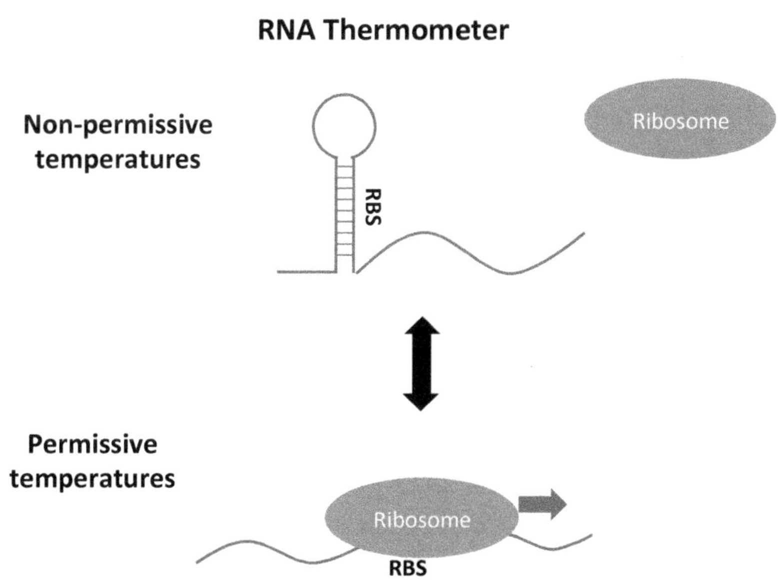

Figure 5.4. General molecular mechanism governing the regulatory activity of a bacterial RNA thermometer. At non-permissive temperatures translation initiation is inhibited by the formation of an inhibitory hairpin that functions to occlude the ribosomal binding site (RBS) within the regulated transcript. At permissive temperatures, the inhibitory hairpin is destabilized, the RBS is exposed, and translation proceeds.

gene expression. Structural changes within a given riboswitch most often modulate target gene expression by altering the efficiency of translation initiation via differential exposure of the ribosomal binding site, or by inducing attenuation of transcript synthesis by the formation of an intrinsic transcriptional terminator (Figure 5.5A-5B) (Smith et al., 2010). In addition to a broader range of characterized regulatory mechanisms, riboswitches differ from RNA thermometers in that the structure of the regulatory element, and thus expression of the regulated gene, is directly influenced by the presence or absence of a specific ligand (Mandal and Breaker, 2004; Smith et al., 2010).

Shigella ribo-regulators

The appreciation and understanding of RNA-based regulation of gene expression has advanced so dramatically that some now suggest ribo-regulators have a greater impact on bacterial gene expression than do their protein counterparts (Storz et al., 2011). A sub-category of the larger field of bacterial ribo-regulation is the study of RNA-based regulation in bacterial pathogens, specifically the role of ribo-regulators in controlling virulence-associated gene expression. The remainder of this chapter will detail the current understanding of RNA-mediated regulation of gene expression in the pathogenic bacteria of the genus *Shigella.* Although relatively few in number, both *trans*- and *cis*-encoded ribo-regulators have been identified in *Shigella* species and, without exception, have been shown to control processes central to the survival and/or virulence of these significant pathogens. Table 5.1 summarizes each *Shigella* ribo-regulator characterized to date, its categorization and, if known, the regulatory targets and cellular processes controlled by each. Each validated *Shigella* ribo-regulator will be discussed below.

Table 5.1. Summary of known ribo-regulators in *Shigella*. * indicates that the regulation of the named target has been experimentally validated. Targets identified by any type of screening approach with no further direct analysis are not included in this table. ^ influenced processes may have been determined experimentally or inferred based on the known function of the validated regulatory target.

Name	Type	Encoded in	Validated Target(s)*	Influenced Process(es)^	Reference(s)
RyhB	Small RNA	*trans*	*evgA* *virB* *sodB*	Acid Resistance Invasion Oxidative Stress	(Africa et al., 2011; Broach et al., 2012; Murphy and Payne, 2007; Oglesby et al., 2005)
CsrB & CsrC	Small RNA	*trans*	CsrA	Invasion	(Gore and Payne, 2010)
RyfA1	Small RNA	*trans*	Unknown	Intercellular spread	(Murphy)
RyfA2	Small RNA	*trans*	Unknown	Invasion	(Murphy)
RnaG	Small RNA	*cis*	*icsA*	Intercellular spread	(Giangrossi et al., 2010; Tran et al., 2011)
shuA	Thermometer	*cis*	*shuA*	Heme iron Utilization	(Kouse et al., 2013)
ompA	Thermometer	*cis*	*ompA*	Unknown	(Murphy)

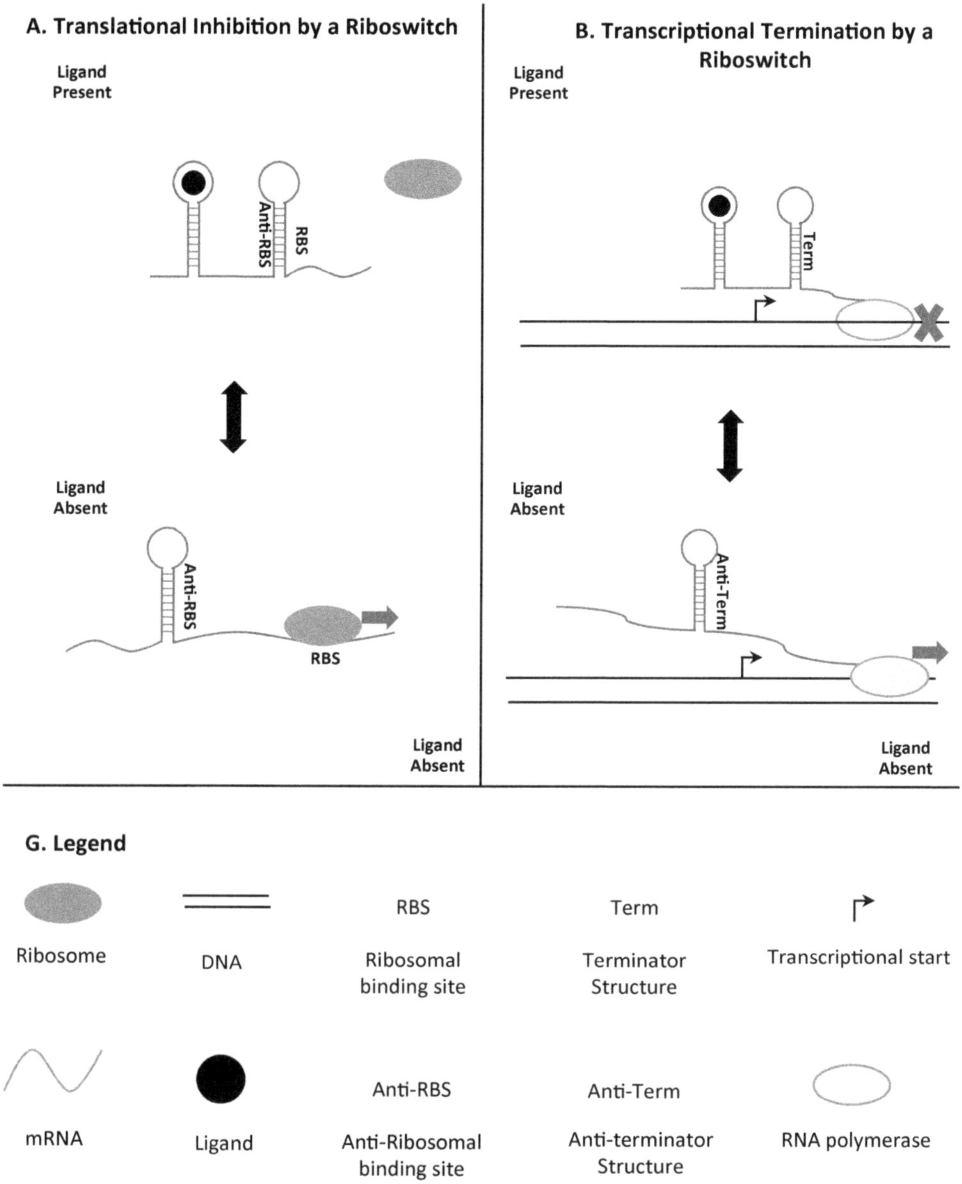

Figure 5.5. General molecular mechanisms underlying transcriptional and translational regulation by bacterial riboswitches. Each panel represents a confirmed mechanism by which bacterial riboswitches have been shown to regulate target gene expression. Each panel shows an example of target gene expression in both the presence (top) and absence (bottom) of the cognate ligand. A) Translational Inhibition. In the presence of its cognate ligand, the riboswitch will form an upstream hairpin that functions to facilitate the formation of a second, downstream hairpin. The second hairpin contains an Anti-ribosomal binding site (Anti-RBS) that sequesters the ribosomal binding site (RBS), thus inhibiting translation initiation. In the absence of the ligand, the upstream hairpin does not form and the Anti-RBS is included in a more stable, alternative structure. In this conformation the RBS is not occluded, the ribosome binds and translation proceeds uninhibited. B) Transcriptional Termination. During transcription, binding of the specific ligand to the growing transcript stabilizes the formation of a termination structure which functions to stop transcription. In the absence of the ligand, an anti-terminator structure forms which functions to inhibit the formation of the terminator structure and thus allows transcription to proceed. C) Figure Key. Each of the symbols used in Figure 5.3 is shown in this panel.

Trans-encoded Shigella Ribo-regulators: RyhB, CsrA, CsrB, RyfA1 and RyfA2

RyhB

RyhB is an iron-regulated *trans*-acting small RNA molecule first identified and characterized in non-pathogenic *E. coli* where it was found control iron homeostasis within the bacterial cell (Jacques et al., 2006; Masse and Gottesman, 2002; Masse et al., 2007; Oglesby-Sherrouse and Murphy, 2013). Maintenance of iron homeostasis is critical to the survival of most bacterial species, as a precise balance must be reached between the absolute requirement for the element, and the toxicity that results when it is present in excess. Iron-responsive regulation of RyhB production is achieved by Fur, a global iron-dependent transcriptional regulator (Masse and Gottesman, 2002). When complexed with iron, Fur functions to repress the expression of most target genes by binding within the regulated promoter and preventing transcription initiation (Escolar et al., 1999). In the case of RyhB, such iron-dependent repression ensures that the regulatory RNA is not produced when environmental iron is abundant. It is under iron-rich conditions that increased production of RyhB-repressed genes, such as those encoding iron containing enzymes and iron storage proteins, is thought to protect the bacterium from iron toxicity (Jacques et al., 2006). During growth of the bacterium under conditions of iron limitation, Fur cannot complex with the element and Fur-mediated repression of target gene expression is relieved (Escolar et al., 1999) . In the case of RyhB, the loss of Fur-mediated repression ensures that the regulatory RNA is produced when the bacterium is growing within an iron-poor environment. It is under these iron-poor conditions that the RyhB-mediated repression of targets genes, such as those encoding non-essential iron containing enzymes and iron storage proteins, is thought to "spare" the element for incorporation into essential iron-containing factors (Jacques et al., 2006). The precise interplay between Fur, RyhB and genes within the RyhB regulon maintains the delicate balance between the absolute requirement for iron and its potent toxicity, and by doing so facilitates bacterial survival.

E. coli RyhB has been shown to control the expression of specific gene targets via multiple regulatory mechanisms including alteration of the stability of, or the translation efficiency from, target transcripts (Masse et al., 2007; Oglesby-Sherrouse and Murphy, 2013; Salvail and Masse, 2012). The first RyhB-dependent regulatory mechanism characterized, and the most extensively studied, is one in which RyhB functions to alter the stability of target mRNA molecules (Figure 5.2A). This first regulatory mechanism is exceptionally well characterized for the RyhB-dependent repression of *E. coli sodB*, a gene encoding an iron-containing super-oxide dismutase. RyhB-dependent repression of *sodB* expression is mediated by a direct interaction between the small RNA and the *sodB* transcript, an interaction that is dependent upon nucleic acid complementarity between the two RNA molecules and that is facilitated by the RNA-binding protein Hfq (Afonyushkin et al., 2005; Desnoyers et al., 2009; Masse and Gottesman, 2002; Vecerek et al., 2003). Following the sequence-specific interaction between RyhB and the *sodB* transcript both RNA molecules are rapidly degraded in a process mediated by RNaseE and RNaseIII (Afonyushkin et al., 2005; Masse et al., 2003; Morita and Aiba, 2011; Prevost et al., 2011; Vecerek et al., 2003).

As an enzyme that detoxifies super-oxide molecules, SodB production is most advantageous to the bacterium when the organism is growing within an iron-rich aerobic environment, as it is within this environment that toxic super-oxide molecules are produced via the Fenton reaction (Imlay et al., 1988; Imlay and Linn, 1988; Maringanti and Imlay, 1999). During growth of the bacterium within an iron-rich environment, Fur-mediated repression results in minimal production of RyhB, and thus maximal production of SodB. Conversely, during growth within iron-poor environments, Fur-mediated repression of RyhB production is relieved and the small RNA molecule functions to decrease *sodB* expression (Masse and Gottesman, 2002). It is under these iron-poor conditions that SodB offers little advantage to a bacterium, and thus that decreased synthesis of the enzyme is advantageous.

In addition to mediating the complete degradation of monocistronic and polycistronic target transcripts, RyhB is also capable of inducing differential degradation of at least one polycistronic transcript (Figure 5.2B); a process revealed while defining the role of *E. coli* RyhB in controlling the production of the Isc Fe-S cluster biosynthesis machinery (Desnoyers et al., 2009). Specifically, RyhB controls the production of the *E. coli* Isc Fe-S cluster synthesis system via differential degradation of a polycistronic transcript, a molecular mechanism that had not been previously characterized for any bacterial small RNA molecule (Desnoyers et al., 2009). The specific target of RyhB-mediated differential degradation is the polycistronic *isc* transcript encoding IscS, IscU and IscA, components of the Isc Fe-S synthesis system, as well as IscR, an Fe-S cluster responsive transcriptional regulator. Studies by Desnoyers *et al.,* demonstrated that during growth of *E. coli* under iron-rich conditions, Fur-mediated repression of RyhB production results in uninhibited expression of each gene within the *isc* operon (Desnoyers et al., 2009). It is under these iron-rich conditions that the Isc Fe-S cluster synthesis machinery functions to synthesize the essential co-factor. Conversely, when *E. coli* is growing within an iron-poor environment where Fur-mediated repression of *ryhB* expression is relieved, the small RNA functions to specifically repress production of the Isc Fe-S synthesis machinery, but not that of the Isc transcriptional regulator (Desnoyers et al., 2009). Specifically, RyhB binds within the large intergenic region separating the *iscR* gene that encodes the transcriptional regulator from the downstream *iscS, iscU* and *iscA* genes encoding components of the Fe-S cluster synthesis machinery (Desnoyers et al., 2009). It is proposed that RyhB-dependent degradation of *iscS, iscU* and *iscA* is initiated at the site of RyhB binding and that the presence of a stable secondary structure located immediately upstream of this binding site prevents degradation towards the 5' end of the molecule, thus sparing *iscR* (Desnoyers et al., 2009). Under these iron-limited conditions, and in the absence of a functional Isc Fe-S cluster building system, IscR functions to activate expression of the *suf* operon. When produced, the Suf system functions to fulfill the Fe-S cluster requirement of the bacterium, and thus facilitate survival under conditions of iron stress (Ayala-Castro et al., 2008). RyhB-dependent modulation of *isc* gene expression represents the first example of differential regulation of genes within a polycistronic transcript by any bacterial small RNA molecule characterized to date.

A second regulatory mechanism by which RyhB functions to control target gene expression is via direct alteration of translation efficiency (Figure 5.2C). RyhB-dependent alteration of target gene translation was first characterized by Prévost *et al.* while investigating the role of *E. coli* RyhB in modulating siderophore production (Prevost et al., 2007). Produced and secreted by bacteria during growth under iron-poor conditions, siderophores are low molecular weight molecules that function to scavenge the limited element from the environment. RyhB was implicated in regulating siderophore production when it was found that the small RNA molecule directly regulates the production of ShiA, a permease required for siderophore biosynthesis (Herrmann and Weaver, 1999; Prevost et al., 2007).

Unlike the RyhB-dependent regulatory mechanisms discussed thus far, RyhB functions to promote ShiA production (Prevost et al., 2007). RyhB-dependent promotion of *shiA* expression is achieved via two distinct, yet complementary, molecular mechanisms. In addition to binding to and stabilizing the *shiA* transcript, RyhB also functions to increase the translational efficiency from the target molecule (Prevost et al., 2007). Specifically, when *E. coli* is growing within an iron-rich environment where the activity of iron scavenging siderophores is less critical, translation from the *shiA* transcript is inefficient due to the presence of an intrinsic inhibitory structure within the *shiA* 5' untranslated region that is predicted to block translation initiation by occlusion of the ribosomal binding site (Prevost et al., 2007). When the bacterium is growing within an iron-poor environment the production of siderophores becomes critical to survival. It is under these iron-poor conditions that Fur-mediated repression of RyhB production is relieved and the small RNA molecule functions to promote *shiA* translation. When present, RyhB binds to complementary sequences within the *shiA* 5' untranslated region and remodels the secondary structure such that the inhibitory structure is eliminated, the ribosomal binding site it exposed, and translation proceeds (Prevost et al., 2007). Investigations characterizing RyhB-dependent regulation of *shiA* expression were the first to experimentally detail a molecular mechanism governing target gene activation by this multifunctional small RNA molecule.

The importance of RyhB to the fitness of *E. coli* is abundantly clear; this single small RNA molecule controls the expression of multiple genes, and by doing so directly influences several key cellular processes. What is remarkable, and to date unique to this small RNA molecule, is the multitude of regulatory mechanisms that RyhB employs to control target gene expression. Whether it is by altering the stability of a mono- or polycistronic mRNA molecule, by inducing differential degradation within a polycistronic transcript, or by altering the translation efficiency from a target transcript, *E. coli* RyhB increases the fitness of the organism by facilitating the survival within, and adaptation to, environments containing differing amounts of the essential yet toxic element, iron.

Analysis of available *Shigella* genome sequences revealed an ortholog of *E. coli* RyhB, faithfully retained from a common microbial ancestor (Oglesby et al., 2005; Skippington and Ragan, 2012). While it is not surprising that the small RNA molecule maintains the ability to regulate the expression of *sodB* and other gene targets conserved between *E. coli* and *Shigella*, it is particularly interesting that *Shigella* RyhB has evolved to regulate the expression of genes required for acid resistance and eukaryotic cell invasion, two processes that differentiate *Shigella* from non-pathogenic *E. coli* and that are essential for *Shigella* virulence (Africa et al., 2011; Broach et al., 2012; Murphy and Payne, 2007; Oglesby et al., 2005).

The first unique target identified for *Shigella* RyhB was that of *S. flexneri evgA*, a gene encoding a transcriptional regulator that promotes the expression of genes required for efficient acid resistance by the pathogen (Masuda and Church, 2002, 2003; Oglesby et al., 2005). The production of *S. flexneri* RyhB results in decreased levels of *evgA* transcript, decreased expression of EvgA-regulated genes and, as expected, decreased acid resistance of the pathogen (Oglesby et al., 2005). Given that the ability of *Shigella* species to tolerate exposure to acid conditions is an essential prerequisite to infection, any factor that facilitates or modifies this process has the potential to impact virulence of these human pathogens.

Although critically important for the virulence of *Shigella,* factors that facilitate acid resistance need not be produced by the bacterium at all times during the course of a natural infection. In fact, when the pathogen is not within the acidic environment of the human stomach it is most likely energetically advantageous for the bacterium to actively repress the production of factors required for acid resistance; RyhB mediates this environment specific regulation. Given the increased solubility of iron at low pH, the stomach represents a unique environment within the infected host where the bioavailability of iron is relatively high. In the relatively iron-rich environment encountered within the human stomach, Fur-mediated repression of *Shigella* RyhB production would result in uninhibited expression of *evgA,* and thus increased production of factors that facilitate acid resistance of the organism. It is within the high-iron, acidic environment of the human stomach that the production of factors required for acid resistance is most advantageous to the survival and pathogenesis of the bacterium. Elsewhere within the infected host the bioavailability of iron is exceptionally low. In these iron-poor environments Fur-mediated repression of RyhB production would be relieved and RyhB would function to inhibit *evgA* expression. The inhibition of EvgA production would in turn result in reduced expression of factors required for acid resistance. It is under these iron-poor, non-acidic conditions that the activity of factors required for acid resistance would not be advantageous for the survival of the pathogen, and thus that their production would be energetically deleterious to the organism. The demonstration that *S. flexneri* RyhB controls the expression of genes required for acid resistance not only revealed that RyhB is a functional ribo-regulator in *Shigella,* but also that RyhB controls a virulence-associated process in this human pathogen (Oglesby et al., 2005).

The molecular mechanism governing RyhB-dependent regulation of *evgA* expression has yet to be experimentally confirmed, however two key findings suggest that this regulation occurs via direct destabilization of the *evgA* message. First, nucleic acid complementarity exists between RyhB and the *evgA* transcript (Oglesby et al., 2005). Second, increased production of *S. flexneri* RyhB leads to a decrease in the overall level of *evgA* transcript present within the bacterial cell (Oglesby et al., 2005). While these data support the hypothesis that RyhB directly regulates *evgA* expression via a sequence specific interaction between the two RNA molecules resulting in destabilization of the target transcript, further studies are required to elucidate the precise molecular mechanism governing this regulation.

The role of RyhB in controlling *Shigella* virulence was expanded by subsequent studies demonstrating that the small RNA molecule modulates expression of *virB* (Murphy and Payne, 2007). Encoded on the large virulence plasmid, the VirB protein directly influences *Shigella* virulence by promoting the expression of several virulence genes including, but not limited to, those encoding components of the Type Three Secretion System, its secreted effectors and their chaperones (Jennison and Verma, 2004). The role of RyhB in controlling the expression of this central regulator was first revealed when it was found that production of *S. dysenteriae* RyhB results in a decrease in the overall amount of *virB* transcript measured within the bacterium (Murphy and Payne, 2007). The observed RyhB-dependent decrease in the abundance of *virB* transcript leads to a decrease in the expression of VirB regulated genes and, as expected, an inhibition of VirB-dependent virulence associated processes (Africa et al., 2011; Murphy and Payne, 2007). In addition to solidifying the role of RyhB in controlling processes essential to *Shigella* virulence, these studies were the first to implicate the conserved, chromosomally encoded small RNA molecule in regulating the expression of a horizontally acquired, plasmid encoded target gene.

The precise molecular mechanism governing RyhB-dependent regulation of *virB* expression remains elusive, however studies to date suggest that this regulation occurs via a mechanism that is unique from those previously described for this small RNA. As is seen for *sodB* and *evgA* transcripts, increased production of RyhB leads to a decrease in the overall levels of *virB* mRNA measured within the bacterium (Broach et al., 2012; Murphy and Payne, 2007; Oglesby et al., 2005). Further investigation into the underlying regulatory mechanism has demonstrated however, that the observed RyhB-dependent reduction in *virB* mRNA levels is not due to destabilization of the target transcript, but instead is mediated by an inhibition of *virB* transcription (Broach et al., 2012). Analyses revealed that RyhB-dependent modulation of *virB* transcription is independent of all factors known to regulate *virB* transcription and that it is mediated, at least in part, by the nucleic acid sequence present in the center of the *virB* open reading frame (Broach et al., 2012). Together these findings support the hypothesis that RyhB-dependent regulation of *virB* transcription is direct, and thus mediated by a novel regulatory mechanism.

Regardless of the regulatory mechanisms employed, it is clear that RyhB regulates the expression of *Shigella evgA* and *virB*, and by doing so plays an important role in controlling virulence of this human pathogen (Africa et al., 2011; Broach et al., 2012; Murphy and Payne, 2007; Oglesby et al., 2005; Payne et al., 2006).

CsrB and CsrC

Originally identified in *E. coli,* the small RNA molecules CsrB and CsrC function by molecular mimicry to modulate the activity of the CsrA protein, a global regulator of carbon metabolism, biofilm formation, quorum sensing and motility (Romeo et al., 2013). CsrA mediates its regulatory function by binding, in the form of a protein dimer, to specific sequences within, or adjacent to, the 5' untranslated region of target transcripts within its regulon. Binding of CsrA to an mRNA molecule results in an alteration in the stability of that transcript and/or in blocking efficient translation from the target molecule (Liu and Romeo, 1997; Romeo et al., 2013). The activity of CsrA is controlled directly by CsrB and CsrC, two small RNA molecules that carry multiple repeats of nucleic acid sequences resembling CsrA binding sites; 18 on the ≈350 nt long CsrB and 9 on the 245 nt long CsrC (Liu et al., 1997; Weilbacher et al., 2003). Binding of multiple CsrA dimers to the repeated sequences within the CsrB and CsrC small RNAs functionally inactivate the regulatory protein by physically sequestering it away from its mRNA targets (Figure 5.2D) (Liu et al., 1997; Weilbacher et al., 2003). The molecular components and regulation mechanism of the Csr regulation systems are highly conserved amongst a variety of Gram-negative bacteria (Romeo et al., 2013).

Studies in *S. flexneri* by Gore and Payne demonstrated not only that *Shigella* CsrB and CsrC are functional ribo-regulators, but also that each of these small RNA molecules controls important virulence-associated processes in this pathogen (Gore and Payne, 2010). The role of the CsrB and CsrC in controlling *Shigella* virulence is rooted in the initial finding that CsrA, the direct regulatory target of each ribo-regulator, promotes the expression of both *virF* and *virB* (Gore and Payne, 2010). VirF and VirB function to promote the production of several virulence-associated factors including the Type Three Secretion system that is required for inter- and intracellular spread of the pathogen (Jennison and Verma, 2004; Schroeder and Hilbi, 2008). Given the established role of CsrB and CsrC in modulating CsrA activity, and the observed impact of CsrA on the expression of *virF* and *virB*, it was predicted that each of these small RNA molecules would function to influence *Shigella* virulence. Consistent with this prediction, analyses demonstrated that increased production of either CsrB or CsrC results in a dramatic reduction in the ability of *S. flexneri* to invade eukaryotic cells (Gore and Payne, 2010). These data are consistent with a model of regulation in which the production of CsrB or CsrC results in decreased availability of CsrA, and thus decreased expression of the CsrA-activated genes *virF* and *virB.* Inhibition of *virF* and *virB* expression, and the subsequent reduction in synthesis of the Type Three Secretion System, explains the observed CsrB- and CsrC-dependent inhibition of virulence-associated processes in *S. flexneri* (Gore and Payne, 2010).

Current studies suggest that the production of CsrB and CsrC is regulated in response to the metabolic state of the host and, interestingly, to the concentration of specific products present in the lumen of the human intestine (Romeo et al., 2013). Modulation of CsrB and CsrC activity in response to host-specific environmental signals is thought to provide a selective advantage by ensuring that their regulatory targets are produced when and where their activity is most advantageous to the invading pathogen. For *Shigella,* Csr-dependent regulation of *virF* and *virB* ensures the production of a wide-range of virulence factors under the conditions that mimic those encountered within the human host, the environment in which they are most advantageous to the invading pathogen.

RyfA1 and RyfA2

Several studies have utilized a range of molecular and *in silico* based approaches to identify or predict novel small RNA molecules in a wide variety of bacterial species. Two independent screens identified RyfA as a putative 300 nt long small RNA molecule encoded on the *E. coli* chromosome (Rivas et al., 2001; Wassarman et al., 2001). Despite evidence that RyfA is highly conserved among strains of pathogenic *E. coli, Salmonella* and *Shigella,* studies designed to validate the existence and/or function of this putative ribo-regulator had not previously been undertaken (Hershberg et al., 2003).

Preliminary studies have demonstrated that *ryfA* is highly conserved across all *Shigella* species and, interestingly, that the gene is present in tandem repeat on the chromosome of *S. dysenteriae* (Murphy). The twin *ryfA* genes of *S. dysenteriae*, designated *ryfA*1 and *ryfA*2, share 95% sequence identity with each other and 90% or 91% sequence identity with the single *ryfA* of *E. coli*, respectively (Murphy). Preliminary analyses suggest that while the overall structure is highly conserved between RyfA1 and RyfA2, variation does exist in nucleic acid sequences contained within a single loop region of each molecule. It is proposed that if this predicted single-stranded loop region is involved in target recognition, this subtle change in structure could mediate the regulation of unique targets by RyfA1 and RyfA2. Despite the high degree of identity documented between RyfA1 and RyfA2, promoter sequences for the genes encoding each are highly variable, suggesting that RyfA1 and RyfA2 are differentially produced by the pathogen. Further differentiating RyfA1 and RyfA2 is the finding that each plays a unique role in controlling *S. dysenteriae* virulence. Specifically, RyfA1 production influences cell-to-cell spread by *S. dysenteriae* while RyfA2 production influences eukaryotic cell invasion by the pathogen (Murphy). While it is clear that RyfA1 and RyfA2 each influence virulence-associated processes in *S. dysenteriae*, the direct regulatory targets and regulatory mechanism(s) of these twin small RNA molecules remain areas of active investigation.

RyfA1 and RyfA2 represent the first set of twin small RNA molecules identified in *Shigella,* and add to the growing body of evidence that small RNA molecules are critical in controlling virulence of these human pathogens. Furthermore, the finding that RyfA1 and RyfA2 control unique virulence-associated processes supports the claim that twin small

RNA molecules, once thought of as simply redundant, can carry out unique functions within the bacterial cell. For these reasons, the continued characterization of RyfA1 and RyfA2 will contribute not only to the understanding of ribo-regulation in *Shigella,* but also to the fundamental understanding of twin small RNA molecules, an increasingly recognized sub-category of ribo-regulators.

Cis-encoded Shigella Ribo-regulators: RnaG, ShuA RNA thermometer and ompA RNA thermometer

RnaG

Unlike RyhB, CsrB and CsrC, which were originally described in *E. coli,* RnaG was first identified and characterized in *Shigella* (Giangrossi et al., 2010). Not only is RnaG the first *cis*-acting small RNA molecule identified in *Shigella,* but it also the first ribo-regulator found to be encoded on the *Shigella* virulence plasmid (Giangrossi et al., 2010). Despite these significant differences, RnaG has one important feature in common with all *Shigella* ribo-regulators characterized thus far; it functions to regulate virulence gene expression.

RnaG, a 450 nt long small RNA molecule, regulates the expression of *Shigella icsA*, a gene encoding an outer-membrane protein essential for intercellular spread of the pathogen (not to be confused with *iscA* discussed above) (Bernardini et al., 1989; Giangrossi et al., 2010; Goldberg and Theriot, 1995; Goldberg et al., 1994). *Shigella* IcsA functions to nucleate host-actin subunits into a tail at the old pole of the bacterium that propels the intracellular pathogen from one eukaryotic cell to a neighboring eukaryotic cell (Bernardini et al., 1989; Buchrieser et al., 2000; Goldberg et al., 1994; Suzuki et al., 1996; Suzuki and Sasakawa, 2001). The ability of *Shigella* to spread from one infected cell within the colonic epithelium to a neighboring cell via actin-based motility is critical to the pathophysiology of shigellosis, marking IcsA and any molecule that influences its expression as virulence factors.

Expression of *icsA* is extensively regulated in response to a variety of environmental conditions by the transcriptional activator VirF, the nucleoid protein H-NS and the small RNA RnaG (Porter and Dorman, 1997; Prosseda et al., 1998; Sakai et al., 1988; Tran et al., 2011). RnaG modulates the expression of *icsA* by two distinct regulatory mechanisms, one of which acts at the level of transcription initiation while the other acts at the level of transcription elongation (Giangrossi et al., 2010). Dual-regulation influencing both transcription initiation and elongation ensures efficient RnaG-dependent regulation of *icsA*, and thus a direct effect on *Shigella* virulence.

Underlying both mechanisms of RnaG-dependent regulation of *icsA* is the genetic arrangement of the genes encoding each RNA molecule. Specifically, *rnaG* is encoded on the DNA strand complementary to that encoding *icsA,* such that the convergent promoters are separated by approximately 120 nucleotides (Giangrossi et al., 2010). Coding of RnaG anti-sense to *icsA* results in an extended region of perfect nucleic acid complementarity between sequences at the 5' end of RnaG and sequences at the 5' end of the *icsA* transcript. It is this perfect nucleic acid complementarity that underlies the first molecular mechanism by which RnaG controls *icsA* expression. When present, RnaG binds specifically to the growing *icsA* transcript, and by doing so facilitates the formation of an intrinsic transcriptional terminator which directly induces premature termination of transcription, and thus reduced expression of the target gene (Figure 5.3A) (Giangrossi et al., 2010; Tran et al., 2011). RnaG represents the first small RNA molecule characterized in any Gram-negative bacterial species shown to repress target gene expression via the induction of transcriptional attenuation (Giangrossi et al., 2010). As such, the potentially expansive impact of this newly recognized mechanism of small RNA mediated regulation on gene expression in Gram-negative bacteria is yet to be realized.

The second molecular mechanism by which RnaG modulates *icsA* expression results directly from the proximity and relative strength of the convergent *rnaG* and *icsA* promoters. The finding that the level of activity of the *icsA* promoter is inversely correlated to the levels of activity of the *rnaG* promoter supports the conclusion that RnaG-dependent regulation of *icsA* expression is mediated by transcriptional interference (Figure 5.3B) (Giangrossi et al., 2010). Specifically, transcription from the relatively strong *rnaG* promoter appears to physically inhibit transcription from the weaker divergent *icsA* promoter (Giangrossi et al., 2010). The documented transcriptional interference is reciprocal as demonstrated by a modest increase *rnaG* promoter activity following inactivation of the *icsA* promoter (Giangrossi et al., 2010). Activity of both the *icsA* and *rnaG* promoters is regulated by two proteins, the temperature-dependent transcriptional silencer H-NS and the transcriptional regulator VirF. The direct interaction of two regulatory proteins with divergent yet proximal promoters is thought to underlie the observed transcriptional interference between *rnaG* and *icsA* (Tran et al., 2011).

The characterization of *Shigella* RnaG has contributed significantly to the field of bacterial small RNA molecules by providing the first evidence for two novel regulatory mechanisms (Giangrossi et al., 2010; Tran et al., 2011). Specifically, RnaG is the first bacterial small RNA molecule shown to mediate regulation of target gene expression by direct attenuation of target gene transcription in Gram-negative bacteria (Figure 5.3A), and the first to demonstrate transcriptional interference at the target gene promoter (Figure 5.3B) (Giangrossi et al., 2010; Tran et al., 2011). Furthermore, it is interesting to note that the first 80 nt of the 450 nt long RnaG small RNA molecule alone are sufficient to induce the termination of *icsA* transcription (Tran et al., 2011). This observation prompts speculation that RnaG may have other regulatory functions independent of that controlling *icsA* transcription. The notion that RnaG may have additional regulatory targets is intriguing given that the specificity of bacterial *cis*-encoded small RNA molecules is generally limited to its single anti-sense target transcript.

ShuA Thermometer

It is well established that iron is essential for the survival of most bacterial species. As part of the innate host immune response, the sequestration of iron within molecules such as ferritin, transferrin, hemoglobin, myoglobin and heme maintains the concentration of free iron within the human body at a level that is too low to support the growth of most invading bacterial pathogens (Nairz et al., 2010; Skaar, 2010). In response to this selective pressure bacteria have evolved multiple high-affinity iron acquisition systems that function to obtain and utilize the iron found within host-associated iron containing molecules (Skaar, 2010; Wandersman and Delepelaire, 2004). Given the abundance of heme bound iron within the human host (95% of all iron) and the importance of iron for survival, it is not surprising that the ability of a pathogenic bacterium to utilize heme as a source of nutrient iron is often correlated with virulence (Otto et al., 1992; Wilks and Burkhard, 2007).

Although heme represents a uniquely rich source of essential iron for an invading bacterium, like iron itself, it is toxic when present in excess (Anzaldi and Skaar, 2010). To both ensure the efficient utilization of heme as source of nutrient iron, and to protect against the toxicity of iron and heme, the production of bacterial heme acquisition systems is highly regulated in response to multiple host-associated conditions including iron limitation, temperature and the presence of heme or heme containing molecules (Battisti et al., 2006; Braun, 2001; Hantke, 2001; Kirby et al., 2001; Murphy et al., 2002; Oh et al., 2009; Rossi et al., 2001; Torres and Payne, 1997). Such complex regulation results in optimal production of heme utilization systems under conditions that mimic those encountered within the host, the only environment in which the pathogen will encounter this rich source of nutrient iron.

For *S. dysenteriae* and *S. sonnei* the ability to utilize host heme as a source of nutrient iron is dependent upon the activity of the Shu (*Shigella* heme uptake) system (Mills and Payne, 1995, 1997; Wilks, 2001; Wyckoff et al., 1998). Studies aimed at elucidating the regulatory mechanisms controlling the production of the Shu system have focused on *S. dysenteriae shuA*, a gene encoding the outer membrane heme and hemoglobin receptor (Burkhard and Wilks, 2007; Mills and Payne, 1995, 1997). Like many genes encoding components of bacterial iron acquisition systems, expression of *shuA* is controlled by the iron-responsive transcriptional regulator Fur (Mills and Payne, 1997). Recent studies have demonstrated that in addition to iron-dependent regulation of *shuA* transcription, the gene is also subject to temperature-dependent post-transcriptional regulation (Kouse et al., 2013). Such multi-level regulation ensures optimal production of ShuA when the bacterium is growing within an iron-limited environment at 37°C, conditions that mimic those encountered within the heme-containing environment of the infected human host.

Subsequent analysis has revealed that an RNA thermometer located within the 5' untranslated region of *shuA* mediates the observed temperature-dependent post-transcriptional regulation of *shuA* expression (Kouse et al., 2013). The *shuA* thermometer was initially discovered in *S. dysenteriae* and represents the first RNA thermometer identified in any *Shigella* species (Kouse et al., 2013). Common to all bacterial RNA thermometers is the presence of an inhibitory structure or structures that blocks translation during growth of the organism at non-permissive temperatures (Figure 5.4) (Kortmann and Narberhaus, 2012; Narberhaus et al., 2006). *In silico* analysis of the *shuA* mRNA molecule revealed the presence of a putative inhibitory structure formed by base pairing between the ribosomal binding site and sequences contained within the 5' untranslated region of the transcript (Kouse et al., 2013). The putative inhibitory structure within the 5' end of the *shuA* transcript shares sequence and structural similarities with a sub-class of RNA thermometers referred to as FourU RNA thermometers (Waldminghaus et al., 2007). The hallmark feature of FourU RNA thermometers is the presence of four consecutive uracil residues that base pair both canonically and non-canonically to sequences contained within the ribosomal binding site of the regulated target (Kortmann and Narberhaus, 2012; Waldminghaus et al., 2007).

The ability to confer temperature-dependent post-transcriptional regulation is an essential feature of any bacterial RNA thermometer (Kortmann and Narberhaus, 2012). Key to the classification of the *shuA* RNA thermometer as a functional ribo-regulator was the finding that sequences constituting the element are sufficient to confer temperature-dependent post-transcriptional regulation onto the expression of a reporter gene (Kouse et al., 2013). The hypothesis that the *shuA* transcript harbors a functional RNA thermometer was further supported by site-directed mutagenesis of the regulatory element that confirmed the functionality of the element (Kouse et al., 2013). Together, these data confirm the presence of a functional RNA thermometer within the 5' untranslated region of the *S. dysenteriae shuA* transcript and represents the first characterization of an RNA thermometer in any *Shigella* species.

The *shuA* RNA thermometer is thought to increase fitness of *S. dysenteriae* by ensuring that the heme binding protein is produced only when its function is most advantageous to the invading pathogen. Underlying the argument that temperature-dependent regulation of ShuA production increases *S. dysenteriae* fitness is the fact that heme is a host-specific iron source. It is proposed that during growth at relatively low non-host associated temperatures (25°-30°C) *shuA* translation is repressed by the formation of the inhibitory structure that prevents binding of the ribosome to the transcript. Given the lack of heme within non-host associated environments, the production of ShuA under such conditions would not be energetically favorable as the protein would not provide *S. dysenteriae* with any selective advantage. Conversely, it is proposed that during growth of *S. dysenteriae* at the relatively high host-associated temperature of 37°C, the inhibitory structure within the *shuA* RNA thermometer is destabilized, the ribosomal binding site is exposed, and translation proceeds. Production of ShuA at host-associated temperatures provides a clear advantage to *S. dysenteriae* as this protein is required to utilize heme, the most abundant source of nutrient iron encountered during infection within the human host. The identification and characterization of a functional RNA thermometer controlling expression of *shuA*, a gene presumed to provide an advantage to *S. dysenteriae* during infection within the human host, represents the first evidence that this class of ribo-regulators plays a direct role in controlling *Shigella* virulence.

ompA Thermometer

The OmpA family of proteins is a collection of highly related porins that are extremely abundant in the outer membrane of many Gram-negative bacterial species (Confer and Ayalew, 2013). Described as the "bacterial Swiss army knife", OmpA carries out a number of functions including ion transport, peptidoglycan binding and attachment in a wide variety of bacterial species (Confer and Ayalew, 2013; Smith et al., 2007). As a direct result of these varied functions, OmpA plays a role in such critical bacterial processes as structural maintenance, stress response and, importantly, virulence (Confer and Ayalew, 2013; Smith et al., 2007). In light of its role in controlling virulence in several bacterial species, it is not surprising that OmpA has recently been identified as a virulence factor in *Shigella* (Ambrosi et al., 2012; Confer and Ayalew, 2013; Smith et al., 2007). Specifically, characterization of *S. flexneri* OmpA has revealed that the protein is essential for the process of actin based cell-to-cell spread by the pathogen (Ambrosi et al., 2012). The finding that OmpA is required for an integral virulence-associated process marks this protein, and any factor that regulates its production, as a *Shigella* virulence factor.

The production of *E. coli* OmpA is highly regulated in response to several specific environmental cues, including temperature (Smith et al., 2007). Many of the complex pathways controlling *ompA* expression have been extensively characterized, but key aspects of the observed temperature-dependent regulation of *ompA* expression have remained elusive. Specifically, it has been observed that growth at relatively high environmental temperatures results in a decrease in the half-life and relative abundance of *ompA* transcript present in the bacterium, yet does not impact the amount of OmpA protein produced. Increased translational efficiency from the lesser amount of *ompA* transcript present during growth of the bacterium at higher temperatures explains the observed differential effect of environmental temperature on *ompA* transcript and OmpA protein levels. Although it has been demonstrated that the *ompA* transcript undergoes temperature-dependent changes in structure, the molecular mechanisms underlying the temperature-dependent regulation of translation efficiency has not yet been elucidated (Afonyushkin et al., 2003).

Preliminary studies have demonstrated that the *E. coli* and *Shigella ompA* transcripts are highly conserved, and that each contains a putative RNA thermometer (Figure 5.4) (Murphy). Key to the classification of the *ompA* RNA thermometer as a functional ribo-regulator was the finding that nucleic acid sequences composing this element are sufficient to confer temperature-dependent post-transcriptional regulation onto the expression of a reporter gene. The model that a conserved RNA thermometer regulates *ompA* expression is consistent with the observed temperature-dependent alteration of *ompA* translational efficiency. Regulation of *Shigella* OmpA production in response to temperature ensures that the protein is made in the precise environment that it is needed to facilitate cell-to-cell spread of the pathogen. The finding that a functional RNA thermometer controls the expression of *ompA* provides the second example of a *Shigella* virulence factor being regulated in this manner, and further validates the importance of this class of ribo-regulators in controlling virulence of this pathogen.

Future Directions in the Field

Continued study of *Shigella* ribo-regulators stands to contribute to our fundamental knowledge of *Shigella* virulence. As a primary pathogen with an infectious dose of as little as 100 organisms, *Shigella* species are highly successful pathogens (Dupont et al., 1989). Despite extensive characterization, we do not yet have a comprehensive roadmap of the many regulated and inter-related networks of protein and RNA molecules that mediate *Shigella* adaptation and virulence (Schroeder and Hilbi, 2008). Without this information, we cannot fully appreciate the processes underlying *Shigella* virulence, and thus the pathogenic success of these species. It has become abundantly clear that ribo-regulators play a key role in controlling processes essential for *Shigella* survival and pathogenesis. To date, *Shigella* ribo-regulators have been implicated in controlling acid resistance, carbon metabolism, nutrient acquisition, eukaryotic cell invasion and cell-to-cell spread by these pathogens. Despite the significant advances made in recent years, investigations of *Shigella* ribo-regulators have likely just scratched the surface of this potentially expansive field of study. With an estimated 2% of the *E. coli* genome dedicated to encoding small RNA molecules alone, it is expected that an equally large pool of ribo-regulators exist in *Shigella* species (Skippington and Ragan, 2012). Only after all *Shigella* ribo-regulators have been identified and characterized will the full impact of these molecules on the physiology and virulence of these important pathogens be elucidated. It is anticipated that the use of emerging technology will facilitate the identification of a multitude of ribo-regulators in each *Shigella* species. Careful characterization of each newly identified ribo-regulator will reveal the impact of these powerful regulators on the survival and pathogenesis of *Shigella* species, while at the same time revealing putative targets for the development of novel therapeutics to treat infections with these and related pathogens.

The identification and characterization of novel ribo-regulators in any bacterial species is a multi-step process that most often utilizes a complex combination of *in silico, in vivo* and *in vitro* analyses. The first required step in the identification and characterization of novel ribo-regulators is to predict their presence within a given bacterial genome or plasmid, a task often achieved using a variety of *in silico* approaches (Acebo et al., 2012; Gomez-Lozano et al., 2012; Liang et al., 2011; Livny and Waldor, 2010; Pichon et al., 2012; Postic et al., 2010; Sharma et al., 2010; Soutourina et al., 2013). One commonly used approach to identify putative ribo-regulators is to combine comparative genomics and structural predictions. This combined *in silico* approach takes advantage of the fact that the conservation of primary sequence and/or secondary structure between transcripts generated from "non-coding" regions on the genomes of related bacteria is highly correlated with the presence of a conserved ribo-regulator (Backofen and Hess, 2010). Specifically, sequence and/or structural conservation within an intergenic region suggest the presence of a *trans*-encoded ribo-regulator, while such conservation within the non-coding portion of a conserved transcript suggests the presence of a *cis*-encoded ribo-regulator. Once the location of putative regulators is identified, the next required step towards the characterization of a novel bacterial ribo-regulator requires experimental proof that the molecule is synthesized. While traditional approaches including northern blot, reverse transcriptase PCR and quantitative Real-time PCR remain effective means by which to validate the production of a given ribo-regulator, RNAseq represents a powerful and increasingly popular approach used to identify all RNA molecules generated by a given species during growth under specific environmental conditions(Acebo et al., 2012; Gomez-Lozano et al., 2012; Sharma et al., 2010; Soutourina et al., 2013). Once a putative ribo-regulator is identified and its production validated, the next required step in characterizing the element is to identify its regulatory targets. In the case of mRNA binding small RNA molecules, regulon identification can be facilitated by software packages designed to identify those transcripts that share a set amount and conformation of sequence complementarity with a given small RNA molecule (Tjaden, 2008, 2012; Tjaden et al., 2006; Wenzel et al., 2012). Despite the availability of target prediction software, direct experimentation remains the gold standard by which to identify the regulon of any given bacterial ribo-regulator. The final and often rate-limiting step in characterizing any bacterial ribo-regulator is to determine if the observed effect on each regulated gene is direct or indirect and to elucidate the molecular mechanism(s) underlying any and all regulatory activity. This final task is becoming increasingly complicated as a growing list of bacterial ribo-regulators has now been shown to control the expression of different targets within their regulon using different molecular mechanisms (Jorgensen et al., 2013; Oglesby-Sherrouse and Murphy, 2013). Systematic analyses will facilitate the identification, characterization and comparison of functional ribo-regulators in each *Shigella* species, an important task that has yet to be completed.

In addition to an increase in number, continued investigations are likely to reveal an increase in the diversity of *Shigella* ribo-regulators. Of note, no riboswitch has been identified in any *Shigella* species to date. Given the importance of riboswitches in controlling gene expression in several related organisms, the identification of this class of ribo-regulator in *Shigella* is extremely likely (Mandal and Breaker, 2004; Smith et al., 2010). Another significant class of ribo-regulators not yet characterized in any *Shigella* species is protein coding small RNA molecules (Ulveling et al., 2011; Vanderpool et al., 2011). A growing number of protein coding small RNA molecules have been characterized recently, yet their impact on bacterial physiology and virulence remains largely unknown. As studies of *Shigella* ribo-regulators advance, the potentially expansive role of riboswitches, coding small RNA molecules and other classes of RNA regulators in controlling gene expression in these pathogens will be revealed.

The identification and characterization of additional *Shigella* ribo-regulators will not only reveal the impact of these molecules on controlling essential processes in the bacterial cell, but may also contribute to the development of novel therapeutics to treat infections with these and related pathogens. An effective anti-shigellosis therapeutic would be one that eliminates the ability of *Shigella* to survive within the human host, or that prevents the production of specific virulence determinants. As regulatory factors that control processes essential for bacterial survival and pathogenesis, ribo-regulators represent potentially potent models for, or targets of, novel antimicrobial agents. A compound that specifically recognizes and stabilizes the inhibitory structure within an RNA thermometer, for example, would prevent translation of the regulated gene at otherwise permissive temperatures. If the stabilizing compound interacts with an RNA thermometer that controls the expression of a *Shigella* virulence-associated gene, it could function to reduce the ability of the pathogen to cause disease. The potential and initial successes of RNA-based antimicrobials are detailed in several recent reviews (Good and Stach, 2011; Hebert et al.; Lioliou et al., 2010; Lloyd, 2012). While the concept of targeting or mimicking ribo-regulators as the basis of a novel anti-*Shigella* therapeutic is sound in theory, what is lacking is a comprehensive understanding of the structure, function and impact of this diverse group of regulators in these important human pathogens; a fact that once again highlights the need to continue and intensify studies aimed at the identification and characterization of *Shigella* ribo-regulators.

Continued investigation of *Shigella* ribo-regulators has the strong potential to advance the broader field of bacterial ribo-regulators. It is interesting to note that several of the known *Shigella* ribo-regulators were first identified and characterized in these species, an observation that highlights the importance of investigating these elements in a variety of bacterial species. *Shigella* represents a particularly unique model system in which to study ribo-regulation, and especially that of virulence-associated gene expression. Specifically, *Shigella* species are closely related to *E. coli* and other enteric pathogens, yet they maintain a collection of unique virulence-associated genes, many of which are encoded on the large virulence plasmid (Dorman and Porter, 1998; Schroeder and Hilbi, 2008). Additionally, each species of *Shigella* is highly related to each other yet none are identical, and each maintains collections of sequences that make it unique. The rare mix of conservation and uniqueness allows investigators characterizing RNA-based regulation in *Shigella* to ask intriguing and important questions that will drive the study of bacterial ribo-regulation forward. Of significance will be the identification and characterization of additional ribo-regulators that are unique to one or more species of *Shigella*, as exemplified by the *trans*-encoded twin small RNA molecules RyfA1 and RyfA2 or the *cis*-encoded *shuA* RNA thermometer, both present only in a subset of *Shigella* species. The study of ribo-regulators unique to a single *Shigella* species may reveal clues about the continued evolution of these strains, by exposing regulatory pathways that are different between these otherwise extremely similar organisms. Another class of *Shigella* ribo-regulators that will be exceptionally interesting to continue to investigate are those that are broadly conserved across bacterial species, but that have evolved to control *Shigella* specific genes and processes. This unique class of regulators is exemplified by regulation of the plasmid born *Shigella* specific virulence-associated gene *virB* by the conserved chromosomally encoded small RNA RyhB (Africa et al., 2011; Broach et al., 2012; Kouse et al., 2013; Murphy and Payne, 2007). Continued study of the interplay between conserved chromosomally encoded ribo-regulators and horizontally acquired plasmid encoded regulatory targets has the potential to provide an evolutionary perspective to the understanding of how bacteria acquire new ribo-regulators and how these elements change in response to the unique selective pressures experienced by pathogenic bacteria.

While relatively few in number, *Shigella* ribo-regulators have proven to be central to gene expression pathways that directly control the physiology and virulence of these significant pathogens. Further studies aimed at identifying novel ribo-regulators, as well as those designed to further characterize the impact and regulatory mechanism(s) of known ribo-regulators in each *Shigella* species will no doubt advance the foundational knowledge of gene expression in these pathogens as well as that of the broader field of bacterial ribo-regulation.

References

Acebo, P., Martin-Galiano, A.J., Navarro, S., Zaballos, A., and Amblar, M. (2012). Identification of 88 regulatory small RNAs in the TIGR4 strain of the human pathogen *Streptococcus pneumoniae*. RNA-A Publication of the RNA Society *18*, 530-546.

Afonyushkin, T., Moll, I., Blasi, U., and Kaberdin, V.R. (2003). Temperature-dependent stability and translation of *Escherichia coli ompA* mRNA. Biochemical and Biophysical Research Communications *311*, 604.

Afonyushkin, T., Vecerek, B., Moll, I., Blasi, U., and Kaberdin, V.R. (2005). Both RNase E and RNase III control the stability of sodB mRNA upon translational inhibition by the small regulatory RNA RyhB. Nucleic Acids Research *33*, 1678-1689.

Africa, L.A.A., Murphy, E.R., Egan, N.R., Wigley, A.F., and Wing, H.J. (2011). The Iron-Responsive Fur/RyhB Regulatory Cascade Modulates the *Shigella* Outer Membrane Protease IcsP. Infection And Immunity *79*, 4543-4549.

Aiba, H. (2007). Mechanism of RNA silencing by Hfq-binding small RNAs. Current Opinion In Microbiology *10*, 134-139.

Ambrosi, C., Pompili, M., Scribano, D., Zagaglia, C., Ripa, S., and Nicoletti, M. (2012). Outer Membrane Protein A (OmpA): A New Player in *Shigella flexneri* Protrusion Formation and Iner-Cellular Spreading. Plos One *7*.

Anzaldi, L.L., and Skaar, E.P. (2010). Overcoming the heme paradox: heme toxicity and tolerance in bacterial pathogens. Infection And Immunity *78*, 4977-4989.

Ayala-Castro, C., Saini, A., and Outten, F.W. (2008). Fe-S cluster assembly pathways in bacteria. Microbiology and molecular biology reviews : MMBR *72*, 110-125, table of contents.

Backofen, R., and Hess, W.R. (2010). Computational prediction of sRNAs and their targets in bacteria. RNA Biol *7*, 33-42.

Battisti, J.M., Sappington, K.N., Smitherman, L.S., Parrow, N.L., and Minnick, M.F. (2006). Environmental signals generate a differential and coordinated expression of the heme receptor gene family of Bartonella quintana. Infection And Immunity *74*, 3251-3261.

Beisel, C.L., and Storz, G. (2010). Base pairing small RNAs and their roles in global regulatory networks. FEMS Microbiology Reviews *34*, 866-882.

Bernardini, M.L., Mounier, J., d'Hauteville, H., Coquis-Rondon, M., and Sansonetti, P.J. (1989). Identification of icsA, a plasmid locus of Shigella flexneri that governs bacterial intra- and intercellular spread through interaction with F-actin. Proc Natl Acad Sci U S A *86*, 3867-3871.

Braun, V. (2001). Iron uptake mechanisms and their regulation in pathogenic bacteria. International Journal of Medical Microbiology: IJMM *291*, 67-79.

Brennan, R.G., and Link, T.M. (2007). Hfq structure, function and ligand binding. Curr Opin Microbiol *10*, 125-133.

Broach, W.H., Egan, N., Wing, H.J., Payne, S.M., and Murphy, E.R. (2012). VirF-Independent Regulation of *Shigella virB* Transcription is Mediated by the Small RNA RyhB. Plos One *7*, 1-11.

Buchrieser, C., Glaser, P., Rusniok, C., Nedjari, H., d'Hauteville, H.l.n., Kunst, F., Sansonetti, P., and Parsot, C. (2000). The virulence plasmid pWR100 and the repertoire of proteins secreted by the type III secretion apparatus of Shigella flexneri. Molecular Microbiology *38*, 760-771.

Burkhard, K.A., and Wilks, A. (2007). Characterization of the outer membrane receptor ShuA from the heme uptake system of Shigella dysenteriae. Substrate specificity and identification of the heme protein ligands. The Journal Of Biological Chemistry *282*, 15126-15136.

Caron, M.P., Lafontaine, D.A., and Masse, E. (2010). Small RNA-mediated regulation at the level of transcript stability. RNA Biology *7*, 140-144.

Confer, A.W., and Ayalew, S. (2013). Review: The OmpA family of proteins: Roles in bacterial pathogenesis and immunity. Veterinary Microbiology *163*, 207-222.

De Lay, N., Schu, D.J., and Gottesman, S. (2013). Bacterial small RNA-based negative regulation: Hfq and its accomplices. J Biol Chem *288*, 7996-8003.

Desnoyers, G., Morissette, A., Prevost, K., and Masse, E. (2009). Small RNA-induced differential degradation of the polycistronic mRNA iscRSUA. The EMBO Journal *28*, 1551-1561.

Dorman, C.J., and Porter, M.E. (1998). The Shigella virulence gene regulatory cascade: a paradigm of bacterial gene control mechanisms. Mol Microbiol *29*, 677-684.

Dupont, H.L., Levine, M.M., Hornick, R.B., and Formal, S.B. (1989). Inoculum size in Shigellosis and implications for expected mode of transmission. Journal of Infectious Dieseases *159*, 1126-1128.

Escolar, L., Perez-Martin, J., and de Lorenzo, V. (1999). Opening the iron box: transcriptional metalloregulation by the Fur protein. Journal Of Bacteriology *181*, 6223-6229.

Frohlich, K.S., and Vogel, J.r. (2009). Activation of gene expression by small RNA. Current Opinion In Microbiology *12*, 674-682.

Geissmann, T., Possedko, M., Huntzinger, E., Fechter, P., Ehresmann, C., and Romby, P. (2006). Regulatory RNAs as mediators of virulence gene expression in bacteria. Handbook Of Experimental Pharmacology, 9-43.

Georg, J., and Hess, W.R. (2011). cis-antisense RNA, another level of gene regulation in bacteria. Microbiology And Molecular Biology Reviews: MMBR *75*, 286-300.

Giangrossi, M., Prosseda, G., Tran, C.N., Brandi, A., Colonna, B., and Falconi, M. (2010). A novel antisense RNA regulates at transcriptional level the virulence gene icsA of Shigella flexneri. Nucleic Acids Research *38*, 3362-3375.

Goldberg, M.B. (2001). Actin-based motility of intracellular microbial pathogens. Microbiology and molecular biology reviews : MMBR *65*, 595-626, table of contents.

Goldberg, M.B., and Theriot, J.A. (1995). Shigella flexneri surface protein IcsA is sufficient to direct actin-based motility. Proceedings Of The National Academy Of Sciences Of The United States Of America *92*, 6572-6576.

Goldberg, M.B., Theriot, J.A., and Sansonetti, P.J. (1994). Regulation of surface presentation of IcsA, a Shigella protein essential to intracellular movement and spread, is growth phase dependent. Infection And Immunity *62*, 5664-5668.

Gomez-Lozano, M., Marvig, R.L., Molin, S., and Long, K.S. (2012). Genome-wide identification of novel small RNAs in Pseudomonas aeruginosa. Environmental Microbiology *14*, 2006-2016.

Good, L., and Stach, J.E.M. (2011). Synthetic RNA silencing in bacteria - antimicrobial discovery and resistance breaking. Frontiers In Microbiology *2*, 185-185.

Gore, A.L., and Payne, S.M. (2010). CsrA and Cra influence Shigella flexneri pathogenesis. Infection And Immunity *78*, 4674-4682.

Gottesman, S., McCullen, C.A., Guillier, M., Vanderpool, C.K., Majdalani, N., Benhammou, J., Thompson, K.M., FitzGerald, P.C., Sowa, N.A., and FitzGerald, D.J. (2006). Small RNA regulators and the bacterial response to stress. Cold Spring Harbor Symposia On Quantitative Biology *71*, 1-11.

Hajnsdorf, E., and Boni, I.V. (2012). Multiple activities of RNA-binding proteins S1 and Hfq. Biochimie *94*, 1544-1553.

Hantke, K. (2001). Iron and metal regulation in bacteria. Current Opinion In Microbiology *4*, 172-177.

Hebert, C.G., Valdes, J.J., and Bentley, W.E. Beyond silencing, engineering applications of RNA interference and antisense technology for altering cellular phenotype. Current Opinion in Biotechnology *19*, 500-505.

Herrmann, K.M., and Weaver, L.M. (1999). The Shikimate Pathway. Annual review of plant physiology and plant molecular biology *50*, 473-503.

Hershberg, R., Altuvia, S., and Margalit, H. (2003). A survey of small RNA-encoding genes in Escherichia coli. Nucleic Acids Res *31*, 1813-1820.

Imlay, J.A., Chin, S.M., and Linn, S. (1988). Toxic DNA damage by hydrogen peroxide through the Fenton reaction in vivo and in vitro. Science *240*, 640-642.

Imlay, J.A., and Linn, S. (1988). DNA damage and oxygen radical toxicity. Science *240*, 1302-1309.

Jacques, J., Jang, S., Prevost, K., Desnoyers, G., Desmarais, M., Imlay, J., and Masse, E. (2006). RyhB small RNA modulates the free intracellular iron pool and is essential for normal growth during iron limitation in Escherichia coli. Molecular Microbiology *62*, 1181-1190.

Jennison, A.V., and Verma, N.K. (2004). Shigella flexneri infection: pathogenesis and vaccine development. FEMS Microbiol Rev *28*, 43-58.

Jorgensen, M.G., Thomason, M.K., Havelund, J., Valentin-Hansen, P., and Storz, G. (2013). Dual function of the McaS small RNA in controlling biofilm formation. Genes Dev *27*, 1132-1145.

Kirby, A.E., Metzger, D.J., Murphy, E.R., and Connell, T.D. (2001). Heme utilization in Bordetella avium is regulated by RhuI, a heme-responsive extracytoplasmic function sigma factor. Infection And Immunity *69*, 6951-6961.

Kortmann, J., and Narberhaus, F. (2012). Bacterial RNA thermometers: molecular zippers and switches.(Report). Nature Reviews Microbiology, 255.

Kouse, A.B., Righetti, F., Kortmann, J., Narberhaus, F., and Murphy, E.R. (2013). RNA-Mediated Thermoregulation of Iron-Acquisition Genes in Shigella dysenteriae and Pathogenic Escherichia coli. Plos One *8*, 1-13.

Liang, H., Zhao, Y.T., Zhang, J.Q., Wang, X.J., Fang, R.X., and Jia, Y.T. (2011). Identification and functional characterization of small non-coding RNAs in Xanthomonas oryzae pathovar oryzae. BMC GENOMICS *12*.

Lioliou, E., Romilly, C.d., Romby, P., and Fechter, P. (2010). RNA-mediated regulation in bacteria: from natural to artificial systems. New Biotechnology *27*, 222-235.

Liu, J.M., and Camilli, A. (2010). A broadening world of bacterial small RNAs. Current Opinion In Microbiology *13*, 18-23.

Liu, M.Y., Gui, G., Wei, B., Preston, J.F., 3rd, Oakford, L., Yuksel, U., Giedroc, D.P., and Romeo, T. (1997). The RNA molecule CsrB binds to the global regulatory protein CsrA and antagonizes its activity in Escherichia coli. J Biol Chem *272*, 17502-17510.

Liu, M.Y., and Romeo, T. (1997). The global regulator CsrA of Escherichia coli is a specific mRNA-binding protein. J Bacteriol *179*, 4639-4642.

Livny, J., and Waldor, M.K. (2010). Mining regulatory 5'UTRs from cDNA deep sequencing datasets. Nucleic Acids Research *38*, 1504-1514.

Lloyd, D.H. (2012). Alternatives to conventional antimicrobial drugs: a review of future prospects. Veterinary Dermatology *23*, 299-e260.

Mandal, M., and Breaker, R.R. (2004). Gene regulation by riboswitches. Nat Rev Mol Cell Biol *5*, 451-463.

Maringanti, S., and Imlay, J.A. (1999). An intracellular iron chelator pleiotropically suppresses enzymatic and growth defects of superoxide dismutase-deficient Escherichia coli. J Bacteriol *181*, 3792-3802.

Masse, E., Escorcia, F.E., and Gottesman, S. (2003). Coupled degradation of a small regulatory RNA and its mRNA targets in Escherichia coli. Genes Dev *17*, 2374-2383.

Masse, E., and Gottesman, S. (2002). A small RNA regulates the expression of genes involved in iron metabolism in Escherichia coli. Proc Natl Acad Sci U S A *99*, 4620-4625.
Masse, E., Salvail, H., Desnoyers, G., and Arguin, M.I. (2007). Small RNAs controlling iron metabolism. Current Opinion In Microbiology *10*, 140-145.
Masuda, N., and Church, G.M. (2002). Escherichia coli gene expression responsive to levels of the response regulator EvgA. J Bacteriol *184*, 6225-6234.
Masuda, N., and Church, G.M. (2003). Regulatory network of acid resistance genes in Escherichia coli. Mol Microbiol *48*, 699-712.
Mills, M., and Payne, S.M. (1995). Genetics and regulation of heme iron transport in Shigella dysenteriae and detection of an analogous system in Escherichia coli O157:H7. J Bacteriol *177*, 3004-3009.
Mills, M., and Payne, S.M. (1997). Identification of shuA, the gene encoding the heme receptor of Shigella dysenteriae, and analysis of invasion and intracellular multiplication of a shuA mutant. Infect Immun *65*, 5358-5363.
Morita, T., and Aiba, H. (2011). RNase E action at a distance: degradation of target mRNAs mediated by an Hfq-binding small RNA in bacteria. Genes Dev *25*, 294-298.
Murphy, E.R. Unpublished data.
Murphy, E.R., and Payne, S.M. (2007). RyhB, an Iron-Responsive Small RNA Molecule, Regulates Shigella dysenteriae Virulence. Infect Immun *75*, 3470-3477.
Murphy, E.R., Sacco, R.E., Dickenson, A., Metzger, D.J., Hu, Y., Orndorff, P.E., and Connell, T.D. (2002). BhuR, a virulence-associated outer membrane protein of Bordetella avium, is required for the acquisition of iron from heme and hemoproteins. Infect Immun *70*, 5390-5403.
Nairz, M., Schroll, A., Sonnweber, T., and Weiss, G.n. (2010). The struggle for iron - a metal at the host-pathogen interface. Cellular Microbiology *12*, 1691-1702.
Narberhaus, F., Waldminghaus, T., and Chowdhury, S. (2006). RNA thermometers. FEMS Microbiology Reviews *30*, 3-16.
Niyogi, S.K. (2005). Shigellosis. Journal Of Microbiology (Seoul, Korea) *43*, 133-143.
Oglesby, A.G., Murphy, E.R., Iyer, V.R., and Payne, S.M. (2005). Fur regulates acid resistance in Shigella flexneri via RyhB and ydeP. Mol Microbiol *58*, 1354-1367.
Oglesby-Sherrouse, A.G., and Murphy, E.R. (2013). Iron-responsive Bacterial Small RNAs: Variations on a Theme. Metallomics *5*, 276-268.
Oh, M.H., Lee, S.M., Lee, D.H., and Choi, S.H. (2009). Regulation of the Vibrio vulnificus hupA gene by temperature alteration and cyclic AMP receptor protein and evaluation of its role in virulence. Infection And Immunity *77*, 1208-1215.
Otto, B.R., Verweij-van Vught, A.M., and MacLaren, D.M. (1992). Transferrins and heme-compounds as iron sources for pathogenic bacteria. Critical Reviews In Microbiology *18*, 217-233.
Papenfort, K., and Vogel, J.r. (2009). Multiple target regulation by small noncoding RNAs rewires gene expression at the post-transcriptional level. Research In Microbiology *160*, 278-287.
Parsot, C. (2009). Shigella type III secretion effectors: how, where, when, for what purposes? Current Opinion In Microbiology *12*, 110-116.
Payne, S.M., Wyckoff, E.E., Murphy, E.R., Oglesby, A.G., Boulette, M.L., and Davies, N.M.L. (2006). Iron and pathogenesis of Shigella: iron acquisition in the intracellular environment. Biometals: An International Journal On The Role Of Metal Ions In Biology, Biochemistry, And Medicine *19*, 173-180.
Philpott, D.J., Edgeworth, J.D., and Sansonetti, P.J. (2000). The pathogenesis of Shigella flexneri infection: lessons from in vitro and in vivo studies. Philos Trans R Soc Lond B Biol Sci *355*, 575-586.

Pichon, C., du Merle, L., Caliot, M.E., Trieu-Cuot, P., and Le Bouguenec, C. (2012). An in silico model for identification of small RNAs in whole bacterial genomes: characterization of antisense RNAs in pathogenic Escherichia coli and Streptococcus agalactiae strains. Nucleic Acids Research *40*, 2846-2861.

Porter, M.E., and Dorman, C.J. (1997). Differential regulation of the plasmid-encoded genes in the Shigella flexneri virulence regulon. Mol Gen Genet *256*, 93-103.

Postic, G., Frapy, E., Dupuis, M., Dubail, I., Livny, J., Charbit, A., and Meibom, K.L. (2010). Identification of small RNAs in Francisella tularensis. BMC Genomics *11*, 625-636.

Prevost, K., Desnoyers, G., Jacques, J.F., Lavoie, F., and Masse, E. (2011). Small RNA-induced mRNA degradation achieved through both translation block and activated cleavage. Genes Dev *25*, 385-396.

Prevost, K., Salvail, H., Desnoyers, G., Jacques, J.-F.o., Phaneuf, E., and Masse, E. (2007). The small RNA RyhB activates the translation of shiA mRNA encoding a permease of shikimate, a compound involved in siderophore synthesis. Molecular Microbiology *64*, 1260-1273.

Prosseda, G., Fradiani, P.A., Di Lorenzo, M., Falconi, M., Micheli, G., Casalino, M., Nicoletti, M., and Colonna, B. (1998). A role for H-NS in the regulation of the virF gene of Shigella and enteroinvasive Escherichia coli. Research In Microbiology *149*, 15-25.

Rivas, E., Klein, R.J., Jones, T.A., and Eddy, S.R. (2001). Computational identification of noncoding RNAs in E. coli by comparative genomics. Curr Biol *11*, 1369-1373.

Romby, P., and Charpentier, E. (2010). An overview of RNAs with regulatory functions in gram-positive bacteria.(Report). Cellular and Molecular Life Sciences (CMLS), 217.

Romeo, T., Vakulskas, C.A., and Babitzke, P. (2013). Post-transcriptional regulation on a global scale: form and function of Csr/Rsm systems.(Report). Environmental Microbiology, 313.

Rossi, M.S., Fetherston, J.D., Letoffe, S., Carniel, E., Perry, R.D., and Ghigo, J.M. (2001). Identification and characterization of the hemophore-dependent heme acquisition system of Yersinia pestis. Infection And Immunity *69*, 6707-6717.

Sakai, T., Sasakawa, C., and Yoshikawa, M. (1988). Expression of four virulence antigens of Shigella flexneri is positively regulated at the transcriptional level by the 30 kiloDalton virF protein. Mol Microbiol *2*, 589-597.

Salvail, H., and Masse, E. (2012). Regulating iron storage and metabolism with RNA: an overview of posttranscriptional controls of intracellular iron homeostasis. Wiley interdisciplinary reviews RNA *3*, 26-36.

Schroeder, G.N., and Hilbi, H. (2008). Molecular pathogenesis of *Shigella* spp.: controlling host cell signaling, invasion, and death by type III secretion. Clinical Microbiology Reviews *21*, 134-156.

Sharma, C.M., Hoffmann, S., and Darfeuille, F. (2010). The primary transcriptome of the major human pathogen Helicobacter pylori. Nature *464*, 250-255.

Skaar, E.P. (2010). The battle for iron between bacterial pathogens and their vertebrate hosts. Plos Pathogens *6*.

Skippington, E., and Ragan, M.A. (2012). Evolutionary Dynamics of Small RNAs in 27 Escherichia coli and Shigella Genomes. Genome Biology and Evolution *4*, 6.

Smith, A.M., Fuchs, R.T., Grundy, F.J., and Henkin, T.M. (2010). Riboswitch RNAs: regulation of gene expression by direct monitoring of a physiological signal. RNA Biology *7*, 104-110.

Smith, S.G.J., Fagan, R.P., Lambert, M.A., and Mahon, V. (2007). A molecular Swiss army knife: OmpA structure, function and expression [electronic resource]. FEMS Microbiology Letters *273*, 1-11.

Soutourina, O.A., Monot, M., Boudry, P., Saujet, L., Pichon, C., Sismeiro, O., Semenova, E., Severinov, K., Le Bouguenec, C., Copp√©e, J.-Y., *et al.* (2013). Genome-Wide Identification of Regulatory RNAs in the Human Pathogen *Clostridium difficile*. PLoS Genetics *9*, 1-18.

Storz, G., Vogel, J.r., and Wassarman, K.M. (2011). Regulation by small RNAs in bacteria: expanding frontiers. Molecular Cell *43*, 880-891.

Suzuki, T., Saga, S., and Sasakawa, C. (1996). Functional analysis of *Shigella* VirG domains essential for interaction with vinculin and actin-based motility. The Journal Of Biological Chemistry *271*, 21878-21885.

Suzuki, T., and Sasakawa, C. (2001). Molecular basis of the intracellular spreading of Shigella. Infect Immun *69*, 5959-5966.

Timmermans, J., and Van Melderen, L. (2010). Post-transcriptional global regulation by CsrA in bacteria. Cellular And Molecular Life Sciences: CMLS *67*, 2897-2908.

Tjaden, B. (2008). TargetRNA: a tool for predicting targets of small RNA action in bacteria. Nucleic Acids Research.

Tjaden, B. (2012). Computational identification of sRNA targets. Methods In Molecular Biology (Clifton, NJ) *905*, 227-234.

Tjaden, B., Goodwin, S.S., Opdyke, J.A., Guillier, M., Fu, D.X., Gottesman, S., and Storz, G. (2006). Target prediction for small, noncoding RNAs in bacteria. Nucleic Acids Res *34*, 2791-2802.

Torres, A.G., and Payne, S.M. (1997). Haem iron-transport system in enterohaemorrhagic Escherichia coli O157:H7. Mol Microbiol *23*, 825-833.

Tran, C.N., Giangrossi, M., Prosseda, G., Brandi, A., Di Martino, M.L., Colonna, B., and Falconi, M. (2011). A multifactor regulatory circuit involving H-NS, VirF and an antisense RNA modulates transcription of the virulence gene icsA of Shigella flexneri. Nucleic Acids Research *39*, 8122-8134.

Ulveling, D., Francastel, C., and Hube, F. (2011). When one is better than two: RNA with dual functions. Biochimie *93*, 633-644.

Vanderpool, C.K., Balasubramanian, D., and Lloyd, C.R. (2011). Dual-function RNA regulators in bacteria. Biochimie *93*, 1943-1949.

Vecerek, B., Moll, I., Afonyushkin, T., Kaberdin, V., and Blasi, U. (2003). Interaction of the RNA chaperone Hfq with mRNAs: direct and indirect roles of Hfq in iron metabolism of Escherichia coli. Mol Microbiol *50*, 897-909.

Waldminghaus, T., Fippinger, A., Alfsmann, J., and Narberhaus, F. (2005). RNA thermometers are common in alpha- and gamma-proteobacteria. Biological Chemistry *386*, 1279-1286.

Waldminghaus, T., Heidrich, N., Brantl, S., and Narberhaus, F. (2007). FourU: a novel type of RNA thermometer in Salmonella. Molecular Microbiology *65*, 413-424.

Wandersman, C.c., and Delepelaire, P. (2004). Bacterial iron sources: from siderophores to hemophores. Annual Review Of Microbiology *58*, 611-647.

Wassarman, K.M. (2002). Small RNAs in bacteria: diverse regulators of gene expression in response to environmental changes. Cell *109*, 141-144.

Wassarman, K.M., Repoila, F., Rosenow, C., Storz, G., and Gottesman, S. (2001). Identification of novel small RNAs using comparative genomics and microarrays. Genes and Development *15*, 1637-1651.

Waters, L.S., and Storz, G. (2009). Regulatory RNAs in bacteria. Cell *136*, 615-628.

Weilbacher, T., Suzuki, K., Dubey, A.K., Wang, X., Gudapaty, S., Morozov, I., Baker, C.S., Georgellis, D., Babitzke, P., and Romeo, T. (2003). A novel sRNA component of the carbon storage regulatory system of Escherichia coli. Mol Microbiol *48*, 657-670.

Wenzel, A., Akbasli, E., and Gorodkin, J. (2012). RIsearch: fast RNA-RNA interaction search useing a simplified nearest-neighbor energy model. Bioinformatics *28*, 9.

Wilks, A. (2001). The ShuS protein of Shigella dysenteriae is a heme-sequestering protein that also binds DNA. Archives Of Biochemistry And Biophysics *387*, 137-142.
Wilks, A., and Burkhard, K.A. (2007). Heme and virulence: how bacterial pathogens regulate, transport and utilize heme. Natural Product Reports *24*, 511-522.
Wyckoff, E.E., Duncan, D., Torres, A.G., Mills, M., Maase, K., and Payne, S.M. (1998). Structure of the Shigella dysenteriae haem transport locus and its phylogenetic distribution in enteric bacteria. Mol Microbiol *28*, 1139-1152.

6

IpaB Ion Channels Induce Pyroptosis in Macrophages

Lidija Senerovic[1], Anna Brotcke Zumsteg[2] and Michael Kolbe*[2]

[1]Institute of Molecular Genetics and Genetic Engineering, University of Belgrade, Vojvode Stepe 444a, P.O.Box 23, 11010 Belgrade, Serbia
[2]*Max-Planck-Institute for Infection Biology, Charitéplatz 1, 10117 Berlin, Germany, kolbe@mpiib-berlin.mpg.de

Abstract
The Gram-negative bacterium *Shigella flexneri* invades the colonic epithelium and causes bacillary dysentery. For successful infection, *S. flexneri* requires many effector proteins secreted via a Type three secretion system (T3SS). The effector protein IpaB is conserved in many medically important Gram-negative bacteria and is an essential virulence factor with multiple roles in *S. flexneri* infection. *S. flexneri* requires IpaB to invade host cells, escape from the phagosome and induce macrophage cell death. The mechanistic details regarding how IpaB functions have recently been elucidated. Secreted IpaB spontaneously oligomerizes and inserts into the plasma membrane of target cells. IpaB oligomers form channels selective for small monovalent cations with optimal activity at an acidic pH. After internalization, IpaB channels permit potassium flux inside macrophage vacuoles causing the disruption of the endolysosomal membranes. Vacuolar leakage is followed by an IPAF-dependent activation of caspase-1 and macrophage death. Mechanistic insights gained from IpaB studies with host cells may be extended to homologues from other medically important enteropathogenic bacteria and could contribute not only to understanding the conserved mechanisms of T3SS dependent infections, but also provide the solutions to fight them.

Introduction
Shigella flexneri is an enteroinvasive bacterium that causes shigellosis. The pathogenicity of *S. flexneri* is based on its capability to invade the epithelium of the colon and to induce a severe inflammatory response in the intestinal tissue. To initiate an infection, *S. flexneri* requires around 30 effector proteins that are secreted through the Type III secretion system (T3SS) needle. Invasion plasmid antigen B (IpaB) is a 62 kDa protein and one of the first effectors secreted by *S. flexneri*. IpaB, a translocator protein, is required for controlled secretion and translocation of other effectors into the host cytosol (Menard et al., 1994a; Veenendaal et al., 2007). The current model of *S. flexneri* infection postulates that after secretion IpaB together with IpaC forms a translocon, which consists of a channel in the host membrane that enables delivery of effector proteins into the host cell

Table 6.1. Comparison of the transmembrane domains of IpaB and its homologues using TopPredII software (Claros and von Heijne, 1994).

Translocator	Bacteria	Length (amino acids)	Predicted transmembrane helices
IpaB	*Shigella flexneri*	580	5
SipB	*Salmonella typhimurium*	593	3
YopB	*Yersinia enterocolitica*	401	5
PopB	*Pseudomonas aeruginosa*	392	4

(Blocker et al., 1999). Although widely accepted, this function of IpaB has not been experimentally demonstrated.

IpaB and IpaC are stored in the bacterial cytoplasm bound to their chaperone IpgC and released after bacterial contact with the host cell (Lunelli et al., 2009; Page et al., 2001; Watarai et al., 1995). The interaction with their chaperone IpgC protects IpaB and IpaC from degradation and keeps them separated and inactive until secretion (Menard et al., 1994b). IpaB can also interact with the needle tip protein IpaD and block secretion of other effectors in the absence of host cells. The structures of the soluble N-terminal regions of IpaB (74-244) and its homolog SipB (82-226) from *Salmonella* were recently reported (Barta et al., 2012). Both IpaB (amino acids 74 to 244) and SipB (amino acids 82 to 226) are almost entirely α-helical and are composed of three antiparallel helices compressed into an intramolecular coiled-coil tertiary structure, which is a conserved feature of these T3SS translocators. Interestingly, the coiled-coil motif is a highly conserved characteristic of the T3SS needle tip proteins from several families of pathogens (Delahay and Frankel, 2002), suggesting its important role in protein-protein interaction. Therefore, IpaD located at the needle tip of *S. flexneri* T3SS apparatus most likely interacts with IpaB within the coiled-coil regions of the both proteins.

The initial site of *S. flexneri* infection is the epithelium where *Shigella* triggers an inflammatory destruction of the gut barrier. *Shigella* is transcytosed from the lumen of the gut via M cells and invades the epithelial cells from the basolateral side (Mounier et al., 1992; Wassef et al., 1989). In the underlying mucosal lymphoid tissue *S. flexneri* is phagocytosed by resident tissue macrophages (Jarry et al., 1989; Soesatyo et al., 1990), but quickly escapes from the phagosome and causes a proinflammatory cell death (Chen et al., 1996; Zychlinsky et al., 1992). Induction of pyroptosis in macrophages is a crucial event during the onset of shigellosis (Sansonetti et al., 2000; Zychlinsky et al., 1994c). To lyse the phagosomal membrane and kill macrophages, *S. flexneri* requires IpaB (Senerovic et al., 2012). We showed recently that IpaB ion channel activity induces destabilization of vacuolar membranes, which is followed by caspase-1 activation and macrophage death (Senerovic et al., 2012). IpaB secreted from the bacteria that escaped to the macrophage cytosol has been suggested to bind to caspase-1 causing its activation (Hilbi et al., 1998). Active caspase-1 induces maturation and release of interleukin 1β (IL-1β) (Zychlinsky et al., 1994a). To invade the epithelium and escape from phagosomes in epithelial cells, *S. flexneri* again requires IpaB (Sansonetti et al., 1986). IpaB targets cholesterol rich membrane rafts for invasion, triggers Golgi fragmentation, induces reorganization of the recycling compartments and contributes to the disruption of the host epithelial barrier (Mounier et al., 2012). IpaB secreted from intracellular bacteria causes epithelial cell-cycle arrest by targeting Mad2L2, an anaphase-promoting complex/ cyclosome inhibitor (Iwai et al., 2007). IpaB-mediated cell-cycle arrest contributes to the efficient colonization of the host cells. Bacterial spreading through the epithelium as well as acute inflammation causes tissue damage and the symptoms of shigellosis. Although many effectors are involved in the pathogenicity of *S. flexneri*, in this chapter we will focus on the mechanism of how IpaB induces macrophage death and describe its critical role for successful infection.

IpaB oligomers spontaneously integrate into host cell membranes

IpaB from *S. flexneri*, as well its homologues SipB from *Salmonella* (Hume et al., 2003), YopB from *Yersinia enterocolitica* (Ryndak et al., 2005), or PopB from *Pseudomonas aerugionsa* (Schoehn et al., 2003), are hydrophobic proteins that directly integrate into host membranes after separation from their chaperones. The membrane topology model obtained using the TopPredII software (Claros and von Heijne, 1994) indicates that IpaB contains five potential transmembrane helices. Although soluble in the presence of its chaperone IpgC, IpaB alone cannot be stably expressed in *E. coli* since it causes cell lysis. When expressed in yeast or other eukaryotic cells, IpaB is rapidly degraded. High amount of pure recombinant IpaB, however, was obtained after coexpression of IpaB with IpgC and then separation of the translocator from the chaperone with mild detergents (De Geyter et al., 2000; Dickenson et al., 2013; Senerovic et al., 2012). Using N, N-Dimethyldodecylamine-N-oxide (LDAO) for purification enabled mechanistic functional studies and biochemical analysis of IpaB (Dickenson et al., 2013; Senerovic et al., 2012). LDAO is a zwitterionic detergent with a small micelle size of 17 kDa, which does not increase with the detergent concentration (Strop and Brunger, 2005). Small micelles enable easy removal of the detergent by dialysis and importantly, avoid concentration of the detergent together with the protein. Although LDAO could destabilize proteins and is assumed relatively harsh (Prive, 2007), IpaB was efficiently solubilized and stayed stable in the presence of this detergent. Limited protease treatment of IpaB in the presence of LDAO shows the stable core of 40 kDa in the region between amino acid 116 and 482 (Senerovic et al., 2012).

Size exclusion chromatography (SEC) of IpaB in the presence of a LDAO concentration above the critical micellar concentration (CMC) shows that LDAO keeps IpaB in monomeric form (Dickenson et al., 2013; Senerovic et al., 2012). However, when the detergent concentration is below the CMC, IpaB forms oligomers of a constant size larger than decamers. Interestingly, as recently shown by Dickenson et al., oligomerization status of IpaB is influenced by the presence of other types of detergents, although the secondary structures stay similar with predominantly α-helical characteristics. Namely, analytical ultracentrifugation (AUC) analysis of crosslinked IpaB solubilized with n-Octyl-oligo-oxyethylene shows that IpaB assembles into homotetramers as a dominant species with a molecular weight of 265 kDa (Dickenson et al., 2013).

IpaB secreted by *S. flexneri* binds to and disrupts membranes (Blocker et al., 1999; De Geyter et al., 2000; High et al., 1992). Purified IpaB also interacts with large unilamellar vesicles (LUVs) composed of typical eukaryotic plasma membrane lipids and red blood cells (Senerovic et al., 2012). The LUVs preloaded with a fluorescent dye at a self-quenching concentration leaked when incubated with IpaB, as demonstrated by an increased flourescence of the dye released from the liposomes. The ability to disrupt phospholipid vesicles depends on the oligomeric state of IpaB (Dickenson et al., 2013). Monomeric IpaB disturbs LUVs with a very low efficiency compared to oligomers which are present in the absence of detergents, which is the situation most similar to *in vivo* conditions. The most efficient membrane disintegration is observed with IpaB tetramers, which is the proposed stoichiometry that exists at the T3SS needle tip just prior translocon formation. IpaB forms oligomers of similar sizes in solution and bound to LUVs, consistent with spontaneous oligomerization of IpaB before insertion into membranes (Senerovic et al., 2012). Crosslinked IpaB oligomers with modified primary amines have reduced ability to disturb liposomes, suggesting that protein flexibility and primary amines of exposed lysine side chains may play a critical role in the observed IpaB-mediated membrane effects (Dickenson et al., 2013). The diameter of the membrane pore formed by IpaB tetramers is estimated to be ≤ 3-5 nm, as judged by the release of the fluorescent dextran with different hydrodynamic diameters from the liposomes. Taken together, these data suggest that oligomerization of IpaB is required for insertion into membranes and pore formation.

Membrane bound IpaB constitutes a channel selective for monovalent cations

The transport properties of the transmembrane pore formed by IpaB oligomers have been addressed using two-electrode voltage clamp (TEVC) technique on *Xenopus laevis* oocytes (Senerovic et al., 2012). When oocytes are treated with exogenous IpaB, recorded electrical currents show a simple almost linear I/V relation, implying that IpaB membrane pores are not voltage-gated.

Lowering the extracellular pH from 9.0 to 4.0 increases the inward-directed current in IpaB-treated cells, suggesting higher activity of IpaB channels in acidic conditions. In the presence of monovalent inorganic cations at neutral pH, large currents are observed decreasing in the order ($K^+ > Na^+ > Li^+$). However, no current is observed for the divalent ions Mg^{2+} and Ca^{2+}, large single-charged ions like tetraethyl ammonium (TEA^+) and N-methylglucamine (NMG^+) or Cl^-. Taken together, TEVC experiments demonstrate that IpaB forms ion channels inside eukaryotic membranes with a pH dependent conductivity and a preference for small monovalent cations.

Endocytosed IpaB channel destabilizes vacuolar membranes

Vacuolar escape is essential in shigellosis because it allows bacteria to survive after phagocytosis and replicate in the cell cytoplasm. Electron microscopy demonstrates that about 15 min after *S. flexneri* is engulfed, *Shigella* destroys the surrounding vacuolar membrane and escapes into the host cell cytosol (Finlay and Falkow, 1988; Maurelli and Sansonetti, 1988; Sansonetti et al., 1986). *S. flexneri* lacking IpaB cannot escape from the phagosome, postulating that IpaB is the factor mediating the phagosomal lysis (Clerc et al., 1987; High et al., 1992). Further studies have proved this hypothesis true using purified IpaB (Senerovic et al., 2012).

Immunofluorescence microscopy with monoclonal antibodies against IpaB shows that exogenosly added IpaB is quickly endocytosed by macrophages (Senerovic et al., 2012). Dynasore, a drug inhibiting endocytosis, blocks all dynamin dependent endocytic pathways including clathrin dependent endocytosis (Macia et al., 2006). Dynasore blocks IpaB internalization and cytotoxicity in macrophages, showing a dependence on dynamin. Vacuolar damage in IpaB-treated macrophages has been detected using the fluorescently labeled fluid-phase endolysosomal marker Dextran. The punctate staining of vacuoles loaded with Dextran changes to diffuse cytosol staining 30 min after exposure of cells to IpaB. Additionally, IpaB permeabilization of vacuoles has been monitored over time with bone marrow-derived macrophages (BMMs) loaded with the pH sensitive fluorescent dye acridine orange (AO) (Senerovic et al., 2012). Using live cell imaging with AO, one can monitor both acidification of the vacuole and lysis events, as AO accumulates in acidic compartments, such as late endosomes and lysosomes, and fluoresces red until the pH decreases, causing a shift of the red fluorescence to orange and green, or the dye become diluted due to phagosomal lysis. Quantification of macrophage vacuoles with decreased AO fluorescence by flow cytometry shows that 30 min of IpaB treatment causes endolysosomal rupture in 40% of the cells. These results together demonstrated that IpaB induces the damage of acidified vacuoles which is followed by the increase of the vacuolar pH and the release of their content into the cytosol.

Quantitative electron probe X-ray microanalysis (EPXMA) can be used to examine changes in the distribution of anorganic ions inside the cells. EPXMA analysis of IpaB-treated macrophages demonstrates that cells treated with IpaB contained significantly higher K^+ and lower Mg^{2+} concentrations inside the vacuoles than the control cells. Since TEVC measurements show that IpaB ion channels exclude divalent cations, the results obtained with EPXMA suggest that IpaB forms ion channels that promote K^+ flux in endosomes. The integrity of endocytic vacuoles depends on tightly regulated osmotic homeostasis, which is maintained by limited permeability for ions and protons. Potassium homeostasis has important role in preserving lysosomal integrity and uncontrolled K^+ influx can lead to increased osmotic pressure and rupture of the vacuoles (Deng et al., 2009; Yao and Zhang, 1997). The data obtained by EPXMA suggest that IpaB mediated potassium flux induce ion disbalance within vacuoles resulting in endolysosomal damage. Taken together, these results support the hypothesis that IpaB is the *S. flexneri* effector responsible for phagosomal rupture. The data also propose a model for IpaB activity, whereby IpaB forms an ion channel that causes endosomal K^+ flux and destabilizes phagosomal membranes.

IpaB ion channels cause macrophage cell death through the IPAF inflammasome

As an intracellular pathogen, *S. flexneri* evolved different mechanisms to ensure survival within the host. For instance, *S. flexneri* induces cell death of macrophages that would otherwise kill bacteria. Macrophages infected with wild type *S. flexneri* undergo pyroptosis, a type of programmed cell death that results in concomitant release of pro-inflammatory cytokines. Cellular characteristics of *S. flexneri*-induced pyroptosis include intense chromatin condensation, DNA cleavage, vacuolization, membrane blebbing, and preservation of organelle structure, followed by plasma membrane damage and release of cellular components into the extracellular medium (Zychlinsky et al., 1992). The type of macrophage-induced cell death depends on bacterial load. At a lower multiplicity of infection (MOI) *S. flexneri* triggers caspase-1 dependent cell death (Zychlinsky et al., 1992), while macrophages infected with high bacterial number die by necrosis (Nonaka et al., 2003; Suzuki et al., 2005). IpaB is required for *S. flexneri*-induced cytotoxicity of macrophages, implicating a role for IpaB in pyroptosis (Zychlinsky et al., 1994b). Subsequently, IpaB is sufficient to induce programmed cell death, as macrophages treated with purified IpaB have a round shape, condensed nuclei and show vacuolization and membrane blebbing. Incorporation of membrane impermeable dye SytoxGreen into macrophage DNA and lactate dehydrogenase release assays demonstrate loss of plasma membrane integrity in IpaB-treated cells and cell death (Senerovic et al., 2012). These data show that treatment of macrophages with IpaB alone is sufficient to induce cellular changes consistent with pyroptosis.

S. flexneri activates caspase-1 in macrophages through an inflammasome containing ICE protease-activating factor (IPAF) and apoptosis-associated speck-like protein containing a caspase recruitment domain (ASC) followed by IL-1β secretion. An IPAF/ASC inflammasome is also formed in response to other intracellular pathogens like *Salmonella*

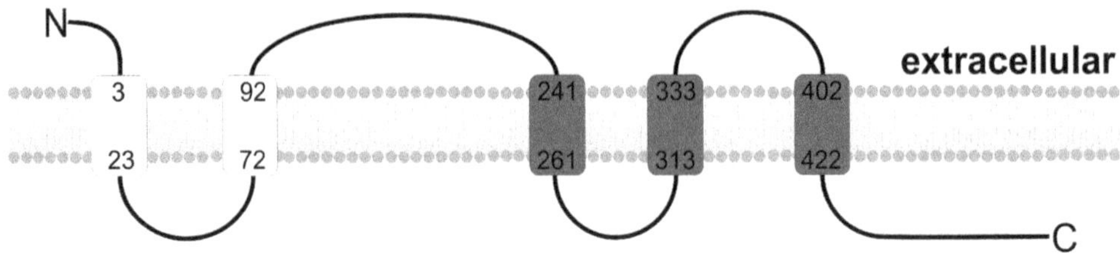

Figure 6.1. Topology of IpaB predicted by TopPredII software with default parameters for prokaryotes (Claros and von Heijne, 1994). Transmembrane domains predicted as certain (cut off 1.0) are gray and potentially assigned domains (cut off 0.6) are white.

typhimurium, *Legionella pneumophila* and *Pseudomonas aeruginosa* (Amer et al., 2006; Franchi et al., 2006; Franchi et al., 2007). Purified IpaB also triggers an IPAF inflammasome to activate caspase-1 and induce IL-1β secretion in macrophages (Senerovic et al., 2012). Like escape from the vacuole, potassium flux mediated by IpaB ion channels is directly involved in caspase-1 activation and pyroptosis of macrophages. Increasing the K^+ concentration in the extracellular medium or blocking K^+ flux with the ATP-sensitive K^+-channel inhibitor glibenclamide significantly decreases IL-1β secretion compared to IpaB-treated cells incubated with physiological K^+ concentration. Interestingly, there are only two signaling molecules known to activate IPAF inflammasome in response to bacterial infection. During *Salmonella*, *Pseudomonas, Legionella* and *Yersinia* infection IPAF is activated by flagellin delivered to the host cytosol by T3SS and T4SS (Miao et al., 2006). The other flagellin-independent signal for IPAF activation involves recognition of T3SS rod protein PrgJ from *Salmonella* or MxiI, from *S. flexneri* (Miao et al., 2010). Since *S. flexneri* lacks flagellin, the finding that IpaB ion channels activate IPAF opens the important question in the immunology field if other signals could be sensed by the IPAF inflammasome and whether caspase-1 activation requires additional upstream partners.

Stimuli like pore forming toxins, silica and uric acid crystals, are known to permeabilize lysosomal membrane (Boya and Kroemer, 2008; Hornung et al., 2008; Kagedal et al., 2001; Zdolsek et al., 1993). Membrane destabilization and the release of lysosomal proteases, like cathepsins, can activate caspase zymogens leading to apoptosis (Hornung et al., 2008; Ishisaka et al., 1998). However, lysosomal leakage and cathepsin release trigger activation of NALP3 inflammasome and the question of whether they can also activate IPAF remains to be answered in the future.

Conclusions

The mechanism of how IpaB induces cell death in macrophages is summarized in the following model shown in Figure 6.2. *S. flexneri* secretes IpaB in the vicinity of targeted cells (I) which spontaneously oligomerizes in solution forming decamers or higher molecular weight structures (II). IpaB oligomers interact with the plasma membrane of macrophages and are quickly internalized in a dynamin-dependent process. After being phagocytosed by macrophages, *S. flexneri* secretes IpaB within the phagolysosome where it oligomerizes and interacts with the vacuolar membrane (III). Within endolysosomal and phagolysosomal compartments, at acidic pH, IpaB oligomers form ion channels selective for monovalent cations (IV). Potassium influx and the consequent increase in osmotic pressure within vacuoles may lead to vacuolar destabilization and membrane rupture (V). In macrophages, damaged endolysosomes can release

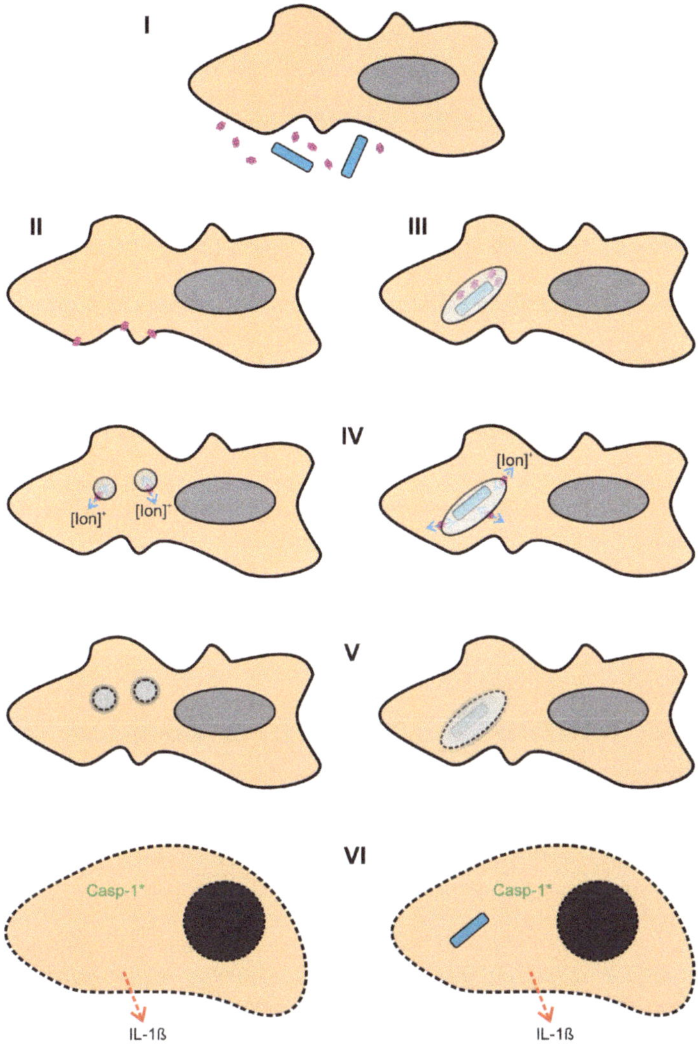

Figure 6.2. Schematic model of IpaB-induced macrophage death. *S. flexneri* secretes IpaB in the vicinity of the host cell (I) or within the phagolysosome after internalization (III). Secreted IpaB oligomerizes in solution (II and III) and then inserts into the membrane (IV). Inside vacuoles at acidic pH, IpaB forms ion channels that mediate K^+ flux disturbing ion equilibrium and leading to the vacuolar damage (V). Permeabilization of the vacuoles is followed by caspase-1 activation causing pyroptosis in macrophages (IV).

cathepsins into the cytosol which in turn activate caspase-1. In addition, ion disbalance at the vacuolar membrane could be sensed by the IPAF inflammasome leading to caspase-1 activation and the release of mature IL-1β. Caspase-1 activation, together with membrane permeabilization and the leakage of cellular content show that purified IpaB causes pyroptosis in macrophages (VI).

IpaB from *S. flexneri* has functional homologues from other enteropathogenic bacteria such as SipB from *Salmonella* (Hume et al., 2003), YopB from *Yersinia* (Ryndak et al., 2005), PopB from *Pseudomonas* (Schoehn et al., 2003), or BipB from *Burkholderia* (Attree and Attree, 2001). These virulence factors are hydrophobic proteins secreted via T3SS with a significant sequence similarity within the hydrophobic region (Buttner and Bonas, 2002; Hakansson et al., 1993). TopPredII software (Claros and von Heijne, 1994) predicts that these translocators contain two conserved transmembrane (tm) helices in the C-teminal region (tm 4 and 5 according to IpaB in Figure 6.1). Tm helix 3 is predicted in IpaB, YopB and PopB, while tm helices 1 and 2 may exist only in IpaB and SipB as this region is missing in the other analyzed translocators. After separation from their chaperones, translocators directly integrate into and permeabilize host membranes. They are all proposed to form translocons that allow transport of other effectors into the host cytosol (Blocker et al., 1999; Hakansson et al., 1993; Neyt and Cornelis, 1999). BipB causes cell fusion in tissue culture models of infection and induces apoptosis in the host cells (Suparak et al., 2005). Secreted SipB activates caspase-1 in macrophages inducing apoptosis (Hersh et al., 1999). YopB is a pore forming protein involved in contact hemolysis (Neyt and Cornelis, 1999). In *Pseudomonas* and *Salmonella* potassium efflux is also important to trigger IPAF inflammasome activation but the proteins forming membrane pores have not been identified yet (Arlehamn et al., 2010). Considering the similar hydrophobic sequence features and related functions, mechanistic details on IpaB activity presented in the model above may be extended to the homologues from other enteropathogenic bacteria that use T3SS for infection. However, in order to be accepted as a common mechanism for how translocators function, this model needs to be experimentally tested for the IpaB homologues.

The methodology to obtain high amounts of pure and active IpaB is well established and has already been applied to different aspects of IpaB function (Dickenson et al., 2013; Mounier et al., 2012; Senerovic et al., 2012). However, some questions still need to be addressed both biochemically and structurally to elucidate molecular details on how IpaB functions. Since attempts to crystallize full length IpaB or other translocators have been so far unsuccessful, *in vitro* experiments with purified protein need to be applied to decipher individual steps in pore formation and vacuolar damage. One interesting open question is: what is the mechanism for inflammasome activation in *S. flexneri* infection? Is ion disbalance and/or vacuolar damage sensed by the IPAF inflammasome or do other released components such as cathepsins induce inflammasome activation? Of course, the main challenge in the field of T3SS is to determine three dimensional structures of IpaB or other translocators either in solution or in the context of membranes.

Acknowledgements

We would like to thank Juana de Diego and Rashmi Gupta for careful reading of the manuscript. The European Research Council has provided financial support to M. Kolbe under the European Community's Seventh Framework Programme (FP7/2007–2013).

References

Amer, A., Franchi, L., Kanneganti, T.D., Body-Malapel, M., Ozoren, N., Brady, G., Meshinchi, S., Jagirdar, R., Gewirtz, A., Akira, S., *et al.* (2006). Regulation of Legionella phagosome maturation and infection through flagellin and host Ipaf. The Journal of biological chemistry *281*, 35217-35223.

Arlehamn, C.S., Petrilli, V., Gross, O., Tschopp, J., and Evans, T.J. (2010). The role of potassium in inflammasome activation by bacteria. The Journal of biological chemistry *285*, 10508-10518.

Attree, O., and Attree, I. (2001). A second type III secretion system in Burkholderia pseudomallei: who is the real culprit? Microbiology (Reading, England) *147*, 3197-3199.

Blocker, A., Gounon, P., Larquet, E., Niebuhr, K., Cabiaux, V., Parsot, C., and Sansonetti, P. (1999). The tripartite type III secreton of Shigella flexneri inserts IpaB and IpaC into host membranes. The Journal of cell biology *147*, 683-693.

Boya, P., and Kroemer, G. (2008). Lysosomal membrane permeabilization in cell death. Oncogene *27*, 6434-6451.

Buttner, D., and Bonas, U. (2002). Port of entry--the type III secretion translocon. Trends in microbiology *10*, 186-192.

Chen, Y., Smith, M.R., Thirumalai, K., and Zychlinsky, A. (1996). A bacterial invasin induces macrophage apoptosis by binding directly to ICE. The EMBO journal *15*, 3853-3860.

Claros, M.G., and von Heijne, G. (1994). TopPred II: an improved software for membrane protein structure predictions. Comput Appl Biosci *10*, 685-686.

Clerc, P.L., Ryter, A., Mounier, J., and Sansonetti, P.J. (1987). Plasmid-mediated early killing of eucaryotic cells by Shigella flexneri as studied by infection of J774 macrophages. Infection and immunity *55*, 521-527.

De Geyter, C., Wattiez, R., Sansonetti, P., Falmagne, P., Ruysschaert, J.M., Parsot, C., and Cabiaux, V. (2000). Characterization of the interaction of IpaB and IpaD, proteins required for entry of Shigella flexneri into epithelial cells, with a lipid membrane. European journal of biochemistry / FEBS *267*, 5769-5776.

Deng, D., Jiang, N., Hao, S.J., Sun, H., and Zhang, G.J. (2009). Loss of membrane cholesterol influences lysosomal permeability to potassium ions and protons. Biochimica et biophysica acta *1788*, 470-476.

Dickenson, N.E., Choudhari, S.P., Adam, P.R., Kramer, R.M., Joshi, S.B., Middaugh, C.R., Picking, W.L., and Picking, W.D. (2013). Oligomeric states of the Shigella translocator protein IpaB provide structural insights into formation of the type III secretion translocon. Protein Sci *22*, 614-627.

Finlay, B.B., and Falkow, S. (1988). Comparison of the invasion strategies used by Salmonella cholerae-suis, Shigella flexneri and Yersinia enterocolitica to enter cultured animal cells: endosome acidification is not required for bacterial invasion or intracellular replication. Biochimie *70*, 1089-1099.

Franchi, L., Amer, A., Body-Malapel, M., Kanneganti, T.D., Ozoren, N., Jagirdar, R., Inohara, N., Vandenabeele, P., Bertin, J., Coyle, A., *et al.* (2006). Cytosolic flagellin requires Ipaf for activation of caspase-1 and interleukin 1beta in salmonella-infected macrophages. Nature immunology *7*, 576-582.

Franchi, L., Stoolman, J., Kanneganti, T.D., Verma, A., Ramphal, R., and Nunez, G. (2007). Critical role for Ipaf in Pseudomonas aeruginosa-induced caspase-1 activation. European journal of immunology *37*, 3030-3039.

Hakansson, S., Bergman, T., Vanooteghem, J.C., Cornelis, G., and Wolf-Watz, H. (1993). YopB and YopD constitute a novel class of Yersinia Yop proteins. Infection and immunity *61*, 71-80.

Hersh, D., Monack, D.M., Smith, M.R., Ghori, N., Falkow, S., and Zychlinsky, A. (1999). The Salmonella invasin SipB induces macrophage apoptosis by binding to caspase-1. Proceedings of the National Academy of Sciences of the United States of America *96*, 2396-2401.

High, N., Mounier, J., Prevost, M.C., and Sansonetti, P.J. (1992). IpaB of Shigella flexneri causes entry into epithelial cells and escape from the phagocytic vacuole. The EMBO journal *11*, 1991-1999.

Hilbi, H., Moss, J.E., Hersh, D., Chen, Y., Arondel, J., Banerjee, S., Flavell, R.A., Yuan, J., Sansonetti, P.J., and Zychlinsky, A. (1998). Shigella-induced apoptosis is dependent on caspase-1 which binds to IpaB. The Journal of biological chemistry *273*, 32895-32900.

Hornung, V., Bauernfeind, F., Halle, A., Samstad, E.O., Kono, H., Rock, K.L., Fitzgerald, K.A., and Latz, E. (2008). Silica crystals and aluminum salts activate the NALP3 inflammasome through phagosomal destabilization. Nature immunology *9*, 847-856.

Hume, P.J., McGhie, E.J., Hayward, R.D., and Koronakis, V. (2003). The purified Shigella IpaB and Salmonella SipB translocators share biochemical properties and membrane topology. Molecular microbiology *49*, 425-439.

Ishisaka, R., Utsumi, T., Yabuki, M., Kanno, T., Furuno, T., Inoue, M., and Utsumi, K. (1998). Activation of caspase-3-like protease by digitonin-treated lysosomes. FEBS letters *435*, 233-236.

Iwai, H., Kim, M., Yoshikawa, Y., Ashida, H., Ogawa, M., Fujita, Y., Muller, D., Kirikae, T., Jackson, P.K., Kotani, S., *et al.* (2007). A bacterial effector targets Mad2L2, an APC inhibitor, to modulate host cell cycling. Cell *130*, 611-623.

Jarry, A., Robaszkiewicz, M., Brousse, N., and Potet, F. (1989). Immune cells associated with M cells in the follicle-associated epithelium of Peyer's patches in the rat. An electron- and immuno-electron-microscopic study. Cell and tissue research *255*, 293-298.

Kagedal, K., Zhao, M., Svensson, I., and Brunk, U.T. (2001). Sphingosine-induced apoptosis is dependent on lysosomal proteases. The Biochemical journal *359*, 335-343.

Lunelli, M., Lokareddy, R.K., Zychlinsky, A., and Kolbe, M. (2009). IpaB-IpgC interaction defines binding motif for type III secretion translocator. Proceedings of the National Academy of Sciences of the United States of America.

Macia, E., Ehrlich, M., Massol, R., Boucrot, E., Brunner, C., and Kirchhausen, T. (2006). Dynasore, a cell-permeable inhibitor of dynamin. Developmental cell *10*, 839-850.

Maurelli, A.T., and Sansonetti, P.J. (1988). Genetic determinants of Shigella pathogenicity. Annual review of microbiology *42*, 127-150.

Menard, R., Sansonetti, P., and Parsot, C. (1994a). The secretion of the Shigella flexneri Ipa invasins is activated by epithelial cells and controlled by IpaB and IpaD. The EMBO journal *13*, 5293-5302.

Menard, R., Sansonetti, P., Parsot, C., and Vasselon, T. (1994b). Extracellular association and cytoplasmic partitioning of the IpaB and IpaC invasins of S. flexneri. Cell *79*, 515-525.

Miao, E.A., Alpuche-Aranda, C.M., Dors, M., Clark, A.E., Bader, M.W., Miller, S.I., and Aderem, A. (2006). Cytoplasmic flagellin activates caspase-1 and secretion of interleukin 1beta via Ipaf. Nature immunology *7*, 569-575.

Miao, E.A., Mao, D.P., Yudkovsky, N., Bonneau, R., Lorang, C.G., Warren, S.E., Leaf, I.A., and Aderem, A. (2010). Innate immune detection of the type III secretion apparatus through the NLRC4 inflammasome. Proceedings of the National Academy of Sciences of the United States of America *107*, 3076-3080.

Mounier, J., Boncompain, G., Senerovic, L., Lagache, T., Chretien, F., Perez, F., Kolbe, M., Olivo-Marin, J.C., Sansonetti, P.J., and Sauvonnet, N. (2012). Shigella effector IpaB-induced cholesterol relocation disrupts the Golgi complex and recycling network to inhibit host cell secretion. Cell Host Microbe *12*, 381-389.

Mounier, J., Vasselon, T., Hellio, R., Lesourd, M., and Sansonetti, P.J. (1992). Shigella flexneri enters human colonic Caco-2 epithelial cells through the basolateral pole. Infection and immunity *60*, 237-248.

Neyt, C., and Cornelis, G.R. (1999). Insertion of a Yop translocation pore into the macrophage plasma membrane by Yersinia enterocolitica: requirement for translocators YopB and YopD, but not LcrG. Molecular microbiology *33*, 971-981.

Nonaka, T., Kuwabara, T., Mimuro, H., Kuwae, A., and Imajoh-Ohmi, S. (2003). Shigella-induced necrosis and apoptosis of U937 cells and J774 macrophages. Microbiology (Reading, England) *149*, 2513-2527.

Page, A.L., Fromont-Racine, M., Sansonetti, P., Legrain, P., and Parsot, C. (2001). Characterization of the interaction partners of secreted proteins and chaperones of Shigella flexneri. Molecular microbiology *42*, 1133-1145.

Prive, G.G. (2007). Detergents for the stabilization and crystallization of membrane proteins. Methods (San Diego, Calif *41*, 388-397.

Ryndak, M.B., Chung, H., London, E., and Bliska, J.B. (2005). Role of predicted transmembrane domains for type III translocation, pore formation, and signaling by the Yersinia pseudotuberculosis YopB protein. Infection and immunity *73*, 2433-2443.

Sansonetti, P.J., Phalipon, A., Arondel, J., Thirumalai, K., Banerjee, S., Akira, S., Takeda, K., and Zychlinsky, A. (2000). Caspase-1 activation of IL-1beta and IL-18 are essential for Shigella flexneri-induced inflammation. Immunity *12*, 581-590.

Sansonetti, P.J., Ryter, A., Clerc, P., Maurelli, A.T., and Mounier, J. (1986). Multiplication of Shigella flexneri within HeLa cells: lysis of the phagocytic vacuole and plasmid-mediated contact hemolysis. Infection and immunity *51*, 461-469.

Schoehn, G., Di Guilmi, A.M., Lemaire, D., Attree, I., Weissenhorn, W., and Dessen, A. (2003). Oligomerization of type III secretion proteins PopB and PopD precedes pore formation in Pseudomonas. The EMBO journal *22*, 4957-4967.

Senerovic, L., Tsunoda, S.P., Goosmann, C., Brinkmann, V., Zychlinsky, A., Meissner, F., and Kolbe, M. (2012). Spontaneous formation of IpaB ion channels in host cell membranes reveals how Shigella induces pyroptosis in macrophages. Cell Death Dis *3*, e384.

Soesatyo, M., Biewenga, J., Kraal, G., and Sminia, T. (1990). The localization of macrophage subsets and dendritic cells in the gastrointestinal tract of the mouse with special reference to the presence of high endothelial venules. An immuno- and enzyme-histochemical study. Cell and tissue research *259*, 587-593.

Strop, P., and Brunger, A.T. (2005). Refractive index-based determination of detergent concentration and its application to the study of membrane proteins. Protein Sci *14*, 2207-2211.

Suparak, S., Kespichayawattana, W., Haque, A., Easton, A., Damnin, S., Lertmemongkolchai, G., Bancroft, G.J., and Korbsrisate, S. (2005). Multinucleated giant cell formation and apoptosis in infected host cells is mediated by Burkholderia pseudomallei type III secretion protein BipB. Journal of bacteriology *187*, 6556-6560.

Suzuki, T., Nakanishi, K., Tsutsui, H., Iwai, H., Akira, S., Inohara, N., Chamaillard, M., Nunez, G., and Sasakawa, C. (2005). A novel caspase-1/toll-like receptor 4-independent pathway of cell death induced by cytosolic Shigella in infected macrophages. The Journal of biological chemistry *280*, 14042-14050.

Veenendaal, A.K., Hodgkinson, J.L., Schwarzer, L., Stabat, D., Zenk, S.F., and Blocker, A.J. (2007). The type III secretion system needle tip complex mediates host cell sensing and translocon insertion. Molecular microbiology *63*, 1719-1730.

Wassef, J.S., Keren, D.F., and Mailloux, J.L. (1989). Role of M cells in initial antigen uptake and in ulcer formation in the rabbit intestinal loop model of shigellosis. Infection and immunity *57*, 858-863.

Watarai, M., Tobe, T., Yoshikawa, M., and Sasakawa, C. (1995). Contact of Shigella with host cells triggers release of Ipa invasins and is an essential function of invasiveness. The EMBO journal *14*, 2461-2470.

Yao, J., and Zhang, G.J. (1997). Lysosomal destabilization via increased potassium ion permeability following photodamage. Biochimica et biophysica acta *1323*, 334-342.

Zdolsek, J., Zhang, H., Roberg, K., and Brunk, U. (1993). H2O2-mediated damage to lysosomal membranes of J-774 cells. Free radical research communications *18*, 71-85.

Zychlinsky, A., Fitting, C., Cavaillon, J.M., and Sansonetti, P.J. (1994a). Interleukin 1 is released by murine macrophages during apoptosis induced by Shigella flexneri. The Journal of clinical investigation *94*, 1328-1332.

Zychlinsky, A., Kenny, B., Menard, R., Prevost, M.C., Holland, I.B., and Sansonetti, P.J. (1994b). IpaB mediates macrophage apoptosis induced by Shigella flexneri. Molecular microbiology *11*, 619-627.

Zychlinsky, A., Perdomo, J.J., and Sansonetti, P.J. (1994c). Molecular and cellular mechanisms of tissue invasion by Shigella flexneri. Annals of the New York Academy of Sciences *730*, 197-208.

Zychlinsky, A., Prevost, M.C., and Sansonetti, P.J. (1992). Shigella flexneri induces apoptosis in infected macrophages. Nature *358*, 167-169.

7

The Cellular Microbiology of *Shigella* Invasion of Epithelial Cells

Guy Tran Van Nhieu[1,2,3] and Philippe Sansonetti[4,5]

[1]Equipe Communication Intercellulaire et Infections Microbiennes, Centre de Recherche Interdisciplinaire en Biologie (CIRB), Collège de France, 75005 Paris, France, guy.tran-van-nhieu@college-de-france.fr
[2]Institut National de la Santé et de la Recherche Médicale U1050, 75005 Paris, France
[3]Centre National de la Recherche Scientifique UMR7241, 75005 Paris, France
[4]Unité de Pathogénie Microbienne Moléculaire. Institut Pasteur. 75015 Paris, France, sansonetti@pasteur.fr
[5]Institut National de la Santé et de la Recherche Médicale U786, 75015 Paris, France

Abstract
In the most impoverished areas of the planet, bacillary dysentery (i.e. shigellosis) remains a major etiology of enteric infections in infants and children below the age of five and still accounts for significant mortality (Kotloff et al., 2013). The subgroup *Shigella* flexneri and *S. sonnei* account for the endemic form of the disease, whereas *Shigella* dysenteriae serotype 1 accounts for the epidemic form that is currently rarely observed (Kotloff et al., 1999). Following ingestion, the bacteria invade the colonic mucosa where they multiply. Under its most severe forms, shigellosis is associated with an intense inflammation of the colonic mucosa leading to its destruction, thus the dysenteric syndrome. An infectious dose as low as a few thousand bacteria has been estimated to be sufficient to confer the disease in humans, indicative of the high virulence of *Shigella*. *Shigella* is a pathogen restricted to humans and higher primates, probably reflecting the specificity of bacterial virulence attributes adapted to cellular invasion and spreading in the colonic mucosa, as well as to the manipulation of host intestinal inflammatory responses. In this chapter, we will focus on our current knowledge of processes involved in the major step of *Shigella* invasion of epithelial cells.

Shigella invasion of the colonic mucosa

Early experiments involving biopsies from infected macaque monkeys indicated that *Shigella* invasion occurs preferentially through M cells, at the level of Peyers patches (Sansonetti and Phalipon, 1999). The intestinal mucosa (IM) consists of several cell types specialized in barrier and absorptive functions. The colonic epithelium, devoid of villi, forms a flat epithelium with deep depressions or crypts along which epithelial cells replicate from precursor/stem cells at the bottom to differentiated cells while migrating towards the apex of the crypts. Cell exfoliation and shedding occurs at the epithelial cell surface in a process that preserves the integrity of the epithelium. The dynamics of this process achieves the high rate of renewal of the epithelium (i.e. 4-5 days). In addition to intestinal epithelial cells, the IM also contains goblet cells that secrete mucus, a gel acting as a first protective layer at the apical surface of the epithelium. Colonic epithelial cells produce beta-defensins that are secreted and embedded in the mucus layer, thus strengthening its antimicrobial barrier function. In addition to their barrier and absorptive functions, intestinal epithelial cells also act as sentinels emitting danger signals upon recognition of virulent microbes, such as IL-8 and other mediators. They are actually on the front line of the innate immune defense and from there orchestrate the occurrence of inflammation as the major mechanism of antibacterial defense, particularly through the recruitment of polymorphonuclear leukocytes (PMN). Inflammation also contributes to establish the proper environment to elicit and orient the adaptive immune response in cooperation with specialized cells of the immune system like dendritic cells and lymphocytes. In the colon, the lymphoid structures that account for the local adaptive immune response are formed in small follicular aggregates or solitary nodules whose overlaying epithelium comprises M cells dedicated to the capture of luminal antigens, allergens and microbes. M cells are devoid of brush border and endowed with phagocytic properties. Unlike the large aggregates found in the distal ileum (i.e. Peyer's patches), lymphoid nodules are diffusely distributed throughout the colonic mucosa, and M cells associated with their epithelium offer multiple possible entry sites for invasive pathogens like *Shigella* (Sansonetti and Phalipon, 1999). Thus, bacterial invasion of the intestinal epithelium implies the co-opting of an integrated system and multiple interactions between epithelial cells and various cell types (Ashida et al., 2011) (Cossart and Sansonetti, 2004). Studies of various enteropathogens that initially breach the intestinal barrier via M cells, indicate that invasion of M cells still requires specific bacterial invasins probably because the basal M cell phagocytic activity is not sufficient to allow efficient bacterial translocation.

As developed elsewhere, *Shigella* uses a type III secretion system to induce its internalization by non-phagocytic cells in a process that, as mentioned above, also appears to occur at the levels of M cells (Tran Van Nhieu et al., 2000). In the rabbit ileal loop model of infection, *Shigella* colonization of the intestinal mucosa is associated with an important inflammatory reaction, leading to the formation of abscesses and the destruction of the villous epithelium, thus recapitulating the major histological lesions observed during the acute phase of bacillary dysentery in the colon. The extent of tissue destruction induced during the invasion of the intestinal mucosa by *Shigella* is essentially mediated by the host inflammatory response (Sansonetti et al., 1983). Rabbit perfusion of with the IL1-R antagonist (IL-1Ra) not only prevented inflammation and tissue destruction during *Shigella* infection, but also bacterial colonization of the intestinal mucosa. These findings stress the importance of the manipulation of inflammatory response by bacteria, with the elicitation of inflammation being required for *Shigella* invasion and dissemination within the epithelium. Because *Shigella* does not invade intestinal epithelial cells very efficiently when challenged apically (Mounier et al., 1992), it is thought that inflammation promotes tissue destabilization and breaking down of the intestinal barrier function, thus favoring bacterial access and invasion at the basolateral side of epithelial cells. Invasion of M cells may ignite inflammatory responses. Indeed, shortly after internalization by M cells, *Shigella* induces the pyroptotic death of the lamina propria macrophages, associated with the release of IL-1 beta and IL-18 (Zychlinsky et al., 1992; Hilbi et al., 1998). These pro-inflammarory cytokines contribute to the recruitment of neutrophils at the site of infection, amplifying tissue damage and epithelial destabilization and in turn bacterial invasion. Shortly following internalization by epithelial cells, the bacteria escape from the phagocytic vacuole and replicate in the cell cytosol (Sansonetti et al., 1986).

During this replication phase, *Shigella* uses actin-based motility to move from cell-to-cell and to disseminate in the epithelium (Bernardini et al., 1989). As opposed to infection of macrophages, *Shigella* invasion of epithelial cells leads to a slow necrotic death. As will be developed further, there is accumulating evidence that *Shigella* actually delays epithelial cells death or exfoliation from the epithelium, and that type III effectors expressed following bacterial-cell contact down-regulate inflammatory responses. Although invasion of epithelial cells may be relatively inefficient through the apical side, it may still provide a discrete route of entry that *Shigella* uses to penetrate epitheliall cells (Phalipon and Sansonetti, 2007). The complexity of the variety of bacterial signals inducing either pro- or anti-inflammatory functions probably reflects the interactions with different sites in the intestinal mucosa, or different stages of the infectious processes.

The T3SS effectors of invasion

Shigella invasion of epithelial cells is triggered by the formation of membrane ruffles localized at the vicinity of bacterial cell contact. Membrane ruffling results from the polymerization of the actin cytoskeleton, in a branched network pushing the plasma membrane outwards. This process is in contrast to the tight apposition of membrane during the induced- "zipper phagocytosis" described for other enteroinvasive pathogens such as *Yersinia* spp or *Listeria monocytogenes* (Cossart and Sansonetti, 2004). *Salmonella* enterica also uses a type III secretion system to invade intestinal epithelial cells, and as *Shigella*, induces localized membrane rufflings. While both *Shigella* and *Salmonella*- triggered phagocytosis present similar macroscopic features, the molecular determinants responsible for invasion, however, differ.

Actin polymerization occurs following activation of "so-called" actin nucleators, including formins involved in the formation of unbranched actin cables, and the Arp2/3 complex, involved in the formation of dendritic branched actin networks (Campellone and Welch, 2010). These nucleators are themselves activated following activation of Rho GTPases. The activated GTP-bound forms of the Cdc42 and Rac GTPases bind to WASP-family proteins and stimulate the actin nucleation activity of the Arp2/3 complex (Hall, 2012). GTP-bound Cdc42 binds to and activates N-WASP, and GTP-bound Rac relieves the inhibition of the Scar/WAVE complex. Activated Rho can lead to polymerization of actin filaments through the activation of formins (Goode and Eck, 2007), but a well-characterized function of Rho is to activate myosin-dependent contractile processes, involved in the formation of stress fibers and focal adhesions during cellular adhesion. While the FHOD3 formin and myosin II have been involved in *Salmonella* invasion, it is unclear whether they also participate in *Shigella* invasion (Truong et al., 2013). Bacterial invasion mediated by the T3SS results from different effectors acting in concert to reorganize the actin cytoskeleton. As for phagocytic uptake of particles by macrophages, actin polymerization is required for the formation of membrane leaflets that surround bacteria, but the completion of the invasion process requires stalling of actin polymerization and the depolymerization of actin filaments (Scott et al., 2005). Furthermore, as opposed to *Salmonella*, *Shigella* does not express known adhesin and does not show constitutive cell binding activity. Thus, besides the T3SS, mechanical links bridging invasive bacteria with the forming phagosomes remain to be characterized.

Membrane leaflets and pseudo-adhesion at *Shigella* invasion sites

Shigella invasion of epithelial cells implicates the joint activation of Rho GTPases and tyrosine kinases (Munter et al., 2006). Inhibition of various Rho GTPases using dominant-negative forms locked in a GDP-bound conformation indicated a critical role for the Cdc42 and Rac GTPases in actin polymerization at invasion sites. However, the use of these constructs has proven misleading in the case of *Salmonella* invasion, because while the role for Rac was confirmed in experiments using small interfering RNAs, Cdc42 did not appear to be required for invasion (Patel and Galan, 2006). These findings can be explained by a lack of selectivity in the dominant-negative form of Cdc42 which, in some cell types, may also titer exchange factors for Rac. Thus, the role of Cdc42 during *Shigella* invasion would probably deserve further investigation. It remains clear, however, that when Cdc42 or Rac are inhibited, actin polymerization and membrane ruffles at *Shigella* invasion sites are severely impaired, with only occasional residual polymerized actin at the intimate bacterial-cell contact sites. In contrast, while interfering with bacterial invasion, Rho inhibition does not drastically impair *Shigella*-induced membrane ruffles. Consistently, the Rho GTPase is recruited after the Rac GTPase at *Shigella* invasion sites and predominantly at the vacuolar membranes sourrounding invading bacteria (Ehsani et al., 2012). Rho allows the organization of *Shigella* invasion sites by allowing the recruitment of the plasma membrane linker ezrin to membrane ruffles at invasion sites, which is not required for the polymerization of actin itself but for later steps of invasion (Skoudy et al., 1999; Dumenil et al., 2000). The role of ezrin during *Shigella* invasion may differ from its canonical role of tethering cell receptors, since it is massively recruited after actin polymerization at bacterial entry sites, at the outskirts of actin foci.

The Arg/ Abl tyrosine kinases have also been implicated in *Shigella* invasion (Burton et al., 2003). These kinases have been shown to phosphorylate the Crk adaptor, known to recruit GEFs for the Cdc42 and Rac GTPases (Vuori and Ruoslahti, 1995). Consistently, expression of a form of Crk mutated for its Abl/Arg phosphorylation site Crk-221F inhibits Cdc42 and Rac activation during *Shigella* invasion (Burton et al., 2003). Thus, by phosphorylating Crk, Arg/Abl contribute to actin polymerization required for membrane ruffling at *Shigella* invasion sites. How Arg/Abl are activated during *Shigella* invasion is not known. A possibility is that such activation occurs downstream of the Src kinase as described in other systems (Plattner et al., 1999; Mader et al., 2011). Along these lines, *Shigella* invasion requires the Src-dependent tyrosyl phosphorylation of the actin-binding protein cortactin (Bougneres et al., 2004). Cortactin can bundle actin filaments via F-actin binding domains, and has been shown to localize and participate in membrane ruffles following Rac activation. In addition, cortactin has also been reported to bind to and to activate the Arp2/3 complex to promote actin polymerization (Weaver et al., 2002; Helgeson and Nolen, 2013). As opposed to WASP/WAVE family proteins, however, cortactin-dependent activation of the Arp2/3 complex is weak and synergizes with the N-WASP protein to nucleate actin filaments, suggesting that cortactin integrates various stimuli that participate in the formation of an actin network at the cell cortex (Helgeson and Nolen, 2013; Han et al., 2014). Src-dependent phosphorylation of cortactin occurs at three tyrosine residues within the proline rich hinge region proximal to the carboxyterminal F-actin binding region (Weed and Parsons, 2001). Tyrosine phosphorylation of cortactin enables its association with the Crk adaptor through its SH2 domain at the plasma membrane (Bougneres et al., 2004). It thus appear that the Crk adaptor coordinates actin polymerization at *Shigella* invasion sites upstream of the Rac GTPase through and Abl/Ark kinases, as well as through the Src-dependent recruitment of cortactin. Consistently, a triple tyrosine mutant or an Arp2/3-binding deficient form of cortactin act in a dominant-negative fashion on *Shigella* induced actin polymerization, indicating that cortactin phosphorylation provides a pathway through which Src activation contributes to *Shigella*-actin foci formation (Bougneres et al., 2004). As observed following inhibition of Src activity, however, the dominant-negative forms of cortactin do not prevent *Shigella*-induced actin polymerization at the bacterial contacts with the cell plasma membranes, but prevent actin polymerization leading to membrane rufflings at the bacterial vicinity. Thus, rather than acting in a linear pathway critical for *Shigella* -induced actin nucleation, Src-mediated signaling may amplify actin polymerization at bacterial entry sites by acting at multiple levels. Such an amplification function may reflect the complex interplay of Crk-Abl/Arg and Crk-cortactin orchestrated by the Src kinase and the Rac GTPase (Weed and Parsons, 2001).

The initial events triggering actin polymerization during *Shigella* invasion are not fully understood, probably because they involve discrete amounts of translocated bacterial effectors and host signaling molecules at the cell plasma membranes. In addition, *Shigella* type III effectors of invasion co-opt tyrosine kinase signaling and Rho GTPases implicated in various pathways leading to cytoskeleton remodeling, further complexifying the picture.

Filopodial capture

Live microscopy experiments indicated that prior to internalization by the cell, *Shigella* was captured by thin cellular extensions, termed filopodia, that retract upon bacterial contact (Romero et al., 2011). Filopodial retraction leads to bacterial contact with the cell body where invasion occurs. Filopodia are sensory organelles expressed on virtually all cell types (Faix et al., 2009). In epithelial cells, filopodia are involved in the initial steps of cell-cell junction formation as well as cell matrix adhesion. Bacterial capture by filopodia is also observed at the apical surface of polarized intestinal epithelial cells, where it amplifies bacterial internalization at entry sites (Romero et al., 2011). This particular mode of cell interaction may allow bacteria to target specific sites of the epithelium, presumably in cells undergoing junctional remodeling. It may also provide an explanation for the ability of *Shigella* to invade cells efficiently in the absence of constitutive cell-binding activity.

Shigella capture by filopodia occurs mainly following contact between one pole of the bacteria and the filopodial tip. As indicated by pre-incubation of bacteria with specific antibodies, bacterial capture by filopodia is dependent on the T3SS tip complex proteins IpaB and IpaD (Romero et al., 2011). Filopodial contact with one bacterial pole is consistent with the observation that translocator components are stored at one bacterial pole prior to cell contact, where active type III secretion takes place. It is unclear if type III secretion at the filopodial tip is required for filopodial retraction. Since particles that establish non-specific interactions with receptor at the cell surface also trigger filopodial retraction, it appears that mere interaction between the tip complex proteins and receptors at the filopodial tip may be sufficient to induce retraction. This process of bacterial capture is in contrast to what has been observed for viral pathogens that glide along the side of filopodia towards the cell body in a myosin II-dependent manner at the onset of cell infection (Sherer and Mothes, 2008). Filopodial retraction induced by bacterial capture is not inhibited by the myosin II inhibitor blebbistatin, but by drugs that interfere with actin polymerization. These seemingly paradoxical findings are explained by measurements of the rates of the actin retrograde flow mediated by the treadmilling of actin filaments in the cell cortex, that pull filopodial actin filaments inwards (Bornschlogl et al.). Capping of actin filaments mediated by receptor engagment following bacterial contact at the filopodia tip results in reduced actin polymerization at the filopodial tip that does not compensate the inward pulling of the cortical retrograde flow. Intriguingly, a discrete number of type III secretion apparatus is sufficient to induce filopodial retraction with forces comparable to those induced by several thousands of beta-1 integrin receptor ligands (Romero et al., 2012). These findings suggest that a discrete number of *Shigella* T3SS tip complex molecules binds to receptors and induces their strong association with actin filaments at the filopodial tip. Among potential filopodial tip receptors, the hyaluronic acid receptor CD44, or beta-1 integrins emerge as putative candidates, since they have been reported to bind to the *Shigella* T3SS tip complex or translocator proteins (Skoudy et al., 2000) (Watarai et al., 1996).

The primary event leading to *Shigella* invasion upon cell contact is the insertion of translocon components into the host cell plasma membrane. This implies the cell-contact dependent secretion of IpaC, which, unlike IpaB and IpaD tip complex proteins, is not exposed at the bacterial surface at the basal state. The precise signal triggering active type III secretion upon cell contact is not understood, but may implicate conformational changes of the needle component MxiH transmitted from tip-to-bottom of the T3SS needle, triggered by release of IpaD and insertion of IpaB into host cell membranes (Roehrich et al., 2013). As for all described T3SSs, in the absence of translocon components the injection of type III effectors is abolished. An *ipaC* mutant, however, still induces bacterial contact-dependent lysis of erythrocytes, albeit at much reduced efficiency compared to wild-type *Shigella* (Blocker et al., 1999). As opposed to contact hemolysis induced by wild-type *Shigella*, the residual erythrocyte lysis activity induced by the *ipaC* mutant is not protected by dextrans, suggesting that in the absence of IpaC, IpaB destabilizes host cell membranes, but does not oligomerize into a bona fide pore (Blocker et al., 1999). These experiments have led to the still holding notion that upon secretion and insertion in host cell membranes, IpaC associates with IpaB and that this association is required for the formation of a pore in the host cell membrane. While pore formation associated with insertion of the T3SS translocon can be detected in red blood cells (RBC), it is not observed in epithelial cells. A likely explanation for this difference is that as opposed to RBCs, epithelial cells repair membrane damage by stimulating endocytosis, potentially permitting the removal of translocon from the plasma membranes, as well as exocytic processes enabling the supply of membranes for patch repair. Also, *Shigella* uptake by epihelial cells into a phagocytic vacuole may contribute to the removal of translocons from the plasma membrane.

Calcium signalling and the T3SS translocon

Shigella induces calcium signals during invasion of epithelial cells (Tran Van Nhieu et al., 2003). These include global Ca^{2+} responses with an atypical pattern of isolated, slow raising peaks within individual cells, reminiscent of responses described in response the the *E. coli* Hly hemolysin (Uhlen et al., 2000) (Tran Van Nhieu et al., 2003). *Shigella* also induces atypical local responses that are confined to the bacterial entry sites and that show abnormally long durations (Tran Van Nhieu et al., 2013). The duration of these *Shigella*-induced local responses is due to the accumulation of inositol (1, 4, 5) triphosphate ($InsP_3$) receptors associated with a restriction of diffusion linked to polymerized actin at invasion sites. Both global and local responses induced by *Shigella* are dependent on the T3SS and on InsP3-mediated signalling (Tran Van Nhieu et al., 2013). While bacterial invasion can occur independently of global Ca^{2+} responses, it requires local Ca^{2+} responses. The precise role of the local Ca^{2+} responses in the invasion process is still not determined, but since they regulate actin polymerization at invasion sites, Ca^{2+} may locally affect early events leading to the activation or organization of actin nucleators at the plasma membranes. *Shigella*-induced local Ca^{2+} responses are observed in the absence of extracellular Ca^{2+} but require $InsP_3$-mediated signalling, indicating that they do not require Ca^{2+} influx, but Ca^{2+} release from internal stores (Tran Van Nhieu et al., 2013). As for Ca^{2+} agonists, *Shigella*-induced actin polymerization at entry sites depends on PLCs that hydrolyze phosphoinositol (4,5) bisphosphate to generare $InsP_3$, as well as $InsP_3$ receptors. Known injected T3SS effectors appear to be dispensible for *Shigella*-induced $InsP_3$-mediated signalling, but as for actin polymerization at invasion sites, $InsP_3$- and Ca^{2+} signalling is totally impaired in mutants lacking the translocon components IpaB or IpaC. These findings suggest that insertion of the T3SS translocon into the host cell plasma membranes is sufficient to elicit the local Ca^{2+} during *Shigella* invasion, most likely through local pertubations in the plasma membrane, since Ca^{2+} influx does not appear necessary. This is a notion that has also been evoked in the case of pore-forming toxin aerolysin, that stimulates Ca^{2+} release from intracellular stores (Krause et al., 1998). Interestingly, the *Listeria* pore-forming toxin LLO was shown to induce local actin polymerization and membrane rufflings that may participate in bacterial internalization (Vadia et al., 2013). It is thus possible that primary signaling events during *Shigella* are linked to the insertion of the T3SS translocon into host cell membranes, further modulated by T3SS effectors.

In addition to its role as a translocon component in pore formation and translocation of injected T3SS effectors upon cell contact, IpaC was also shown to induce actin polymerization in a cell-permeabilized assay (Tran Van Nhieu et al., 1999). Using monoclonal antibodies that recognize various domains of IpaC, the region of IpaC required for this process was localized to the last 72 amino-acid residues. The role of this IpaC carboxyterminal domain during *Shigella* invasion is supported by the inhibitory activity if these antibodies on bacterial-induced actin foci in this permeabilized cell assay (Tran Van Nhieu et al., 1999). Following cell microinjection, IpaC induces microspikes that are reminiscent of those induced upon activation of the Cdc42 GTPase. As for *Shigella*-induced invasion foci formation, dominant-negative forms of the Cdc42 and Rac GTPases inhibit IpaC-induced actin reorganization and membrane protrusions (Tran Van Nhieu et al., 1999). More recent studies indicated that the Src tyrosine kinase, which has been implicated in *Shigella* invasion, is also recruited and activated at bacterial invasion sites in an IpaC-dependent manner (Mounier et al., 2009). The effects of small epitope insertions in various domains of IpaC indicated that while N-terminal insertions did not affect bacterial invasion, an insertion at residue 351 abolished prevented *Shigella*-induced actin polymerization at cell contact sites and bacterial internalization. The IpaC351 insertion did not prevent injection of T3SS effectors, although it occured at reduced levels compared to wid-type IpaC. Reduced levels of T3SS injection may result from a indirect effect of the C-terminal insertion on IpaC-mediated signalling and actin polymerization (Mounier et al., 2009). Indeed, similar reduced levels of injected T3SS effectors are also observed when cells are treated with inhibitors of actin polymerization (Mounier et al., 2009). The IpaC351 insertion, or the presence of actin inhibitors, result in an injection of the IpaA effector restricted to a cell area contacting a bacterial pole, as opposed to an IpaA distribution in membrane leaflets and around the phagocytic cup surrounding the invading bacteria in untreated cells. These findings are consistent for a role of actin polymerization in bacterial T3SS injection (Figure 7.1). Thus, during early translocation events following bacterial contact at discrete sites, IpaC, through its C-terminal domain, may initiate signaling leading to actin polymerization. Actin polymerization, in turn, may allow further membrane apposition to the bacterial surface following initial contacts and other injection events. In this model, efficient T3SS injection would require IpaC-mediated signaling and *Shigella*-invasion would not only need to "inject to signal" but also "signal to inject" to promote its internalization by epithelial cells.

How IpaC induces actin polymerization through its carboxy-terminal domain is not entirely understood. The *Salmonella* orthologue SipC has been shown to directly induce actin polymerization by activating the Arp2/3 complex (Hayward and Koronakis, 1999) (Chang et al., 2005). However, mutations in the SipC domains that prevent actin nucleation in vitro only have partial effects on bacterial-induced actin polymerization and invasion (Chang et al., 2005). Dominant-negative forms of Cdc42 and Rac strongly inhibit IpaC-dependent actin polymerization and *Shigella* invasion suggesting that the implication of endogenous GEFs upstream of these GTPases (Mounier et al., 1999) (Tran Van Nhieu et al., 1999).

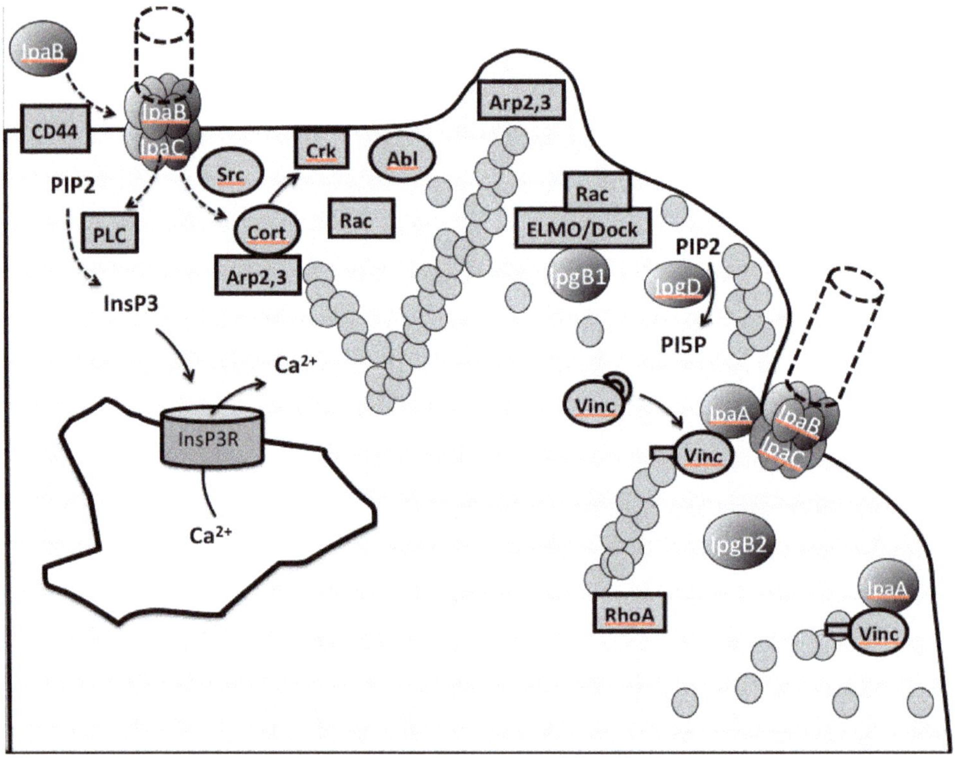

Figure 7.1. Type III effectors of Shigella invasion of epithelial cells. Upon cell contact, the *Shigella* T3SS inserts the translocon components IpaB and IpaC into the host cell plasma membrane. The cylinder in dotted line represents the T3SS needle. The initial steps of host membrane sensing by the Shigella T3SS may involve interaction between IpaB and the hyaluronan receptor CD44 and targeting of cholesterol microdomains. Translocon insertion into the plasma membranes triggers the activation of PLCs, leading to InsP3-dependent Ca^{2+} release from intracellular pools and local Ca^{2+} signals, required for actin polymerization and membrane rufflings at bacterial invasion sites. IpaC, through its carboxyterminal domain, induces the recruitment and activation of the Src tyrosine kinase. Src-dependent phosphorylation of cortactin allows its recruitment at the plasma membrane through the Crk adaptor. Crk may also be phosphorylated by the Abl/Arg tyrosine kinases, permitting the recruitment of a GEF for Rac and activation of this GTPase. The Src and Abl/Arg tyrosine kinases thus participate in the amplification of Arp2/3 mediated actin nucleation and membrane rufflings. Injected type III effectors also reorganize the actin cytoskeleton. IpgB1, acting as Rac GEF or by recruiting the ELMO/Dock180 complex, further enhances Rac activation and Arp2/3-dependent actin polymerization. IpgD, by hydrolyzing PI(4,5)P2 at the plasma membranes, release cortical tension thus favoring de novo actin polymerization at invasion sites. IpgB2, acting as a RhoGEF, coordinates with IpgB1 to allow bacterial invasion, presumaby through myosin activation. IpaA, acting a a talin mimic, activates vinculin to tether the invading bacteria to actin filaments. IpaA in complex with vinculin caps the barbed ends of actin filaments to limit actin polymerization and favor depolymerization, required for the completion of the invasion process.

The IpgB1 and IpgB2 type III effectors and the Rac and Rho GTPases

An *ipgB1* mutant still induces actin polymerization foci but of smaller amplitude, and shows a fifty percent invasion defect compared to wild-type (Ohya et al., 2005). IpgB1 and IpgB2 belong to the class of type III effectors containing the "WXXXE" motif (Alto et al., 2006). Such effectors were previously described as mimics of the Rac and Rho GTPases, respectively, based on their interaction with targets of GTP-bound GTPases. The situation is more complex for IpgB1, which has been proposed to act as a mimic of RhoG, rather than Rac (Handa et al., 2007). In this study, it was shown that as activated RhoG, IpgB1 binds to the ELMO protein, that recruits Dock180 acting as a GEF for Rac. Along these lines, ELMO/Dock 180 is recruited at *Shigella* invasion sites in an IpgB1-dependent manner and inhibition of ELMO prevented IpgB1-induced actin polymerization (Handa et al., 2007). Also, and as opposed to what was expected if IpgB1 acted as a mimic of activated Rac, it was found that a dominant-negative form of the Rac inibited IpgB1-mediated actin polymerization. However, the structural characterization indicates that "WXXXE effectors" correspond to GEFs for Rho GTPases, rather than mimics, with the WXXXE motif representing a motif involved in a structural loop conserved in GEFs (Orchard and Alto, 2012). Consistently, using in vitro assays, purified IpgB1 and IpgB2 proteins were shown to harbor a GEF activity towards Rac and RhoA, respectively (Huang et al., 2009) (Klink et al., 2010). It is possible that IpgB1 effector employs different mechanisms to activate actin polymerization at *Shigella* invasion site, which would explain the seemingly paradoxical published findings. Also, the mode of action of IpgB1 will also require further clarification, because its link with the Abl/Arg and Src tyrosine kinases, also required for actin polymerization leading to membrane ruffling at *Shigella* invasion sites, has not been established. By inducing membrane ruffling during *Shigella* invasion, IpgB1 appears to increase the kinetics of bacterial uptake (Ehsani et al., 2012).

As opposed to an *ipgB1* mutant, an *ipgB2* mutant is not deficient in inducing actin polymerization, membrane rufflings at bacterial entry sites and invasion of epithelial cells. The absence of effects for IpgB2 in inducing actin foci is not unexpected from its function in mimicking / activating the Rho GTPase, which typically leads to the stimulation of acto-myosin contraction (Hall, 2012). In many cell processes, Rac and Rho activation are antagonistic, with the down-regulation of one favoring the activation of the other (Iden and Collard, 2008). IpgB2 may induce an acto-myosin-dependent entry, that may be masked by IpgB1-dependent ruffles. This has been proposed for *Salmonella*, where a myosin pathway of invasion dependent on the SopB effector has been described in addition to the SopE1/2-pathway dependent on the activation of Rac and membrane ruffles (Hanisch et al., 2011). These alternate modes of entry may be used during invasion of different cell types or perhaps, particular loci on the epithelial cells where the context would be more favorable to membrane rufflings or acto-myosin contraction. Another possibility is that both IpgB1 and IpgB2 coordinate their activity to promote *Shigella* invasion. This hypothesis is supported by the obervation that a double *ipgB1 ipgB2* mutant, but not single *ipgB1* or *ipgB2* mutants, shows a defect in invasion of polarized cultured intestinal cells (Hachani et al., 2008). Such concerted activity suggest a precise spatio-temporal control of the Rac and Rho GTPases by IpgB1 and IpgB2, that is revealed in the context of invasion at the apical surface of polarized intestinal epithelial cells.

IpaA as a focal adhesion protein activator

The type III invasion effector IpaA is not required to induce actin polymerization during the primary step of bacterial interaction with the epithelial cell. Instead, IpaA induces bacterial anchoring into invasion foci, since an *ipaA* mutant induces actin projections that appear to repell bacteria upon cell contact (Tran Van Nhieu et al., 1997). Furthermore, in the absence of IpaA, *Shigella*-induced actin polymerization at invasion sites is not down-regulated, suggesting that IpaA promotes actin depolymerization required for completion of the entry process (Tran Van Nhieu et al., 1997).

IpaA is a 633 amino-acid protein shown to bind to vinculin, a focal adhesion protein involved, through its association with talin, in the anchoring of the actin filaments to beta-1 integrin receptors at the plasma membrane (Peng et al., 2011; Calderwood et al., 2013). Vinculin-dependent cytoskeletal anchorage, however, is tightly regulated, since the bulk of the vinculin pool is maintained inactive by intramolecular interaction between the F-actin binding carboxyterminal tail and the vinculin globular head, whose N-terminal domain is involved in talin association (Peng et al., 2011). It is generally admitted that upon integrin binding to extracellular matrix proteins, talin association with the cytoplasmic tail of the beta-1 integrin subunit triggers binding and activation of vinculin, required for the strengthening of adhesion (Calderwood et al., 2013). In vitro, however, vinculin unfolding is only observed in the presence of high concentrations of talin peptides containing a vinculin bindin site (VBS), suggesting that other processes occur in the cell to favor vinculin activation. Such processes may involve phosphatidylinositol (4,5) bisphosphate (PI(4,5)P_2) binding to vinculin, or its phosphorylation by the Src kinase (Peng et al., 2011).

IpaA is the only protein reported to date to activate vinculin and promote its association with actin filaments in stoechiometric amounts (Bourdet-Sicard et al., 1999). This property makes IpaA a particularly interesting tool to probe vinculin functions. Vinculin binding occurs through the last 150 carboxyterminal residues of IpaA. This domain contains three vinculin binding sites (VBSs) that appear to play different role in vinculin association (Park et al., 2011). The IpaA carboxyterminal VBS act as a mimic of talin by promoting the unfurling of the vinculin D1 domain and helical-bundle conversion observed during vinculin activation (Izard et al., 2006). IpaA VBS1 binding to vinculin occurs with high affinity, with an estimated dissociation constant in the picomolar range, that is two orders of magnitude lower than those determined for physiological VBSs (Izard et al., 2006). The second proximal IpaA VBS2 binds in an additional manner to another part of the vinculin D1 domain, thereby stabilizing the IpaA interaction (Tran Van Nhieu and Izard, 2007). Such mode of IpaA VBS2-vinculin interaction may be representative of a class of endogenous VBSs, that may allow the recruitment of vinculin without promoting its activation. An IpaA polypeptide containing both VBS1 and VBS2 activates vinculin and shows a vinculin-dependent capping activity of actin filaments at their barbed ends, revealing that in addition to its role in anchoring actin cytosleleton, vinculin also regulates the dynamics of actin polymerization / depolymerization (Ramarao et al., 2007). The IpaA VBS1/2 -vinculin capping activity is partial, however, suggesting that it is regulated by other interactions. By inhibiting actin polymerization, full capping activity mediated by an IpaA-vinculin complex may account for the depolymerization of actin filaments during *Shigella* invasion. The third IpaA VBS3, located upstream of IpaVBS1 and VBS2, discovered more recently, acts as a more "classical" endogenous VBS (Park et al., 2011). In functional assays, IpaAVBS3 appears redundant to IpaAVBS1 and VBS2, since their simultaneous removal leads to full IpaA inactivation in terms of *Shigella* invasion of epithelial cells. The reason for this functional redundancy is unclear but probably relates to a specific mode of vinculin activation by IpaA. The role of the first 483 IpaA residues is not well characterized, but there is evidence that it is implicated in disassembly of actin filaments. Expression of this domain in cells leads to their retraction with the disassembly of focal adhesions (Demali et al., 2006). Such disassembly has been associated with the IpaA aminoterminal domain joint abilities to activate the Rho GTPase by a yet undefined mechanism, and to inhibit the association of talin to the cytoplasmic tail of beta 1-integrins (Demali et al., 2006).

By associating peripherally with the IpaB-C translocon components recruiting vinculin, IpaA appears to promote the formation of a transient bacterial adhesion structure required for efficient *Shigella* invasion. The induction of such structure, similar to those observed following engagement of integrin receptors, could provide the bacteria with the means to anchor to the actin cytoskeleton and to be subjected to cytoskeletal retraction forces similar to those exerted during cell attachment to the extracellular matrix, following injection into host cells.

IpgD: a multifunctional phosphatidyl-inositol (4, 5)phosphatase

The function of IpgD as a lipid phosphatase was first hinted by the presence in its sequence of 2 conserved motives, found in various eurkaryotic phosphatases, as well as in the *Salmonella* orthologie SopB type III effector (Norris et al., 1998; Niebuhr et al., 2000). Previous works on SopB had reported that this effector hydrolized the inositol (1, 3, 4, 5, 6) pentaphosphate ($InsP_5$) into inositol (1, 4, 5, 6)tetraphosphate (InsP4). Cell transfection with SopB leads to the formation of membrane rufflings, which are presumably induced by InsP4 and its propensity to activate a GEF for the Rac GTPase, indicating that SopB may participate in bacterial invasion (Zhou et al., 2001). While a *sopB* mutant only shows a modest defect in bacterial invasion, the combination of *sopB sopE1/2* mutations results in a non-invasive *Salmonella* (Zhou et al., 2001). More recent works indicate that SopB also hydrolyzes phosphatidylinositol polyphosphates to produce phosphatodylinositol triphosphate (PIP_3) and that this activity enables it together with the *Salmonella* GEF SopE, to trigger the recruitment and activation of the WAVE complex downstream of Rac (Humphreys et al., 2012). The *Shigella* orthologs of SopB and SopE are IpgD and IpgB1, respectively. A concerted action of IpgD and IpgB1 during *Shigella* invasion, however, has not yet been shown. An *ipgD* mutant is partially deficient at inducing actin polymerization at bacterial invasion sites, although this defect is of lesser amplitude than for the *ipgB1* mutation, and may be cell type dependent (Niebuhr et al., 2000). This partial *ipgD* phenotype suggests that this type III effector plays a role in bacterial invasion in a specific context that may not be faithfully represented by standard cultured cell system.

IpgD hydrolyzes $PIP(4,5)P_2$ in the 4-position to generate PI(5)P, and has been implicated in various cellular processes that depend on its PI(4,5)P2 hydrolytic activity, or on the synthesis of PI(5)P (Niebuhr et al., 2002). $PI(4,5)P_2$ is present on endosomal and plasma membranes and also binds to proteins containing a so-called "plextrin-homology" (PH)-domain and contributes to the regulation of their activities. $PI(4,5)P_2$ also binds to the focal adhesion proteins such as vinculin, and regulates their association with these adhesion structures (Palmer et al., 2009). Through its $PI(4,5)P_2$ hydrolytic activity, and as indicated by force measurements using beads coated with integrin-ligands, IpgD appears to disconnect cortical actin from the plasma membranes (Niebuhr et al., 2002). Such deconnexion is presumed to favor the depolymerization of actin filaments, thus dynamizing the de novo polymerization of actin filaments at the bacterial-cell contact sites, required for actin foci formation. PI(5)P, generated by the hydrolysis of $PI(4,5)P_2$ by IpgD, also plays a role in signaling during *Shigella* infection. In addition to its role in invasion, through PI(5)P production, IpgD regulates various cell signals. IpgD activates the Akt / PI3 Kinase cell survival pathway (Pendaries et al., 2006). PI(5)P produced by IpgD also affects cell trafficking by impairing lysosomal degradation, which, in combination with the activation of the Akt/PI3 kinase pathway, could benefit the pathogen by retarding cell death and favoring bacterial intracellular replication to allow cell-to-cell spread (Ramel et al., 2011). Finally, extracellular ATP is an important mediator of paracrine cell-cell signaling, that can act as a danger signal to locally amplify inflammatory responses. While *Shigella* activates cell-to-cell signalling through connexin hemichanels and extracellular ATP during the early steps of cell invasion, later stages of bacterial intracellular replication are associated with the closing of these hemichanels (Tran Van Nhieu et al., 2003; Puhar et al., 2013). Hemichannels closure is regulated by PI(5)P through IpgD by a still undefined mechanism. Thus, IpgD serves as a multifunctional effector, which contributes to invasion by remodeling the actin cytoskeleton, impairs cell trafficking, promotes cell survival to favor bacterial intracellular replication and cell-to-cell spreading, and inhibits signaling through connexin hemichannels to prevent cell-to-cell communication and dampen inflammation.

On the versatility of the *Shigella* type III effectors of invasion and beyond.
IpgD is not the only example of a multifunctional effector. Multi-tasking emerges as a general rule for *Shigella* type III effectors of invasion. IpgB2 mediates the activation of RhoA in a process implicating GEF-H1, which in addition to its role in invasion, leads to the stimulation of NF-kB dependent signaling and increases pro-inflammatory signals (Fukazawa et al., 2008). The translocon component IpaB, in addition to regulating the activity of the type III secretory apparatus, has been reported to bind to CD44, the hyaluronic acid receptor, thereby allowing targeting of cholesterol microdomains during *Shigella* invasion (Skoudy et al., 2000; Lafont et al., 2002). Furthermore, effectors within these groups may act in concert to divert cell processes, as shown for the *Salmonella* SopB SopE proteins, or SipA SipC proteins during bacterial invasion (McGhie et al., 2001; Humphreys et al., 2012). In addition, the activity of a given type III effector can be regulated by intracellular responses, as illustrated by the caspase-3 dependent cleavage of the *Salmonella* SptP type III effector (Srikanth et al., 2010). Finally, type III effectors may have different functions in various cell types. For example, IpaB has also been reported to directly bind to caspase 1, which together with the activation of the NlrC4, inflammasome, regulates pyroptotic cell death in the macrophage (Hilbi et al., 1998; Yang et al., 2013). Also, during interaction with T-lymphocytes, the IpgD-mediated hydrolytic activity on PI(4, 5)P2 on ERM (ezrin-radixin-moesin) proteins prevents their subsequent activation by phosphorylation and their anchoring at the plasma membrane (Konradt et al., 2011). The inhibition of the ERM cortical link with the membranes is thought to prevent their chemokine-induced migration, an effect that could explain the lack of priming of short-lasting protective immunity (Konradt et al., 2011). The diversity of roles of each effector renders complex the precise identification of their contribution in interdependent processes.

Penetrating into intestinal epithelial cells is only the initial step of the *Shigella* infectious process. Proficient invasion and dissemination within the epithelium also requires the co-opting of intracellular processes, such as autophagy. Autophagy is a degradation process involved in the maintenance of cell homeostasis, that can also be activated to control bacterial infection. The IcsB type III effector was shown to bind to the autophagy component Atg5, to prevent the recognition of bacteria by the autophagic machinery and favor bacterial intracellular replication (Ogawa et al., 2005). The induction of autophagic responses, however, were shown to control the activation of the inflammasome by remnants of the *Shigella* vacuolar membranes, following release of the bacteria in the cell cytosol (Dupont et al., 2009). Thus, through its effects on the autophagic responses, and beyond its role in favoring bacterial intracellular replication, IcsB may also regulate inflammatory responses and the cell fate. As opposed to pathogens that persist or replicate inside a vacuole, *Shigella* rapidly escapes from the phagocytic vacuole following internalization and, until recently, little attention was paid to its interference with trafficking processes. *Shigella* type III effectors that are up-regulated upon cell contact, however, were shown to interfere with trafficking pathways. These include the VirA protein that, by acting as a GAP for Rab1, inhibits recycling (Dong et al., 2012) and IpaJ, which, through its demyristoylation activity of ARF1p and ARF2p interferes with Cargo transport to the Golgi and triggers its desorganization during infection, perhaps in combination with IpaB (Mounier et al., 2012; Burnaevskiy et al., 2013). The combined action of these effectors interfering with trafficking probably accounts for the inhibition of secretory pathway that modulates the release of pro-inflammatory mediators during *Shigella* infection.

Important advances have been made in recent years in understanding the function of isolated type III effectors. The cellular microbiology of *Shigella* invasion gives us the view of host molecular processes being diverted, and the integrated view of the combined action of *Shigella* effectors. In future works, it will be important to pursue our efforts to further characterize the molecular function of the *Shigella* type III effectors as well as their combined action as a "functional package" during *Shigella* invasion.

References

Alto, N. M., F. Shao, et al. (2006). Identification of a bacterial type III effector family with G protein mimicry functions. Cell 124(1): 133-45.
Ashida, H., M. Ogawa, et al. (2011). Shigella deploys multiple countermeasures against host innate immune responses. Curr Opin Microbiol 14(1): 16-23.
Bernardini, M. L., J. Mounier, et al. (1989). Identification of icsA, a plasmid locus of Shigella flexneri that governs bacterial intra- and intercellular spread through interaction with F-actin. Proc Natl Acad Sci U S A 86(10): 3867-71.
Blocker, A., P. Gounon, et al. (1999). The tripartite type III secreton of Shigella flexneri inserts IpaB and IpaC into host membranes. J Cell Biol 147(3): 683-93.
Bornschlogl, T., S. Romero, et al. (2013). Filopodial retraction force is generated by cortical actin dynamics and controlled by reversible tethering at the tip. Proc Natl Acad Sci U S A 110(47): 18928-33.
Bougneres, L., S. E. Girardin, et al. (2004). Cortactin and Crk cooperate to trigger actin polymerization during Shigella invasion of epithelial cells. J Cell Biol 166(2): 225-35.
Bourdet-Sicard, R., M. Rudiger, et al. (1999). Binding of the Shigella protein IpaA to vinculin induces F-actin depolymerization. Embo J 18(21): 5853-62.
Burnaevskiy, N., T. G. Fox, et al. (2013). Proteolytic elimination of N-myristoyl modifications by the Shigella virulence factor IpaJ. Nature 496(7443): 106-9.
Burton, E. A., R. Plattner, et al. (2003). Abl tyrosine kinases are required for infection by Shigella flexneri. Embo J 22(20): 5471-9.
Calderwood, D. A., I. D. Campbell, et al. (2013). Talins and kindlins: partners in integrin-mediated adhesion. Nat Rev Mol Cell Biol 14(8): 503-17.
Campellone, K. G. and M. D. Welch (2010). A nucleator arms race: cellular control of actin assembly. Nat Rev Mol Cell Biol 11(4): 237-51.
Chang, J., J. Chen, et al. (2005). Delineation and characterization of the actin nucleation and effector translocation activities of Salmonella SipC. Mol Microbiol 55(5): 1379-89.
Cossart, P. and P. J. Sansonetti (2004). Bacterial invasion: the paradigms of enteroinvasive pathogens. Science 304(5668): 242-8.
Demali, K. A., A. L. Jue, et al. (2006). IpaA targets beta1 integrins and rho to promote actin cytoskeleton rearrangements necessary for Shigella entry. J Biol Chem 281(51): 39534-41.

Dong, N., Y. Zhu, et al. (2012). Structurally Distinct Bacterial TBC-like GAPs Link Arf GTPase to Rab1 Inactivation to Counteract Host Defenses. Cell 150(5): 1029-41.

Dumenil, G., P. Sansonetti, et al. (2000). Src tyrosine kinase activity down-regulates Rho-dependent responses during Shigella entry into epithelial cells and stress fibre formation. J Cell Sci 113 (Pt 1): 71-80.

Dupont, N., S. Lacas-Gervais, et al. (2009). Shigella phagocytic vacuolar membrane remnants participate in the cellular response to pathogen invasion and are regulated by autophagy. Cell Host Microbe 6(2): 137-49.

Ehsani, S., J. C. Santos, et al. (2012). Hierarchies of host factor dynamics at the entry site of Shigella flexneri during host cell invasion. Infect Immun 80(7): 2548-57.

Faix, J., D. Breitsprecher, et al. (2009). Filopodia: Complex models for simple rods. Int J Biochem Cell Biol 41(8-9): 1656-64.

Fukazawa, A., C. Alonso, et al. (2008). GEF-H1 mediated control of NOD1 dependent NF-kappaB activation by Shigella effectors. PLoS Pathog 4(11): e1000228.

Goode, B. L. and M. J. Eck (2007). Mechanism and function of formins in the control of actin assembly. Annu Rev Biochem 76: 593-627.

Hachani, A., L. Biskri, et al. (2008). IpgB1 and IpgB2, two homologous effectors secreted via the Mxi-Spa type III secretion apparatus, cooperate to mediate polarized cell invasion and inflammatory potential of Shigella flexenri. Microbes Infect 10(3): 260-8.

Hall, A. (2012). Rho family GTPases. Biochem Soc Trans 40(6): 1378-82.

Han, S. P., Y. Gambin, et al. (2014). Cortactin scaffolds Arp2/3 and WAVE2 at the epithelial zonula adherens. J Biol Chem.

Handa, Y., M. Suzuki, et al. (2007). Shigella IpgB1 promotes bacterial entry through the ELMO-Dock180 machinery. Nat Cell Biol 9(1): 121-8.

Hanisch, J., R. Kolm, et al. (2011). Activation of a RhoA/myosin II-dependent but Arp2/3 complex-independent pathway facilitates Salmonella invasion. Cell Host Microbe 9(4): 273-85.

Hayward, R. D. and V. Koronakis (1999). Direct nucleation and bundling of actin by the SipC protein of invasive Salmonella. Embo J 18(18): 4926-34.

Helgeson, L. A. and B. J. Nolen (2013). Mechanism of synergistic activation of Arp2/3 complex by cortactin and N-WASP. Elife 2: e00884.

Hilbi, H., J. E. Moss, et al. (1998). Shigella-induced apoptosis is dependent on caspase-1 which binds to IpaB. J Biol Chem 273(49): 32895-900.

Huang, Z., S. E. Sutton, et al. (2009). Structural insights into host GTPase isoform selection by a family of bacterial GEF mimics. Nat Struct Mol Biol 16(8): 853-60.

Humphreys, D., A. Davidson, et al. (2012). Salmonella virulence effector SopE and Host GEF ARNO cooperate to recruit and activate WAVE to trigger bacterial invasion. Cell Host Microbe 11(2): 129-39.

Iden, S. and J. G. Collard (2008). Crosstalk between small GTPases and polarity proteins in cell polarization. Nat Rev Mol Cell Biol 9(11): 846-59.

Izard, T., G. Tran Van Nhieu, et al. (2006). Shigella applies molecular mimicry to subvert vinculin and invade host cells. J Cell Biol 175(3): 465-75.

Klink, B. U., S. Barden, et al. (2010). Structure of Shigella IpgB2 in complex with human RhoA: implications for the mechanism of bacterial guanine nucleotide exchange factor mimicry. J Biol Chem 285(22): 17197-208.

Konradt, C., E. Frigimelica, et al. (2011). The Shigella flexneri type three secretion system effector IpgD inhibits T cell migration by manipulating host phosphoinositide metabolism. Cell Host Microbe 9(4): 263-72.

Kotloff, K. L., J. P. Nataro, et al. (2013). Burden and aetiology of diarrhoeal disease in infants and young children in developing countries (the Global Enteric Multicenter Study, GEMS): a prospective, case-control study. Lancet 382(9888): 209-22.

Kotloff, K. L., J. P. Winickoff, et al. (1999). Global burden of Shigella infections: implications for vaccine development and implementation of control strategies. Bull World Health Organ 77(8): 651-66.

Krause, K. H., M. Fivaz, et al. (1998). Aerolysin induces G-protein activation and Ca2+ release from intracellular stores in human granulocytes. J Biol Chem 273(29): 18122-9.

Lafont, F., G. Tran Van Nhieu, et al. (2002). Initial steps of Shigella infection depend on the cholesterol/sphingolipid raft-mediated CD44-IpaB interaction. Embo J 21(17): 4449-57.

Mader, C. C., M. Oser, et al. (2011). An EGFR-Src-Arg-cortactin pathway mediates functional maturation of invadopodia and breast cancer cell invasion. Cancer Res 71(5): 1730-41.

McGhie, E. J., R. D. Hayward, et al. (2001). Cooperation between actin-binding proteins of invasive Salmonella: SipA potentiates SipC nucleation and bundling of actin. Embo J 20(9): 2131-9.

Mounier, J., G. Boncompain, et al. (2012). Shigella Effector IpaB-Induced Cholesterol Relocation Disrupts the Golgi Complex and Recycling Network to Inhibit Host Cell Secretion. Cell Host Microbe 12(3): 381-9.

Mounier, J., V. Laurent, et al. (1999). Rho family GTPases control entry of Shigella flexneri into epithelial cells but not intracellular motility. J Cell Sci 112 (Pt 13): 2069-80.

Mounier, J., M. R. Popoff, et al. (2009). The IpaC carboxyterminal effector domain mediates Src-dependent actin polymerization during Shigella invasion of epithelial cells. PLoS Pathog 5(1): e1000271.

Mounier, J., T. Vasselon, et al. (1992). Shigella flexneri enters human colonic Caco-2 epithelial cells through the basolateral pole. Infect Immun 60(1): 237-48.

Munter, S., M. Way, et al. (2006). Signaling during pathogen infection. Sci STKE 2006(335): re5.

Niebuhr, K., S. Giuriato, et al. (2002). Conversion of PtdIns(4,5)P(2) into PtdIns(5)P by the S.flexneri effector IpgD reorganizes host cell morphology. Embo J 21(19): 5069-78.

Niebuhr, K., N. Jouihri, et al. (2000). IpgD, a protein secreted by the type III secretion machinery of Shigella flexneri, is chaperoned by IpgE and implicated in entry focus formation. Mol Microbiol 38(1): 8-19.

Norris, F. A., M. P. Wilson, et al. (1998). SopB, a protein required for virulence of Salmonella dublin, is an inositol phosphate phosphatase. Proc Natl Acad Sci U S A 95(24): 14057-9.

Ogawa, M., T. Yoshimori, et al. (2005). Escape of intracellular Shigella from autophagy. Science 307(5710): 727-31.

Ohya, K., Y. Handa, et al. (2005). IpgB1 is a novel Shigella effector protein involved in bacterial invasion of host cells. Its activity to promote membrane ruffling via Rac1 and Cdc42 activation. J Biol Chem 280(25): 24022-34.

Orchard, R. C. and N. M. Alto (2012). Mimicking GEFs: a common theme for bacterial pathogens. Cell Microbiol 14(1): 10-8.

Palmer, S. M., M. P. Playford, et al. (2009). Lipid binding to the tail domain of vinculin: specificity and the role of the N and C termini. J Biol Chem 284(11): 7223-31.

Park, H., C. Valencia-Gallardo, et al. (2011). Novel vinculin binding site of the IpaA invasin of Shigella. J Biol Chem 286(26): 23214-21.

Patel, J. C. and J. E. Galan (2006). Differential activation and function of Rho GTPases during Salmonella-host cell interactions. J Cell Biol 175(3): 453-63.

Pendaries, C., H. Tronchere, et al. (2006). PtdIns5P activates the host cell PI3-kinase/Akt pathway during Shigella flexneri infection. Embo J 25(5): 1024-34.

Peng, X., E. S. Nelson, et al. (2011). New insights into vinculin function and regulation. Int Rev Cell Mol Biol 287: 191-231.

Phalipon, A. and P. J. Sansonetti (2007). Shigella's ways of manipulating the host intestinal innate and adaptive immune system: a tool box for survival? Immunol Cell Biol 85(2): 119-29.

Plattner, R., L. Kadlec, et al. (1999). c-Abl is activated by growth factors and Src family kinases and has a role in the cellular response to PDGF. Genes Dev 13(18): 2400-11.

Puhar, A., H. Tronchere, et al. (2013). A Shigella effector dampens inflammation by regulating epithelial release of danger signal ATP through production of the lipid mediator PtdIns5P. Immunity 39(6): 1121-31.

Ramarao, N., C. Le Clainche, et al. (2007). Capping of actin filaments by vinculin activated by the Shigella IpaA carboxyl-terminal domain. FEBS Lett 581(5): 853-7.

Ramel, D., F. Lagarrigue, et al. (2011). Shigella flexneri infection generates the lipid PI5P to alter endocytosis and prevent termination of EGFR signaling. Sci Signal 4(191): ra61.

Roehrich, A. D., E. Guillossou, et al. (2013). Shigella IpaD has a dual role: signal transduction from the type III secretion system needle tip and intracellular secretion regulation. Mol Microbiol 87(3): 690-706.

Romero, S., G. Grompone, et al. (2011). ATP-mediated Erk1/2 activation stimulates bacterial capture by filopodia, which precedes Shigella invasion of epithelial cells. Cell Host Microbe 9(6): 508-19.

Romero, S., A. Quatela, et al. (2012). Filopodium retraction is controlled by adhesion to its tip. J Cell Sci 125(Pt 21): 4999-5004.

Sansonetti, P. J., T. L. Hale, et al. (1983). Alterations in the pathogenicity of Escherichia coli K-12 after transfer of plasmid and chromosomal genes from Shigella flexneri. Infect Immun 39(3): 1392-402.

Sansonetti, P. J. and A. Phalipon (1999). M cells as ports of entry for enteroinvasive pathogens: mechanisms of interaction, consequences for the disease process. Semin Immunol 11(3): 193-203.

Sansonetti, P. J., A. Ryter, et al. (1986). Multiplication of Shigella flexneri within HeLa cells: lysis of the phagocytic vacuole and plasmid-mediated contact hemolysis. Infect Immun 51(2): 461-9.

Scott, C. C., W. Dobson, et al. (2005). Phosphatidylinositol-4,5-bisphosphate hydrolysis directs actin remodeling during phagocytosis. J Cell Biol 169(1): 139-49.

Sherer, N. M. and W. Mothes (2008). Cytonemes and tunneling nanotubules in cell-cell communication and viral pathogenesis. Trends Cell Biol 18(9): 414-20.

Skoudy, A., J. Mounier, et al. (2000). CD44 binds to the Shigella IpaB protein and participates in bacterial invasion of epithelial cells. Cell Microbiol 2(1): 19-33.

Skoudy, A., G. T. Nhieu, et al. (1999). A functional role for ezrin during Shigella flexneri entry into epithelial cells. J Cell Sci 112 (Pt 13): 2059-68.

Srikanth, C. V., D. M. Wall, et al. (2010). Salmonella pathogenesis and processing of secreted effectors by caspase-3. Science 330(6002): 390-3.

Tran Van Nhieu, G., A. Ben-Ze'ev, et al. (1997). Modulation of bacterial entry into epithelial cells by association between vinculin and the Shigella IpaA invasin. Embo J 16(10): 2717-29.

Tran Van Nhieu, G., R. Bourdet-Sicard, et al. (2000). Bacterial signals and cell responses during Shigella entry into epithelial cells. Cell Microbiol 2(3): 187-93.

Tran Van Nhieu, G., E. Caron, et al. (1999). IpaC induces actin polymerization and filopodia formation during Shigella entry into epithelial cells. Embo J 18(12): 3249-62.

Tran Van Nhieu, G., C. Clair, et al. (2003). Connexin-dependent inter-cellular communication increases invasion and dissemination of Shigella in epithelial cells. Nat Cell Biol 5(8): 720-6.

Tran Van Nhieu, G. and T. Izard (2007). Vinculin binding in its closed conformation by a helix addition mechanism. Embo J 26(21): 4588-96.

Tran Van Nhieu, G., B. Kai Liu, et al. (2013). Actin-based confinement of calcium responses during Shigella invasion. Nat Commun 4: 1567.

Truong, D., D. Brabant, et al. (2013). Formin-mediated actin polymerization promotes Salmonella invasion. Cell Microbiol 15(12): 2051-63.

Uhlen, P., A. Laestadius, et al. (2000). Alpha-haemolysin of uropathogenic E. coli induces Ca2+ oscillations in renal epithelial cells. Nature 405(6787): 694-7.

Vadia, S., E. Arnett, et al. (2013). The pore-forming toxin listeriolysin O mediates a novel entry pathway of L. monocytogenes into human hepatocytes. PLoS Pathog 7(11): e1002356.

Vuori, K. and E. Ruoslahti (1995). Tyrosine phosphorylation of p130Cas and cortactin accompanies integrin-mediated cell adhesion to extracellular matrix. J Biol Chem 270(38): 22259-62.

Watarai, M., S. Funato, et al. (1996). Interaction of Ipa proteins of Shigella flexneri with alpha5beta1 integrin promotes entry of the bacteria into mammalian cells. J Exp Med 183(3): 991-9.

Weaver, A. M., J. E. Heuser, et al. (2002). Interaction of cortactin and N-WASp with Arp2/3 complex. Curr Biol 12(15): 1270-8.

Weed, S. A. and J. T. Parsons (2001). Cortactin: coupling membrane dynamics to cortical actin assembly. Oncogene 20(44): 6418-34.

Yang, J., Y. Zhao, et al. (2013). Human NAIP and mouse NAIP1 recognize bacterial type III secretion needle protein for inflammasome activation. Proc Natl Acad Sci U S A 110(35): 14408-13.

Zhou, D., L. M. Chen, et al. (2001). A Salmonella inositol polyphosphatase acts in conjunction with other bacterial effectors to promote host cell actin cytoskeleton rearrangements and bacterial internalization. Mol Microbiol 39(2): 248-59.

Zychlinsky, A., M. C. Prevost, et al. (1992). Shigella flexneri induces apoptosis in infected macrophages. Nature 358(6382): 167-9.

8

The Molecular Role of the Polar Outer Membrane Protein IcsA (VirG) in *Shigella* Actin-based Motility

Anuradha Janakiraman

Department of Biology, City College and Graduate Center of CUNY, 160 Convent Avenue, MR 616, New York, NY 10031, USA, Anuj@ccny.cuny.edu

Abstract

Central to the cytosolic life-cycle of the Gram-negative bacterial pathogen, *Shigella*, is the ability to polymerize host cell actin at the bacterial pole to generate actin comet tails. The actin tails generate a propulsive force that aid in intracellular and intercellular bacterial motility and is an essential feature of *Shigella* virulence. The bacterial outer membrane protein, IcsA (VirG), plays a critical role in *Shigella* actin-based motility. Here we review key findings and discuss recent developments that aid in our understanding of the polar localization of IcsA and IcsA-mediated *Shigella* actin-tail assembly.

Introduction

"The discovery of the dysentery bacillus stirred my young heart with hopes of eradicating the disease." Dr. Kiyoshi Shiga, discoverer of *Shigella dysenteriae,* 1936.

The Gram-negative bacteria *Shigella* spp. are the leading cause of bacillary dysentery in humans and monkeys leading to morbidity and mortality worldwide; 165 million cases occur annually with approximately 1 million associated deaths (Kotloff et al., 1999). A central feature of *Shigella* pathogenesis involves bacterial invasion of the colonic epithelium and bacterial spread through the colonic mucosal layer. *Shigella* is unique among Gram-negative pathogens in that it accesses the host cell cytoplasm and utilizes the host cell machinery to spread within tissues. The critical steps in the intracellular life-cycle of *Shigella*, depicted schematically in Figure 8.1A, is the topic of several excellent reviews (Ashida et al., 2012; Haglund and Welch, 2011; Ireton, 2013; Marteyn et al., 2012; Ray et al., 2009; Sasakawa, 2010). *Shigella* enters colonic mucosal cells by bacterium-induced macropinocytosis, after which it lyses the endocytic vacuole and is released into the host cell cytoplasm. In the cytoplasm, the bacterium recruits host cell actin to form a tail at one end of the cell (Figure 8.1A). Active assembly of this actin tail generates a force that is sufficient to propel the bacterium through the cytoplasm and into adjacent cells. *Shigella* mutants unable to undergo actin-based motility are attenuated in mouse experimental models, in humans and monkeys, indicating the essentiality of actin assembly in *Shigella* pathogenesis.

The process of actin tail assembly is mediated by the *Shigella* outer membrane protein IcsA (*i*ntra*c*ellular *s*pread A) also known as VirG (*vir*ulence G); disruption of *icsA* leads to loss of actin tail formation, IcsA-mediated actin assembly is key in disease progression as *icsA* mutants are markedly attenuated in humans and animal models. In addition to a major role in the pathology of *Shigella*, unipolar localization of IcsA has served as a model system to examine the molecular basis of spatiotemporal protein targeting in bacteria. Over the past 25 years much progress has been made in elucidating the genetic, molecular and cellular mechanisms that are critical to the intracellular life-cycle of *Shigella*. The focus of this chapter is on significant discoveries and the current understanding of the polar localization of IcsA and its role in *Shigella* actin-based motility.

Overview of *Shigella* pathogenesis

Shigella was first isolated and described by Dr. Kiyoshi Shiga in 1898 (Trofa et al., 1999). His discovery was the first demonstration of a bacterium being the causal agent of non-amoebic dysentery. Initially named *Bacillus dysenteriae*, the genus was first termed *Shigella* in the 1930 edition of Bergey's Manual of Determinative Bacteriology (Trofa et al., 1999). Today *Shigella* spp. are designated into 4 subgroups: *S. dysenteriae*, *S. flexneri*, *S. boydii* and *S. sonnei*. Shigellosis occurs endemically but certain populations such as immunocompromised individuals, residents of refugee camps, military field units, are at increased risk (Hale, 1991). It is also the most common cause of disease in children of developing nations (Sansonetti, 2001). The infection usually spreads through a fecal-oral route by contaminated food and water supplies and/or person-to-person contact (Kotloff et al., 1999). As few as 10 bacterial cells can cause Shigellosis, a disease that is manifested by bloody diarrhea and dysentery. The reason for the low dose of infectivity is suggested to be increased resistance of *Shigella* to stomach acids compared to other enterics such as *E. coli* or *Salmonella* (Hale, 1991). While *Shigella* infections are self-limiting, the standard recommendation in the US is that anyone infected with *Shigella* should be treated with antibiotics; this is for public health reasons given the low infectious inoculum of the bacterium (Kotloff et al., 1999).

It was conclusively demonstrated in the mid 1960's that *S. flexneri* causes disease by invading the intestinal mucosa. The bacteria breach the intestinal barrier primarily through the antigen sampling M cells and are engulfed by macrophages (Sansonetti and Phalipon, 1999). However, the bacteria are able to evade the macrophage by triggering apoptosis gaining access to the basolateral surface of the epithelium (Mounier et al., 1992; Zychlinsky et al., 1992). Invasion of the non-phagocytic epithelial cells is facilitated by a type III secretion system which helps induce uptake of the bacterium into an endocytic vacuole (Bergman et al., 1994; Clerc and Sansonetti, 1987; Hale and Bonventre, 1979; Mounier et al., 1992; Sansonetti et al., 1982b; Vasselon et al., 1992). Access to the cell cytoplasm is gained by lysing the vacuole and once in the cytoplasm the bacteria rapidly disseminate into adjacent tissues (Sansonetti et al., 1986). The intracellular motility of *Shigella* was first described by Ogawa in the late 1960's (Ogawa et al., 1968) but the movement was attributed to actin-based motility only in the 1980's. The bacterial encoded outer membrane protein IcsA is essential for actin-based motility (Bernardini et al., 1989; Lett et al., 1989; Makino et al., 1986; Pal et al., 1989). The continuous assembly of an actin tail generates a propulsive force such that the bacterium pushes out into the adjacent cell via peripheral membrane protrusions, which are internalized resulting in bacteria existing in double-membrane vacuoles (Bernardini et al.,

1989; Kadurugamuwa et al., 1991; Prevost et al., 1992). The bacterium lyses the double membrane vacuole with the aid of the *icsB* gene product and escapes into the cytoplasm of the adjacent cell continuing its life-cycle (Allaoui et al., 1992). Actin-based motility of *Shigella* is essential to its pathogenesis. The infected epithelial cells release cytokines and attract polymorphonuclear cells exacerbating the inflammation resulting from the initial infection (Sansonetti, 2001; Sasakawa, 2010). This leads to degeneration of the epithelium and eventual disintegration of the mucosa leading to blood and mucus in the diarrhea (Hale, 1991).

Conducting genetic and molecular analysis of *Shigella* virulence was hampered by the innate resistance of most animals to oral infections of *Shigella* (Hale, 1991). However, some animal models of bacterial uptake do exist: oral infections of starved guinea pigs and rhesus monkeys with *Shigella* established and subsequently confirmed the invasive properties of *Shigella* species (LaBrec et al., 1964; Takeuchi et al., 1968). Furthermore, ligated rabbit ileal loops provided a useful mechanism to study *Shigella* invasion and inflammation (Gots et al., 1974). An infant mouse model exists but is of limited use because the bacterium is not exposed to the full spectrum of host defenses (Fernandez et al., 2003). Alongside these intestinal models of *Shigella* infections exists the Sereny test - invasion of *Shigella* in corneal epithelial cells of guinea pig leads to keratoconjunctivitis and a measure of virulence (Murayama et al., 1986). A pulmonary model of *Shigella* infection was used extensively in the mid to late 1990s (Mallett et al., 1995). Finally, *Shigella* are capable of invading and spreading in tissue culture monolayers leading to plaque formation, wherein dead host cells are surrounded by adjacent infected cells (Gerber and Watkins, 1961; Hale and Bonventre, 1979; Hale et al., 1979; LaBrec et al., 1964; Oaks et al., 1985). Recently a guinea pig model of colonic infection has been developed and although it does not allow for an oral infection it shows a lot of promise in terms of optimal characterization of *Shigella* infections with respect to environmental adaptation in the colon (Shim et al., 2007).

The *Shigella* Virulence Plasmid

The chromosomal virulence determinants can generally be classified as gene products that affect the ability of *Shigella* to survive in the intestinal tract or in the mucosa, cytotoxins that increase the virulence of the bacterium, and regulatory factors that affect the expression of the plasmid virulence genes (Hale, 1991). However, intergeneric hybrid studies revealed that the virulence of *Shigella* could not be reconstituted from transferring the complete chromosome of *S. flexneri* to *E. coli* indicative of the fact that a *Shigella* specific element was required for virulence (Falkow et al., 1963; Formal et al., 1965a; Formal et al., 1965b). Later, Sansonetti and colleagues identified that *Shigella spp* virulence is dependent upon the presence of a family of large plasmids (Sansonetti et al., 1982a; Sansonetti et al., 1981; 1982b, Maurelli, 1988 #109). The genes encoded on this large, approximately 220 KB plasmid, include adhesins that help the bacteria adhere to host cell surface, type III secreted proteins involved in bacterial uptake by host cells, disruption of endocytic vesicles, gene products involved in intracellular and intercellular spread, and genes involved in the regulation of plasmid encoded genes (Dorman, 2008). The complete sequence analyses of large virulence plasmids from various *Shigella* strains are mostly identical except for minor differences in their insertion sequence (IS) elements; the high proportion of IS elements in the virulence plasmids (acquired likely through horizontal gene transfer) have implications in the structural instability within the plasmid (Dorman, 2008). Studies examining the relationship between the evolution of *Shigella* virulence plasmids and *E. coli* have revealed that in fact *Shigella* are within the genera of *E. coli* much like enteroinvasive *E. coli* (EIEC) and several lateral DNA acquisition events from various ancestral virulence plasmids have been key to the modern day *Shigella* species and its pathogenicity (Pupo et al., 2000; Venkatesan et al., 2001; Wei et al., 2003).

The icsA locus

A critical locus involved in *Shigella* cell-to-cell spread is located on the virulence plasmid. Sasakawa and colleagues were initially able to identify various virulence associated regions on the plasmid by combining data collected from independent Tn5 insertion mutants isolated on the 230Kb pMYSH6000 plasmid, a SalI restriction endonuclease fragment digestion map of the same plasmid, and four virulence associated phenotypes – the Sereny test (mentioned earlier), invasion of a cultured epithelial cell line, ability to bind the dye Congo Red, and faster growth of avirulent mutants (Sasakawa et al., 1986). The *virG* locus was first identified by transposon insertion within the SalI fragment "G" located 20 kb from the *mxi-spa* type III secretion system involved in invasion (Lett et al., 1989). The homologous locus in *S. flexneri* 5a is designated as *icsA* (Bernardini et al., 1989). Bacteria that lack a functional IcsA are unable to spread in the corneal epithelium in a Sereny test and in cultured monolayers (Bernardini et al., 1989; Lett et al., 1989; Makino et al., 1986; Pal et al., 1989). IcsA mediates actin tail assembly in *Shigella*; inert particles bound to purified IcsA or *E. coli* expressing IcsA are able to form actin tails in cytoplasmic extracts indicative of IcsA being both necessary and sufficient for actin-tail assembly by *Shigella* (Magdalena and Goldberg, 2002; Goldberg and Theriot, 1995).

IcsA is an 1102 amino acid, 110 kDa outer membrane protein that belongs to the autotransporter family of proteins. The domain organization of IcsA is characteristic of the AIDA-I Va family of autotransporters and has an N-terminal signal peptide (1-52) followed by the passenger or catalytic domain (alpha domain residues 53-758) and the conserved C-terminal translocation domain (beta domain residues 759-1102) (Figure 8.1B) (Dautin and Bernstein, 2007; Goldberg, 2001). Translocation across the outer membrane for members of this family occurs by the insertion of the beta-domain into the outer membrane to form an amphipathic beta-barrel and passage of the mature amino-terminal alpha-domain though the beta-barrel to the bacterial surface (Dautin and Bernstein, 2007). In addition to the three major structural and functional domains, IcsA has an intramolecular autochaperone (AC) domain (residues 591-758) situated within the C-terminal end of the alpha domain (Figure 8.1B) (May and Morona, 2008). Amino acids 634-735 within this region, known as the self-associating autotransporter (SAAT) domain, are implicated in homo-aggregation and biofilm formation (Teh and Morona, 2013; Teh et al., 2012). Biochemical characterization of IcsA has been impeded because of the inability to purify high enough quantities of stable IcsA for structural analysis. However, a recent breakthrough came in the form of solving the structure of the autochaperone domain of IcsA (PDB3ML3) which reveals two coils of a right handed parallel beta helix (Figure 8.1B) (Kuhnel and Diezmann, 2011).

Secretion of IcsA

The first indication that IcsA was a surface exposed protein came from immunofluorescence studies where non-permeabilized bacteria were labeled with anti-IcsA anti-serum (Goldberg et al., 1993). Lett and colleagues confirmed that IcsA was indeed exposed on the surface by using radiolabeled ^{125}I to externally label *Shigella* (Lett, 1995). Over the past decade several groups have contributed towards our understanding of how IcsA traverses the Gram-negative cell envelope. IcsA belongs to the type Va family of classical autotransporters, the largest family of translocating outer membrane proteins in the Gram-negatives and thought to have all the functional elements required for secretion of the proteins within the protein itself (Dautin and Bernstein, 2007). However, recent studies reveal that accessory factors play critical roles in optimal IcsA secretion and insertion across the Gram-negative envelope.

The N-terminal signal peptide targets IcsA to the inner membrane SecYEG translocon, which catalyzes its rapid energy driven transport across the inner membrane into the periplasm (Brandon et al., 2003). The precise folded nature of the protein in the cytoplasm is unclear although it has been established that proteins secreted through the Sec machinery need to be unfolded to transit across the inner membrane (Fisher and DeLisa, 2004). Once in the periplasm, the passenger domain of IcsA is indicated to be at least partially folded (Brandon and Goldberg, 2001). Furthermore, periplasmic

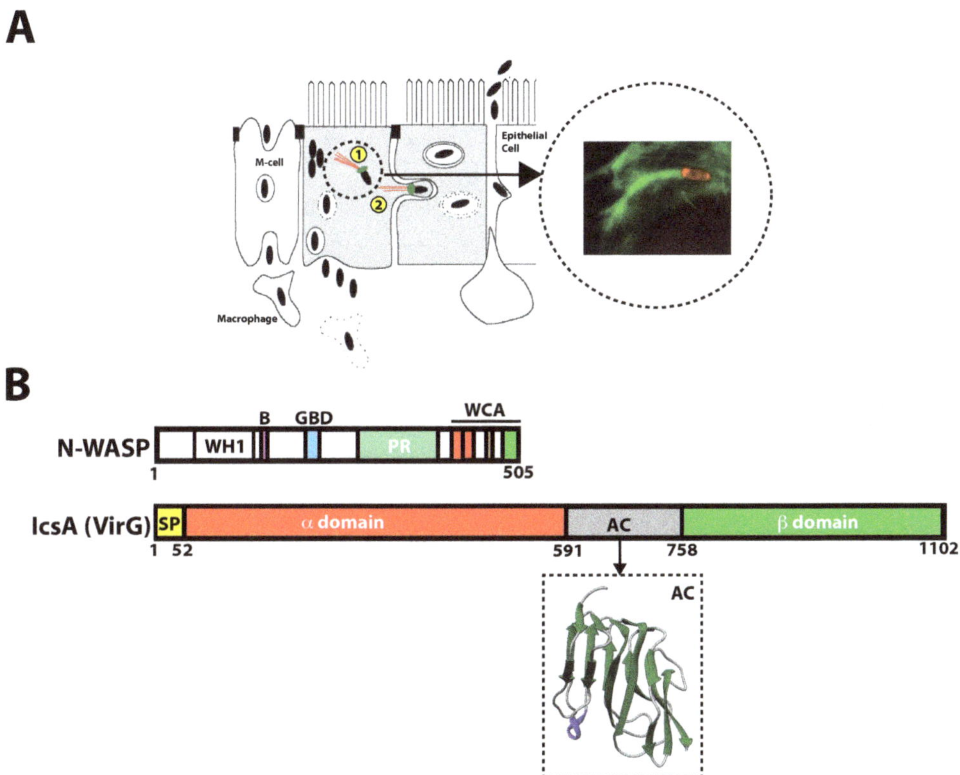

Figure 8.1. (A) The intracellular life-cycle of *Shigella*. *Shigella* (black rods) gain access to the basolateral surface and induce invasion into host epithelial cells. They lyse the membrane-bound vacuole and enter the host cytosol where they proliferate and assemble an actin tail at the old pole of the bacterium. Actin-based motility allows bacterial spread within (1) and between cells (2) and is a hallmark of *Shigella* pathogenicity. Figure 8 has been adapted from (Brandon and Goldberg, 2000). The inset shows an image of a host cell infected with *Shigella*. Intact bacteria with IcsA visualized by immunofluorescence (red) and actin-tail stained with phalloidin (green) is shown; the image has been adapted from (Goldberg, 2001). (B) Domain organization of N-WASP and IcsA proteins. The GenBank accession numbers for N-WASP and IcsA are NP_003932 and M22802 respectively. WH1, WIP interacting domain; B, basic region; GBD, GTP binding domain; PR, proline-rich sequences; WCA, verprolin, cofilin, acidic domain; SP, signal peptide; AC, autochaperone. The crystal structure of IcsA AC region (PDB3LM3) is shown below.

chaperones such as DegP, SurA, and Skp are required for optimal IcsA biogenesis (Hong and Payne, 1997; Purdy et al., 2007; Purdy et al., 2002; Wagner et al., 2009). While the IcsA passenger domain is presented on the bacterial surface less efficiently and impacts bacterial intercellular spread in chaperone mutants, the beta-barrel is folded and inserted in its native conformation and IcsA processing by the protease IcsP (discussed below) is preserved. The proper folding of IcsA beta-barrel in a *skp* mutant recapitulates studies with translocation of other autotransporters in periplasmic chaperone mutants and taken together these data suggest that the role of the periplasmic chaperones is to likely retain passenger domains in an extended or partially folded, translocation-competent form protected from degradative periplasmic proteases (Wagner et al., 2009).

Similar to other autotransporters, IcsA biogenesis requires the BAM complex on the Gram-negative envelope (Jain and Goldberg, 2007). Upon insertion into the outer membrane, the beta-domain of the IcsA autotransporter adopts a beta-barrel conformation, which is evolutionarily conserved based on reported structures of other beta-barrel outer membrane proteins. After insertion of the beta-barrel into the outer membrane, the passenger domain is translocated into the exterior and folds into a characteristic beta helix. The autochaperone region of IcsA is implicated in proper folding of the passenger domain as the alpha domain is not presented extracellularly when the autochaperone region is disrupted (Klemm et al., 2006; May and Morona, 2008; Teh et al., 2012). Another conserved feature in the primary sequence of these autotransporters, including IcsA, is the presence of a linker alpha helix in the beta-barrel domain (Kuhnel and Diezmann, 2011). Since the extracellular IcsA alpha domain is cleaved, the beta domain alpha helix, which spans the length of the beta-barrel lumen is proposed to plug the pore and play a role in the selective permeability of the outer membrane protein (Kuhnel and Diezmann, 2011). The precise molecular nature of how the BAM complex facilitates the incorporation of the beta-barrel into the outer membrane and how this is coordinated with the translocation of the alpha domain across the outer membrane are not completely understood as yet.

IcsP mediated cleavage of IcsA extracellular domain

The folded IcsA passenger domain is slowly cleaved extracellularly at the R758-R759 (SS**RR**ASS) bond position by a *Shigella* outer membrane protease, IcsP (SopA), which is encoded on the virulence plasmid (Egile et al., 1997; Shere et al., 1997). IcsP bears 60% identity at the primary sequence level to *E. coli* OmpT and is a functional homolog (Egile et al., 1997; Shere et al., 1997). Cleavage of ~20% of the membrane-bound IcsA by IcsP results in a ~95 kDa product, $IcsA_{53\text{-}757}$ containing the actin polymerizing motif, which is released into the extracellular milieu. However, IcsP mediated cleavage of IcsA is not critical for IcsA mediated cell-to-cell spread as *Shigella icsP* mutants still support actin based motility and show similar plaquing abilities relative to the wild type in cultured cells (Egile et al., 1997; Shere et al., 1997). While IcsP mediated cleavage may have a niche-specific role *in vivo*, perhaps the normal function of cleavage is more of a "maintenance" role in processing inefficiently incorporated outer membrane IcsA molecules. IcsA is distributed circumferentially in cells lacking IcsP, leading to the prediction that IcsP may itself be circumferentially distributed (Steinhauer et al., 1999). However, a very recent study has demonstrated an asymmetrical localization pattern for IcsP; IcsP is concentrated in the septa and new pole in bacterial cells lacking O antigen (Tran et al., 2013). The implications of the cellular localization of IcsP and IcsP-mediated cleavage of IcsA on the polar localization of IcsA are discussed below.

Polar localization of IcsA

Using affinity purified IcsA antiserum, Goldberg and colleagues first demonstrated that IcsA is localized predominantly on the old pole of the bacteria (Goldberg et al., 1993). As the bacterial cells begin to divide, IcsA begins to be expressed on the second pole as well. Early and more recent studies have all reported that drastic surface modifications such as the removal of O antigen side chains play a critical role in the establishment and maintenance of IcsA polar localization; aberrant IcsA localization is observed in a group of lipopolysaccharide (LPS) mutants, secretion of IcsA alpha domain and actin based motility are also altered in these rough mutants, and finally these cells lose their ability to form plaques or provide a positive outcome in Sereny tests (Morona and Van Den Bosch, 2003; Sandlin et al., 1995). Taken together these data underscore the importance of IcsA unipolar localization in *Shigella* virulence.

What bacterial-specific factors mediate such preferential accumulation of IcsA at the old pole of the bacterium? Several studies have attempted to address this question by examining the molecular mechanisms of polar localization of IcsA. The prevailing model is that targeting of IcsA to the pole occurs in the cytoplasm such that secretion across the cytoplasmic membrane via the Sec translocon and subsequently across the outer membrane occurs at the pole, leading to the polar display of the protein on the bacterial cell surface. IcsP slowly cleaves IcsA at all sites but as the insertion of IcsA is predominantly occurring at one pole it results in the unipolar distribution of IcsA. The bacterial factors implicated in polar localization of IcsA either directly or indirectly are depicted in Figure 8.2. The salient studies leading to the development of the above model are discussed below.

IcsP mediated cleavage maintains polar IcsA

Early work suggested that cleavage of IcsA is likely to be critical in the localization of IcsA. Mutational analysis of the IcsA SSRRASS cleavage site prevented IcsA phosphorylation by protein kinase A (PKA), led to a loss of IcsA cleavage, a loss of unipolar IcsA localization and directional actin assembly, as well as a "super-*icsA*" phenotype – formation of larger plaques indicative of greater spread in cultured monolayers (d'Hauteville and Sansonetti, 1992). A later study identified that inactivation of *sopA* (*icsP*), the virulence plasmid encoded gene product responsible for IcsA cleavage at the R-R bond, did indeed lead to a loss of IcsA polar localization and aberrant actin polymerization (Egile et al., 1997; Shere et al., 1997). However, in contrast to the d'Hauteville study, Egile *et al.* noted reduced plaque formation and a decrease in virulence in the *icsP* mutant. Another study mutagenizing the R758-R759 cleavage site demonstrated a lack of cleavage and secretion of IcsA, but virulence, plaque formation, IcsA localization or actin polymerization were unaltered compared to wild type (Fukuda et al., 1995). In their work Shere *et al.* reported data consistent with both Fukuda *et al.* and d'Hauteville *et al.* observations (Shere et al., 1997). Also while IcsA continued to be localized in a polar manner more circumferential distribution of IcsA was noted consistent with reduced cleavage. Consistent with the Fukuda *et al.* study, Shere *et al.*, demonstrated that reduced IcsA cleavage and localization with altered actin tail formation did not result in reduced plaquing ability in cultured monolayers. In fact, similar to results reported by d'Hauteville *et al.*, they noted faster intracellular movement in *icsP* mutants. To clarify these conflicting reports, Sandlin and colleagues determined localization of IcsA in a virulence plasmid cured derivative of *Shigella* as also in *E. coli* cells with an intact O antigen but lacking the protease OmpT, a homolog of IcsP (Sandlin and Maurelli, 1999). In both instances they were able to demonstrate unipolar localization of IcsA indicating that virulence plasmid determinants including IcsP are not essential in the establishment of IcsA localization and that a conserved pathway of preferentially localizing proteins such as IcsA to one pole of the bacterium exist in both *Shigella* and *E. coli*. A recent study shows that less IcsP relative to IcsA is present at the old pole in bacteria lacking O antigen side chains and this contributes to the polar localization of IcsA (Tran et al., 2013). Taken together these studies suggest that while there is evidence for a role for IcsP in maintaining the preferential accumulation of IcsA at the old pole, IcsP, is not essential for IcsA polarity.

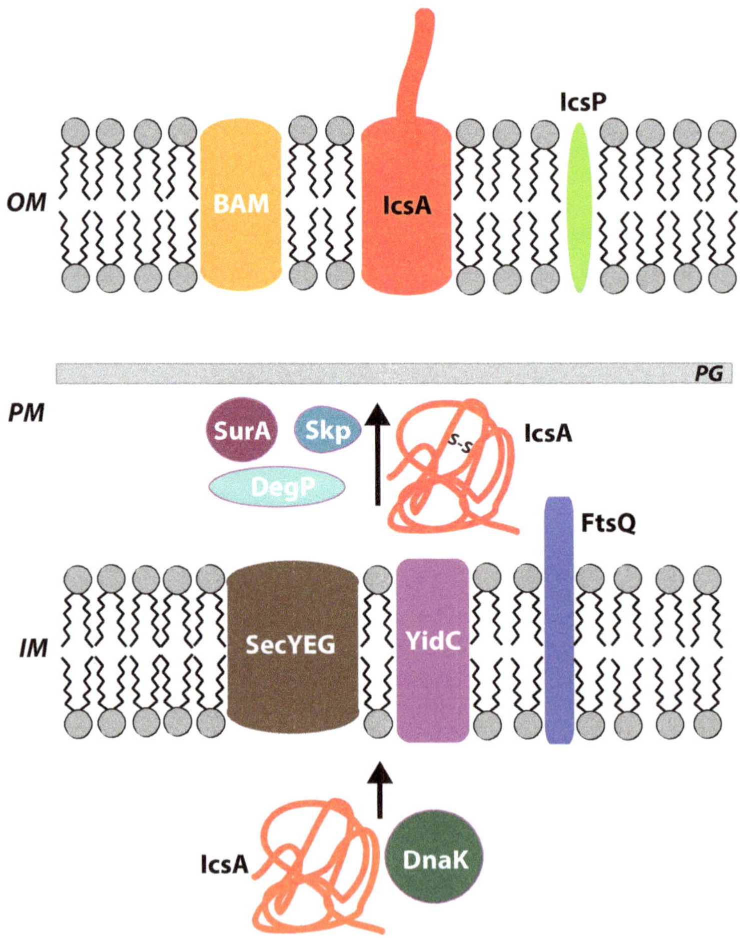

Figure 8.2. Localization of IcsA at the old pole of *Shigella*. Schematic representation of proteins identified in various studies as promoting polar localization and secretion of IcsA is depicted. The folded state of IcsA in the cytoplasm is currently unknown. In the periplasm, IcsA is thought to be partially folded containing at least one disulfide bond. RpoS, the stationary phase sigma factor, which inhibits polar localization during stationary phase, has not been included in the cartoon. OM, outer membrane; PM, periplasm, PG, peptidoglycan; IM, inner membrane.

Polar targeting of IcsA is conserved amongst Shigella and related bacteria

Using translations fusions of various regions of IcsA to green fluorescent protein (GFP) it was first revealed that localization of IcsA is independent of the signal sequence and therefore of signal peptide mediated secretion (Charles et al., 2001). Disruption of the Sec machinery did not disrupt polar targeting of IcsA confirming that secretion is not required for IcsA targeting to the pole (Charles et al., 2001). These results indicate that the positional information recognized by IcsA is present in the cytoplasm or the cytoplasmic face of the inner membrane at the bacterial old pole. Moreover, amino acid sequences 1-104 and 507-620 were each necessary and sufficient for polar localization of IcsA not only in *Shigella* but in other Enterobacteriacae and *Vibrio* (Charles et al., 2001). The localization of IcsA-GFP fusions can be disrupted in the presence of wild type IcsA suggesting a competition by the tagged and untagged versions of the protein for the same target (Charles et al., 2001). These data considered with those from experiments demonstrating that several other classical autotransporters may also be asymmetrically distributed (Jain et al., 2006) suggest that the overall molecular mechanism that allows targeting of specific proteins to the old pole may be conserved in these bacterial subgroups. It is important to distinguish here that while the molecular mechanism of targeting to the pole may be conserved in different bacteria, given the diverse sequence and functions of autotransporters, the biological significance of such asymmetric localization is still unclear.

Polar positional information is controlled in part by cell wall modifying enzymes

Following inhibition of the cell-division-specific transpeptidase FtsI, or in the absence of functional FtsZ, the essential tubulin-like GTPase division protein, a cytoplasmic derivative of IcsA, $IcsA_{507-620}$-GFP fusion, localizes at or near the future poles along filamentous *E. coli* cells (Janakiraman and Goldberg, 2004). Furthermore, the localization of IcsA to these sites is not determined by the Min proteins, the topological specificity factors for division. Nor is it determined by nucleoid occlusion, a molecular mechanism preventing division atop segregating chromosomes (Janakiraman and Goldberg, 2004). These data indicate that positional information recognized by the polar protein IcsA is independent of both known spatial-control factors of the cell-division machinery. A transposon screen conducted to identify the polar positional cues, indicated that the placement of the polar material recognized by IcsA appears to be in part controlled by MreB, an actin-like cell shape determinant, and low molecular weight penicillin binding proteins, a set of enzymes that make minor modifications to the peptidoglycan (Nilsen et al., 2004; Nilsen et al., 2005).

Role for division protein FtsQ in IcsA targeting

Using a genetic reporter screen for identifying factors that may serve as IcsA polar targets, a recent report indicated that elevated levels of RpoS, the stationary phase sigma factor, block IcsA polar localization (Fixen et al., 2012). This result was consistent with a prior study, which noted that IcsA localization is growth-phase dependent (Goldberg et al., 1994). The polar localization of IcsA in exponentially growing cells is compatible with the notion that during infection *Shigella* are rapidly multiplying and spreading to adjacent cells and likely never quite reach stationary phase. In the same screen, and independent of the RpoS mediated role on IcsA targeting, the cell division protein FtsQ was implicated in IcsA and perhaps more generally in targeting other autotransporters to the pole by contributing to the positional cues (Fixen et al., 2012). FtsQ is an inner membrane essential cell division protein that requires YidC and the Sec translocase for insertion into the inner membrane (Urbanus et al., 2001; van der Laan et al., 2001). In *E. coli*, FtsQ forms a complex with other essential cell division proteins FtsB and FtsL and is thought to link the early division proteins, which are mostly cytoplasmic, with the transmembrane or periplasmic components of the multiprotein division complex that are recruited later (Buddelmeijer and Beckwith, 2004). The periplasmic C-terminal domain of FtsQ plays a necessary and sufficient role in localization of FtsQ to the midcell septal site (Guzman et al., 1997; Scheffers et al., 2007). While the precise function of FtsQ in cell division is incompletely understood, the periplasmic region of FtsQ is postulated to play a role in peptidoglycan synthesis during cell division (van den Ent et al., 2008). IcsA localization was circumferential in a conditional *ftsQ* mutant grown at the non-permissive temperature and polar localization could be reconstituted by expressing wild type FtsQ *in trans* (Fixen et al., 2012). The extracytoplasmic region of FtsQ was also found to play a role in targeting an $IcsA_{507-620}$-mCherry fusion to the pole suggesting that the periplasmic domain of FtsQ contributes to establishing bacterial polarity at the cytoplasmic face of the inner membrane (Fixen et al., 2012).

Role for membrane insertase YidC in IcsA localization

In *E. coli*, the biogenesis of 17-32% of all cytoplasmic membrane proteins inserted into the membrane via the Sec translocon requires the widely conserved essential cytoplasmic membrane insertase YidC (Xie and Dalbey, 2008). The aforementioned cell division factor FtsQ, which contributes to polar positional information crosslinks to YidC suggesting that FtsQ may itself be a YidC substrate (Urbanus et al., 2001). A recent study explored the role of YidC-dependent biogenesis of FtsQ and perhaps other cytoplasmic membrane proteins in the cytoplasmic localization of IcsA. In filamentous cells depleted of YidC, a cytoplasmic derivative of IcsA, $IcsA_{507-620}$-GFP, localized mostly to the poles and not to future division sites (Gray et al., 2014). However, FtsQ was properly inserted into the membrane in these filaments (Gray et al., 2014), indicating that FtsQ is not a substrate of YidC. These data predict a model wherein depletion of YidC leads to a specific disruption in the expression, positioning or maturation of a cytoplasmic membrane protein factor that is independent of FtsQ and that provides IcsA positional information.

Role for cytoplasmic chaperone DnaK in IcsA localization

Whether IcsA is targeted passively through diffusion to be eventually captured upon contact with a polar target, or is targeted actively through a specific delivery mechanism in the cytoplasm is unclear. The molecular chaperone DnaK was implicated in the proper targeting and secretion of IcsA and another *Shigella* autotransporter, SepA, in the cytoplasm, suggesting that perhaps large multidomain autotransporters are retained in a "localization competent" state in the cytoplasm by chaperones such as DnaK, albeit DnaK is not directly involved in the polar localization of IcsA (Janakiraman et al., 2009).

IcsA and *Shigella* actin-based motility

After lysis of the phagocytic vacuole, *Shigella* assembles an actin tail at one end of the cell that aids in intracellular and intercellular movement. Such host-actin facilitated microbial movement has been termed "actin-based motility." Other intracellular pathogens such as *Listeria, Mycobacterium, Burkholderia, Vaccinia* and *Rickettsia* also share actin-based motility with *Shigella* but possess distinct mechanisms (Ireton, 2013). The proteins that are essential for *Shigella*-based actin mobility *in vivo* are outlined in Figure 8.3.

About 45 years ago Ogawa and colleagues first reported that intracellular *Shigella* are motile and that this motility could be abolished by treating host cells with tetracycline suggesting the requirement of bacterial protein synthesis in bacterial movement (Ogawa et al., 1968). Previous studies have demonstrated that an *icsA* mutant does not polymerize actin on its surface, does not move intracellularly, does not form protrusions from the cell surface, and does not spread from cell to cell, indicating that *Shigella* actin-based motility is mediated by IcsA alone (Bernardini et al., 1989; Lett et al., 1989; Makino et al., 1986; Sansonetti et al., 1991). Consistent with the polar localization of IcsA, *Shigella* actin tails, which are ~7 µm long and 0.7 µm wide are restricted to the old pole of the bacterium and are generally absent at the lateral ends of the cell (Goldberg, 2001). Early evidence suggested that the surface exposed alpha-domain of IcsA contains all the elements required for actin-assembly; inert particles coated with purified alpha domain polypeptide form actin tails in cytoplasmic extracts, and amino acids 53-508 which lie within the alpha domain increase actin polymerization in *in vitro* assays (Magdalena and Goldberg, 2002). However, IcsA could not directly induce actin nucleation, indicating the recruitment of one or more host cytosolic components.

Role of N-WASP in IcsA mediated Shigella actin-based motility

In infected human cells, the Neuronal Wiskott-Aldrich Syndrome Protein (N-WASP) accumulates at the bacterial pole that produces the F-actin tail (Figure 8.1B) (Egile et al., 1999; Rohatgi et al., 1999; Suzuki et al., 1998). N-WASP is a member of the WASP/Scar family of proteins that connect several signaling pathways to the actin cytoskeleton (Snapper et al., 2001). A large proportion of cellular N-WASP is complexed with WIP

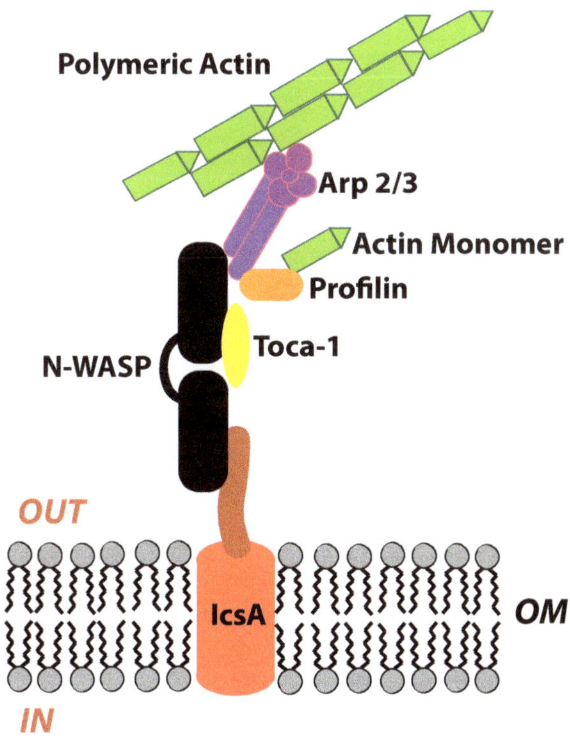

Figure 8.3. *Shigella* IcsA mediated actin-tail assembly. A schematic representation of host actin cytoskeletal modulators that mediate actin-tail polymerization by *Shigella*. Vinculin and Cdc42 whose roles in *Shigella* actin assembly are not fully clarified are not shown.

(WASP-interacting protein), which stabilizes the autoinhibited form of N-WASP wherein the N-terminal regulatory domain of N-WASP blocks the C-terminal Arp2/3 activating segment (Rohatgi et al., 1999). N-WASP binds to the activated form of the small Rho GTPase Cdc42, or phosphatidylinositol 4,5-bisphosphate (PIP2) or Toca-1 (transducer of Cdc42-dependent actin assembly) a highly conserved member of *S. pombe* Cdc15 homology (PCH) family, and is released from the folded autoinhibited state to an unfolded active state (Ho et al., 2004). The activation of N-WASP allows the unmasking of the C-terminal WCA domain that binds and stimulates the Arp2/3 complex, a set of proteins that promote actin nucleation (Figure 8.1B) (Rohatgi et al., 1999).

The essentiality of N-WASP for actin based intracellular and intercellular motility of *Shigella*, but not bacterial entry, was confirmed by experiments involving dominant negative N-WASP alleles or mouse cell lines depleted for N-WASP (Snapper et al., 2001). Reconstitution experiments with purified proteins demonstrate that IcsA activates N-WASP resulting in Arp2/3-dependent actin polymerization (Egile et al., 1999). IcsA uses an amino-terminal domain with glycine-rich repeats (residues 103-433) to bind directly to N-WASP protein (Egile et al., 1999). More recent studies with linker mutational analyses identified residues within the IcsA SAAT domain in interacting with N-WASP (May and Morona, 2008; Teh and Morona, 2013). The domain of N-WASP that binds IcsA lies in the N-terminal region which includes the WH1 domain of N-WASP (Figure 8.1B)(Suzuki et al., 2002). IcsA is functionally analogous to Cdc42 in retaining N-WASP in an open conformation and in agreement with this notion, actin based motility of *Shigella* is largely independent of Cdc42 (Shibata et al., 2002). Mutant N-WASP constructs in constitutively open conformations showed an increase in plaquing efficiency suggesting a role in intercellular spread (Adamovich et al., 2009). Once in an open conformation N-WASP can be modulated by phosphoregulation but analyses of the phosphoregulation status of N-WASP and its role on *Shigella* actin based motility have generated conflicting data. While Burton *et al.* suggested a role for Abl kinase mediated phosphorylation in promoting intracellular and intercellular dissemination (Burton et al., 2005), Adamovich *et al.* reported no changes in *Shigella* induced actin tail formation in phophomicking or phosphodisruptive N-WASP mutants (Adamovich et al., 2009).

Actin is present in the monomeric (G-actin) or polymeric filamentous form of several repeating monomers (F-actin). For a review of actin filament assembly and function in intracellular motility of pathogens the reader is referred to several excellent reviews (Campellone and Welch, 2010; Cossart, 2000; Goldberg, 2001; Haglund and Welch, 2011; Ireton, 2013; Stevens et al., 2006; Welch and Way, 2013). Nucleation of actin filament assembly occurs *de novo* by the formation of short actin oligomers made up of 2 or 3 repeating subunit of actin monomers and is aided by several proteins including Arp2/3 and formins (Campellone and Welch, 2010), classes of proteins relevant to the discussion of *Shigella* intracellular and intercellular spread respectively. N-WASP in the open conformation binds Arp2/3 complex through the WCA domain and activates Arp2/3 mediated actin nucleation and polymerization. Arp2 and Arp3 structurally resemble monomeric actin and are thought to mimic an actin dimer to stimulate polymerization of branched actin filaments off of preexisting actin polymers (Ireton, 2013). Arp2/3 is required for IcsA mediated *Shigella* actin based motility as indicated by the following experimental results - Arp2/3 colocalizes along the length of *Shigella* actin tails, *E. coli* cells expressing IcsA and coated with N-WASP fail to show movement in cells lacking Arp2/3 but motility can be partially restored upon addition of Arp2/3, and in *in vitro* polymerization assays, in the presence of N-WASP an IcsA concentration dependent increase in Arp2/3 mediated actin polymerization in noted (Egile et al., 1999; Frischknecht et al., 1999; Gouin et al., 1999). Taken, together these results initially indicated that the formation of a ternary complex between IcsA, N-WASP, and Arp2/3 at the bacterial surface alone is sufficient to cause actin nucleation/polymerization in the presence of actin monomers.

Role for Toca-1 in Shigella actin-based motility

However, most of cellular N-WASP exists in a complex with WIP, which stabilizes the N-WASP autoinhibitory conformation state (Ho et al., 2004). The autoinhibitory interaction of N-WASP is alleviated by a number of upstream activators, including Cdc42 and Toca-1, an essential component of the Cdc42 signaling cascade (Ho et al., 2004). Consistent with the necessity of *Shigella* to circumvent N-WASP autoinhibition, results indicate a critical role for Toca-1 in actin-based motility of *Shigella* (Leung et al., 2008). Host Toca-1 associates with intracellular bacteria prior to actin-based movement and independent of IcsA. Instead, Toca-1 recruitment is mediated by an as yet unidentified *Shigella* factor secreted through the type III secretion system (Leung et al., 2008). Depletion of Toca-1 leads to the reduction of bacterial-induced actin assembly and polymerization. Actin based motility is restored upon expression of a constitutively active form of N-WASP suggesting that Toca-1 contributes to the relief of N-WASP autoinhibition. These findings indicate that while *in vitro Shigella* IcsA alone is sufficient to activate N-WASP, *in vivo Shigella* stimulates host Arp2/3-dependent actin polymerization by recruiting N-WASP through the interaction with IcsA but also requires Toca-1, which is recruited by an as yet unidentified mechanism (Leung et al., 2008).

Roles of Profilin, Cofilin, Vinculin, VASP, and Mena in Shigella actin based-motility

The roles of a number of additional host proteins including profilin, cofilin, vinculin and vasodilator-stimulated phosphoprotein (VASP), have been examined in *Shigella* actin-based motility largely due to colocalization of these cytoskeletal proteins with *Shigella* actin tails (Ally et al., 2004; Gouin et al., 1999; Mimuro et al., 2000; Suzuki et al., 1996). Further studies with some of the host proteins such as vinculin have led to conflicting roles in *Shigella* actin based motility. Nonetheless, the current understanding of the functions of these host proteins in *Shigella* motility is briefly summarized below.

Profilin can interact with monomeric actin and plays a modulatory role in maintaining the monomer-filament actin dynamics (Cossart, 2000). Profilin can act as both an actin-polymerizing factor or as an actin-sequestering factor based on the nature of the barbed ends of filamentous actin. Profilin is required for maximal rates of *Shigella* movement but is not essential for actin tail assembly. In cell extracts depleted of profilin, *E. coli* expressing IcsA display a significant decrease in the rate of motility but not in actin tail assembly; the rate of motility can be restored by expressing wild type profilin *in trans* suggesting a role for this host protein in sustaining maximal rates of *Shigella* spread (Loisel et al., 1999; Mimuro et al., 2000). Profilin can bind proline-rich sequences and while N-WASP has a poly L-proline patch near the WCA domain at the C-terminal end conflicting data exist regarding the requirement of profilin interaction with N-WASP in *Shigella* actin-based motility. While a profilin mutant (H133S) that is unable to interact with a proline-rich sequence was shown to enhance movement of *E. coli* expressing IcsA to the same extent as wild type profilin in cells depleted of profilin (Egile et al., 1999), Mimuro *et al.* indicated that expression of H133S mutant in *Xenopus* extracts that are depleted of profilin continued to exhibit significantly reduced bacterial motility compared to extracts expressing wild type profilin (Mimuro et al., 2000). Furthermore, these authors showed that expression of a N-WASP derivative lacking the proline-rich sequences in N-

WASP-depleted cytoplasmic extracts reduced actin tail formation by ~90% suggesting that profilin association with N-WASP was essential for spread (Mimuro et al., 2000).

Cofilin is an actin-associated protein that plays a role in both actin severing and depolymerization (Cossart, 2000). Cofilin colocalizes with *Shigella* actin tails and is required for maximal *Shigella* motility based upon actin-based motility reconstitution assays with purified proteins (Loisel et al., 1999). Given the biological role of cofilin, the requirement of cofilin in bacterial motility may likely be restricted to the maintenance of actin dynamics *in vivo* rather than a more direct involvement in *Shigella* actin tail assembly.

Actin cytoskeletal proteins such as vinculin, a component of focal contacts and intercellular adhesions, and VASP, member of a family of proteins that modulate actin dynamics, colocalize with *Shigella* actin tails (Kadurugamuwa et al., 1991; Suzuki et al., 1996). Since IcsA has no proline-rich sequences, it was hypothesized that vinculin, which contains proline-rich sequences and binds VASP, may serve as the intermediary in linking IcsA to profilin mediated actin motility (Laine et al., 1997). VASP binds profilin, which as discussed earlier promotes the rate of *Shigella* intracellular and intercellular motility (Reinhard et al., 1995). Moreover, a VASP proline peptide in the presence of profilin significantly increased *Shigella* motility within cells (Zeile et al., 1996). Cellular vinculin is maintained in an autoinhibited folded form and is cleaved upon *Shigella* infection (Laine et al., 1997). The cleaved amino-end of vinculin binds IcsA *in vitro*, and colocalizes with the actin tail (Laine et al., 1997; Suzuki et al., 1996). Furthermore, vinculin is recruited to *Shigella* in the absence of N-WASP confirming a direct interaction with IcsA (Ally et al., 2004). However, vinculin is not required in reconstituted actin-motility assays nor is motility affected in cell lines, which have reduced levels of vinculin (Goldberg, 1997; Loisel et al., 1999). In their 2004 study, Ally *et al.* clarified some of these conflicting data by showing that *Shigella* actin-tail assembly and motility remained unaltered in VASP deficient cells suggesting that a VASP-profilin interaction was not required for the role of profilin in enhancing *Shigella* spread speeds (Ally et al., 2004). Moreover, they also showed that although vinculin is recruited to *Shigella*, and very likely by a direct interaction with IcsA as vinculin could be detected in the absence of both N-WASP and VASP, it failed to recruit Arp2/3 to *Shigella*. Taken together, these data indicate that the precise role of vinculin in *Shigella* actin based movement is not effected through Arp2/3 actin nucleation and continues to be unclear.

Finally, Mena, a VASP family protein, is also recruited to the *Shigella* pole in the presence of IcsA (Ally et al., 2004; Gertler et al., 1996). Mena associated with *Shigella* actin tails in a pattern similar to that observed for VASP (Frischknecht et al., 1999; Gouin et al., 2005). These results suggest that while VASP family proteins are recruited to *Shigella* poles and actin tails by binding to vinculin, they do not appear to have a function in intracellular spread. Whether the localization of Mena protein has a role in intercellular spread has not been tested. In summary, although several actin cytoskeletal proteins localize to *Shigella* actin tails, conclusive evidence only points to the essentiality of host proteins N-WASP, Arp2/3 and Toca-1 in *Shigella* actin-based assembly and motility.

Intercellular spread of Shigella and roles for formins and myosin-X

Motile intracellular *Shigella* engages components of the cell intermediate junctions to form a membrane bound cell extension known as 'protrusion' that is internalized by the adjacent cell leaving the bacterium transiently placed in a double-membrane vacuole. The bacterium lyses the double-membrane vacuole thereby entering the adjacent cell cytoplasm where it multiplies and exhibits actin-based motility leading to the spread of infection. In cultured cells, *Shigella* often form protrusions at the site of focal adhesions (Prevost et al., 1992). Cadherin expression is necessary for the proper structure, dynamics and internalization of the protrusions (Sansonetti et al., 1994). Other host factors involved in the formation of protrusions and intercellular spread, are beginning to be identified and understood. Focal adhesion sites are rich in the mammalian Diaphanous-related formins Dia1 and Dia2. Like Arp2/3 complexes, formins are nucleators of actin polymerization and play roles in a multitude of cellular functions including promoting the formation of membrane extensions, cell polarity, cell migration, cytokinesis and intracellular trafficking (Campellone and Welch, 2010; Pruyne et al., 2002). While *Shigella* actin tails that generate a propulsive force for intracellular motility are arranged in branched arrays, typical of Arp2/3 mediated actin polymerization, actin tails in bacteria in protrusions are arranged in parallel arrays (Gouin et al., 1999). Cells depleted of Dia show a reduction in formation of protrusions and in intercellular spread (Heindl et al., 2010). Significantly no reduction in actin-tail assembly is noted in the cell body (Heindl et al., 2010).

Most recently, a role for Myosin-X (Myo10) in connecting *Shigella* actin tails to the peripheral membranes has been unearthed indicating a significant role for Myo10 in *Shigella* intercellular spread (Bishai et al., 2013). Depletion of myosin X has no discernable effect on the number of *Shigella* protrusions. However, a reduction in the length of *Shigella* protrusions as well as an increase in the width of the protrusions that make contact with the main body of the cell is seen, suggesting a role for Myo10 in maintaining the architecture of the protrusions.

Conclusions

Analysis of the localization of *Shigella* IcsA to the bacterial pole and IcsA-mediated actin tail assembly have served as model systems and revealed novel insights into spatiotemporal control of protein localization in bacteria, overall pathogen biology, and the regulation of host-cell actin assembly. Key questions that remain unanswered concern the nature of the IcsA polar localization factor in *Shigella* and the molecular mechanism that allows IcsA to sense its target. Additionally, the influences of fundamental forces such as membrane potential or cell geometry on IcsA localization have not been investigated as yet. These basic cellular properties have been shown to govern protein localization in cell division and sporulation in bacteria (Ramamurthi and Losick, 2009; Strahl and Hamoen, 2010). Substantial progress has been made in terms of our understanding of *Shigella* actin-based motility. While the role of IcsA in promoting *Shigella* motility within cells is reasonably well understood, aspects of *Shigella* cell-to-cell spread are much less clear. How *Shigella* switches from Arp2/3 mediated branched actin tail polymerization to Dia-mediated bundled actin tail assembly in protrusions is not known. Also unknown are the molecular basis of protrusion formation, dynamics and internalization. While some host factors that play key roles in bacterial spread between cells are beginning to be identified and characterized, very little to nothing is understood of the bacterial factors that are likely to maintain active control over the process. A recent report suggests that IcsA is able to self-associate on the bacterial surface and that changes in IcsA oligomerization status lead to inefficient intercellular spread (May et al., 2012). Further investigations are required to elaborate whether such IcsA self-association has physiological relevance in interactions with N-WASP and actin tail assembly in protrusions. Lastly, while the focus of this chapter has been on the role of IcsA in actin-based motility, it is worth mentioning that IcsA induces autophagy by binding the autophagy protein 5 (ATG5), an E3 ubiquitin ligase (Ogawa et al., 2005). The type III secreted factor IcsB enable *Shigella* to escape autophagy by competitively binding to IcsA and preventing its association with ATG5 (Ogawa et al., 2005).

Acknowledgements
I thank Marcia Goldberg for comments on the manuscript. Work in my laboratory is supported by a grant from NSF (MCB1158059) and core facility support for CCNY comes from NIH/NIMHD/RCMI 8G12MD007603-29.

References

Adamovich, D., Nakamura, F., Worth, A., Burns, S., Thrasher, A., Hartwig, J., and Snapper, S. (2009). Activating mutations of N-WASP alter *Shigella* pathogenesis. Biochem Biophys Res Commun *384*, 284-289.

Allaoui, A., Mounier, J., Prevost, M.-C., Sansonetti, P.J., and Parsot, C. (1992). *icsB*: a *Shigella flexneri* virulence gene necessary for the lysis of protrusions during intercellular spread. Mol Microbiol *6*, 1605-1616.

Ally, S., Sauer, N.J., Loureiro, J.J., Snapper, S.B., Gertler, F.B., and Goldberg, M.B. (2004). *Shigella* interactions with the actin cytoskeleton in the absence of Ena/VASP family proteins. Cell Microbiol *6*, 355-366.

Ashida, H., Ogawa, M., Kim, M., Mimuro, H., and Sasakawa, C. (2012). Bacteria and host interactions in the gut epithelial barrier. Nat Chem Biol *8*, 36-45.

Bergman, T., Erickson, K., Galyov, E., Persson, C., and Wolf-Watz, H. (1994). The lcrB (*yscN/U*) gene cluster of *Yersinia pseudotuberculosis* is involved in Yop secretion and shows high homology to the *spa* gene clusters of *Shigella flexneri* and *Salmonella typhimurium*. J Bacteriol *176*, 2619-2626.

Bernardini, M.L., Mounier, J., d'Hauteville, H., Coquis-Rondon, M., and Sansonetti, P.J. (1989). Identification of *icsA*, a plasmid locus of *Shigella flexneri* that governs bacterial intra- and intercellular spread through interaction with F-actin. Proc Natl Acad Sci *86*, 3867-3871.

Bishai, E.A., Sidhu, G.S., Li, W., Dhillon, J., Bohil, A.B., Cheney, R.E., Hartwig, J.H., and Southwick, F.S. (2013). Myosin-X facilitates *Shigella*-induced membrane protrusions and cell-to-cell spread. Cell Microbiol *15*, 353-367.

Brandon, L., and Goldberg, M. (2001). Periplasmic Transit and Disulfide Bond Formation of the Autotransported *Shigella* Protein IcsA. J Bacteriol *183*, 951-958.

Brandon, L.D., Goehring, N., Janakiraman, A., Yan, A.W., Wu, T., Beckwith, J., and Goldberg, M.B. (2003). IcsA, a polarly localized autotransporter with an atypical signal peptide, uses the Sec apparatus for secretion, although the Sec apparatus is circumferentially distributed. Mol Microbiol *50*, 45-60.

Brandon, L.D., and Goldberg, M.B. (2000). Exploitation of mammalian host cell function by *Shigella* spp. In Virulence Mechanisms of Bacterial Pathogens, K.A. Brogden, ed. Washington, D.C. ASM Press, pp. 175-187.

Buddelmeijer, N., and Beckwith, J. (2004). A complex of the *Escherichia coli* cell division proteins FtsL, FtsB and FtsQ forms independently of its localization to the septal region. Mol Microbiol *52*, 1315-1327.

Burton, E., Oliver, T., and Pendergast, A. (2005). Abl Kinases Regulate Actin Comet Tail Elongation via an N-WASP-Dependent Pathway. Mol Cell Biol *25*, 8834-8843.

Campellone, K.G., and Welch, M.D. (2010). A nucleator arms race: cellular control of actin assembly. Nat Rev Mol Cell Biol *11*, 237-251.

Charles, M., Perez, M., Kobil, J.H., and Goldberg, M.B. (2001). Polar targeting of *Shigella* virulence factor IcsA in Enterobacteriacae and *Vibrio*. Proc Natl Acad Sci *98*, 9871-9876.

Clerc, P., and Sansonetti, P.J. (1987). Entry of *Shigella flexneri* into HeLa cells: evidence for directed phagocytosis involving actin polymerization and myosin accumulation. Infect Immun *55*, 2681-2688.
Cossart, P. (2000). Actin-based motility of pathogens: the Arp2/3 complex is a central player. Cell Microbiol *2*, 195-205.
d'Hauteville, H., and Sansonetti, P.J. (1992). Phosphorylation of IcsA by cAMP-dependent protein kinase and its effect on intracellular spread of *Shigella flexneri*. Mol Microbiol *6*, 833-841.
Dautin, N., and Bernstein, H. (2007). Protein Secretion in Gram-Negative Bacteria via the Autotransporter Pathway. Annu Rev Microbiol *61*, 89-112.
Dorman, C. (2008). The virulence plasmids of *Shigella flexneri*. In Microbial Megaplasmids, E. Schwartz, ed. Heidelberg. Springer, pp. 151-170.
Egile, C., d'Hauteville, H., Parsot, C., and Sansonetti, P.J. (1997). SopA, the outer membrane protease responsible for polar localization of IcsA in *Shigella flexneri*. Mol Microbiol *23*, 1063-1073.
Egile, C., Loisel, T.P., Laurent, V., Li, R., Pantaloni, D., Sansonetti, P.J., and Carlier, M.-F. (1999). Activation of the CDC42 effector N-WASP by the *Shigella flexneri* IcsA protein promotes actin nucleation by Arp2/3 complex and bacterial actin-based motility. J Cell Biol *146*, 1319-1332.
Falkow, S., Schneider, H., Baron, L.S., and Formal, S.B. (1963). Virulence of *Escherichia-Shigella* Genetic Hybrids for the Guinea Pig. J Bacteriol *86*, 1251-1258.
Fernandez, M.I., Thuizat, A., Pedron, T., Neutra, M., Phalipon, A., and Sansonetti, P.J. (2003). A newborn mouse model for the study of intestinal pathogenesis of Shigellosis. Cell Microbiol *5*, 481-491.
Fisher, A.C., and DeLisa, M.P. (2004). A little help from my friends: quality control of presecretory proteins in bacteria. J Bacteriol *186*, 7467-7473.
Fixen, K.R., Janakiraman, A., Garrity, S., Slade, D.J., Gray, A.N., Karahan, N., Hochschild, A., and Goldberg, M.B. (2012). Genetic reporter system for positioning of proteins at the bacterial pole. mBio *3*, e00251-00211.
Formal, S.B., Labrec, E.H., Kent, T.H., and Falkow, S. (1965a). Abortive Intestinal Infection with an *Escherichia coli-Shigella flexneri* Hybrid Strain. J Bacteriol *89*, 1374-1382.
Formal, S.B., LaBrec, E.H., Palmer, A., and Falkow, S. (1965b). Protection of monkeys against experimental Shigellosis with attenuated vaccine. J Bacteriol *90*, 63-68.
Frischknecht, F., Cudmore, S., Moreau, V., Reckmann, I., Röttger, S., and Way, M. (1999). Tyrosine phosphorylation is required for actin-based motility of vaccinia but not *Listeria* or *Shigella*. Curr Biol *9*, 89-92.
Fukuda, I., Suzuki, T., Munakata, H., Hayashi, N., Katayama, E., Yoshizawa, M., and Sasakawa, C. (1995). Cleavage of *Shigella* surface protein VirG occurs at a specific site, but the secretion is not essential for intracellular spreading. J Bacteriol *177*, 1719-1726.
Gerber, D.F., and Watkins, H.M. (1961). Growth of *Shigellae* in monolayer tissue cultures. J Bacteriol *82*, 815-822.
Gertler, F.B., Niebuhr, K., Reinhard, M., Wehland, J., and Soriano, P. (1996). Mena, a relative of VASP and Drosophila Enabled, is implicated in the control of microfilament dynamics. Cell *87*, 227-239.
Goldberg, M.B. (1997). *Shigella* actin-based motility in the absence of vinculin. Cell Motil Cytoskeleton *37*, 44-53.
Goldberg, M.B. (2001). Actin-based motility of intracellular microbial pathogens. Microbiol Mol Biol Rev *65*, 595-626.

Goldberg, M.B., Bârzu, O., Parsot, C., and Sansonetti, P.J. (1993). Unipolar localization and ATPase activity of IcsA, a *Shigella flexneri* protein involved in intracellular movement. J Bacteriol *175*, 2189-2196.

Goldberg, M. B., and J. A. Theriot. (1995). *Shigella flexneri* surface protein IcsA is sufficient to direct actin-based motility. Proc. Natl. Acad. Sci. USA

Goldberg, M.B., Theriot, J.A., and Sansonetti, P.J. (1994). Regulation of surface presentation of IcsA, a *Shigella* protein essential to intracellular movement and spread, is growth phase-dependent. Infect Immun *62*, 5664-5668.

Gots, R.E., Formal, S.B., and Giannella, R.A. (1974). Indomethacin inhibition of *Salmonella typhimurium*, *Shigella flexneri*, and cholera-mediated rabbit ileal secretion. J Infect Dis *130*, 280-284.

Gouin, E., Gantelet, H., Egile, C., Lasa, I., Ohayon, H., Villiers, V., Gounon, P., Sansonetti, P.J., and Cossart, P. (1999). A comparative study of the actin-based motilities of the pathogenic bacteria *Listeria monocytogenes*, *Shigella flexneri* and *Rickettsia conorii*. J Cell Sci *112*, 1697-1708.

Gouin, E., Welch, M.D., and Cossart, P. (2005). Actin-based motility of intracellular pathogens. Curr Opin Microbiol *8*, 35-45.

Gray, A.N., Li, Z., Henderson-Frost, J., and Goldberg, M.B. (2014). Biogenesis of YidC Cytoplasmic Membrane Substrates Is Required for Positioning of Autotransporter IcsA at Future Poles. J Bacteriol *196*, 624-632.

Guzman, L.M., Weiss, D.S., and Beckwith, J. (1997). Domain-swapping analysis of FtsI, FtsL, and FtsQ, bitopic membrane proteins essential for cell division in *Escherichia coli*. J Bacteriol *179*, 5094-5103.

Haglund, C.M., and Welch, M.D. (2011). Pathogens and polymers: microbe-host interactions illuminate the cytoskeleton. J Cell Biol *195*, 7-17.

Hale, T.L. (1991). Genetic basis of virulence in *Shigella* species. Microbiol Rev *55*, 206-224.

Hale, T.L., and Bonventre, P.F. (1979). *Shigella* infection of Henle intestinal epithelial cells: role of the bacteria. Infect Immun *24*, 879-886.

Hale, T.L., Morris, R.E., and Bonventre, P.F. (1979). *Shigella* infection of Henle intestinal epithelial cells: role of the host cell. Infect Immun *24*, 887-894.

Heindl, J., Saran, I., Yi, C., Lesser, C., and Goldberg, M. (2010). Requirement for Formin-Induced Actin Polymerization during Spread of *Shigella flexneri*. Infect Immun *78*, 193-203.

Ho, H.Y., Rohatgi, R., Lebensohn, A.M., Le, M., Li, J., Gygi, S.P., and Kirschner, M.W. (2004). Toca-1 mediates Cdc42-dependent actin nucleation by activating the N-WASP-WIP complex. Cell *118*, 203-216.

Hong, M., and Payne, S.M. (1997). Effect of mutations in *Shigella flexneri* chromosomal and plasmid-encoded lipopolysaccharide gene on invasion and serum resistance. Mol Microbiol *24*, 779-791.

Ireton, K. (2013). Molecular mechanisms of cell-cell spread of intracellular bacterial pathogens. Open Biol *3*, 130079.

Jain, S., and Goldberg, M. (2007). Requirement for YaeT in the Outer Membrane Assembly of Autotransporter Proteins. J Bacteriol *189*, 5393-5398.

Jain, S., van Ulsen, P., Benz, I., Schmidt, M.A., Fernandez, R., Tommassen, J., and Goldberg, M.B. (2006). Polar localization of the autotransporter family of large bacterial virulence proteins. J Bacteriol *188*, 4841-4850.

Janakiraman, A., Fixen, K.R., Gray, A.N., Niki, H., and Goldberg, M.B. (2009). Genome-scale proteomic screen identifies role for DnaK in chaperoning of polar autotransporters in *Shigella*. J Bacteriol *191*, 6300-6311.

Janakiraman, A., and Goldberg, M.B. (2004). Evidence for polar positional information independent of cell division and nucleoid occlusion. Proc Natl Acad Sci *101*, 835-840.

Kadurugamuwa, J.L., Rohde, M., Wehland, L., and Timmis, K.N. (1991). Intercellular spread of *Shigella flexneri* through a monolayer mediated by membraneous protrusions and associated with reorganization of the cytoskeletal protein vinculin. Infect Immun *59*, 3463-3471.

Klemm, P., Vejborg, R.M., and Sherlock, O. (2006). Self-associating autotransporters, SAATs: functional and structural similarities. Int J Med Microbiol *296*, 187-195.

Kotloff, K.L., Winickoff, J.P., Ivanoff, B., Clemens, J.D., Swerdlow, D.L., Sansonetti, P.J., Adak, G.K., and Levine, M.M. (1999). Global burden of *Shigella* infections: implications for vaccine development and implementation of control strategies. Bull WHO *77*, 651-666.

Kuhnel, K., and Diezmann, D. (2011). Crystal Structure of the Autochaperone Region from the *Shigella flexneri* Autotransporter IcsA. J Bacteriol *193*, 2042-2045.

LaBrec, E.H., Schneider, H., Magnani, T.J., and Formal, S.B. (1964). Epithelial cell penetration as an essential step in the pathogenesis of bacillary dysentery. J Bacteriol *88*, 1503-1518.

Laine, R.O., Zeile, W., Kang, F., Purich, D.L., and Southwick, F.S. (1997). Vinculin proteolysis unmasks an ActA homolog for actin-based *Shigella* motility. J Cell Biol *138*, 1255-1264.

Lett, M. (1995). Extracellular Transport of VirG Protein in *Shigella*. J Biol Chem *270*, 30874-30880.

Lett, M.-C., Sasakawa, C., Okada, N., Sakai, T., Makino, S., Yamada, M., Komatsu, K., and Yoshikawa, M. (1989). *virG*, a plasmid-coded virulence gene of *Shigella flexneri*: identification of the *virG* protein and determination of the complete coding sequence. J Bacteriol *171*, 353-359.

Leung, Y., Ally, S., and Goldberg, M.B. (2008). Bacterial actin assembly requires Toca-1 to relieve N-wasp autoinhibition. Cell Host Microbe *3*, 39-47.

Loisel, T.P., Boujemaa, R., Pantaloni, D., and Carlier, M.-F. (1999). Reconstitution of actin-based motility of *Listeria* and *Shigella* using pure proteins. Nature *401*, 613-616.

Magdalena, J., and Goldberg, M.B. (2002). Quantification of *Shigella* IcsA required for bacterial actin polymerization. Cell Motil Cytoskeleton *51*, 187-196.

Makino, S., Sasakawa, C., Kamata, K., Kurata, T., and Yoshikawa, M. (1986). A genetic determinant required for continuous reinfection of adjacent cells on large plasmid in *S. flexneri 2a*. Cell *46*, 551-555.

Mallett, C.P., Hale, T.L., Kaminski, R.W., Larsen, T., Orr, N., Cohen, D., and Lowell, G.H. (1995). Intransal or intragastric immunization with proteosome-*Shigella* lipopolysaccharide vaccines protects against lethal pneumonia in a murine model of *Shigella* infection. Infect Immun *63*, 2382-2386.

Marteyn, B., Gazi, A., and Sansonetti, P. (2012). *Shigella*: a model of virulence regulation *in vivo*. Gut Microbes *3*, 104-120.

May, K., and Morona, R. (2008). Mutagenesis of the *Shigella flexneri* Autotransporter IcsA Reveals Novel Functional Regions Involved in IcsA Biogenesis and Recruitment of Host Neural Wiscott-Aldrich Syndrome Protein. J Bacteriol *190*, 4666-4676.

May, K.L., Grabowicz, M., Polyak, S.W., and Morona, R. (2012). Self-association of the *Shigella flexneri* IcsA autotransporter protein. Microbiology *158*, 1874-1883.

Mimuro, H., Suzuki, T., Suetsugu, S., Miki, H., Takenawa, T., and Sasakawa, C. (2000). Profilin is required for sustaining efficient intra- and intercellular spreading of *Shigella flexneri*. J Biol Chem *275*, 28893-28901.

Morona, R., and Van Den Bosch, L. (2003). Lipopolysaccharide O antigen chains mask IcsA (VirG) in *Shigella flexneri*. FEMS Microbiol Lett *221*, 173-180.

Mounier, J., Vasselon, T., Hellio, R., Lesourd, M., and Sansonetti, P. (1992). *Shigella flexneri* enters human colonic Caco-2 epithelial cells through the basolateral pole. Infect Immun *60*, 237-248.

Murayama, S.Y., Sakai, T., Makino, S., Kurata, T., Sasakawa, C., and Yoshikawa, M. (1986). The use of mice in the Sereny test as a virulence assay of Shigellae and enteroinvasive *Escherichia coli*. Infect Immun *51*, 696-698.

Nilsen, T., Ghosh, A.S., Goldberg, M.B., and Young, K.D. (2004). Branching sites and morphological abnormalities behave as ectopic poles in shape-defective *Escherichia coli*. Mol Microbiol *52*, 1045-1054.

Nilsen, T., Yan, A.W., Gale, G., and Goldberg, M.B. (2005). Presence of multiple sites containing polar material in spherical *Escherichia coli* cells that lack MreB. J Bacteriol *187*, 6187-6196.

Oaks, E.V., Wingfield, M.E., and Formal, S.B. (1985). Plaque formation by virulent *Shigella flexneri*. Infect Immun *48*, 124-129.

Ogawa, H., Nakamura, A., and Nakaya, R. (1968). Cinemicrographic study of tissue cell cultures infected with *Shigella flexneri*. Jpn J Med Sci Biol *21*, 259-273.

Ogawa, M., Yoshimori, T., Suzuki, T., Sagara, H., Mizushima, N., and Sasakawa, C. (2005). Escape of intracellular *Shigella* from autophagy. Science *307*, 727-731.

Pal, T., Newland, J.W., Tall, B.D., Formal, S.B., and Hale, T.L. (1989). Intracellular spread of *Shigella flexneri* associated with the *kcpA* locus and a 140-kilodalton protein. Infect Immun *57*, 477-486.

Prevost, M.C., Lesourd, M., Arpin, M., Vernel, F., Mounier, J., Hellio, R., and Sansonetti, P.J. (1992). Unipolar reorganization of F-actin layer at bacterial division and bundling of actin filaments by plastin correlate with movement of *Shigella flexneri* within HeLa cells. Infect Immun *60*, 4088-4099.

Pruyne, D., Evangelista, M., Yang, C., Bi, E., Zigmond, S., Bretscher, A., and Boone, C. (2002). Role of formins in actin assembly: nucleation and barbed-end association. Science *297*, 612-615.

Pupo, G.M., Lan, R., and Reeves, P.R. (2000). Multiple independent origins of *Shigella* clones of *Escherichia coli* and convergent evolution of many of their characteristics. Proc Natl Acad Sci *97*, 10567-10572.

Purdy, G., Fisher, C., and Payne, S. (2007). IcsA Surface Presentation in *Shigella flexneri* Requires the Periplasmic Chaperones DegP, Skp, and SurA. J Bacteriol *189*, 5566-5573.

Purdy, G., Hong, M., and Payne, S. (2002). *Shigella flexneri* DegP Facilitates IcsA Surface Expression and Is Required for Efficient Intercellular Spread. Infect Immun *70*, 6355-6364.

Ramamurthi, K.S., and Losick, R. (2009). Negative membrane curvature as a cue for subcellular localization of a bacterial protein. Proc Natl Acad Sci *106*, 13541-13545.

Ray, K., Marteyn, B., Sansonetti, P.J., and Tang, C.M. (2009). Life on the inside: the intracellular lifestyle of cytosolic bacteria. Nat Rev Microbiol *7*, 333-340.

Reinhard, M., Giehl, K., Abel, K., Haffner, C., Jarchau, T., Hoppe, V., Jockusch, B.M., and Walter, U. (1995). The proline-rich focal adhesion and microfilament protein VASP is a ligand for profilins. EMBO J *14*, 1583-1589.

Rohatgi, R., Ma, L., Miki, H., Lopez, M., Kirchhausen, T., Takenawa, T., and Kirschner, M.W. (1999). The interaction between N-WASP and the Arp2/3 complex links Cdc42-dependent signals to actin assembly. Cell *97*, 221-231.

Sandlin, R.C., Lampel, K.A., Keasler, S.P., Goldberg, M.B., Stolzer, A.L., and Maurelli, A.T. (1995). Avirulence of rough mutants of *Shigella flexneri*: requirement of O antigen for correct unipolar localization of IcsA in bacterial outer membrane. Infect Immun *63*, 229-237.

Sandlin, R.C., and Maurelli, A.T. (1999). Establishment of unipolar localization of IcsA in *Shigella flexneri* 2a is not dependent on virulence plasmid determinants. Infect Immun *67*, 350-356.

Sansonetti, P.J. (2001). Microbes and microbial toxins: paradigms for microbial-mucosal interactions III. Shigellosis: from symptoms to molecular pathogenesis. Am J Physiol Gastrointest Liver Physiol *280*, G319-323.

Sansonetti, P.J., Arondel, J., Fontaine, A., d'Hauteville, H., and Bernardini, M.L. (1991). *OmpB* (osmo-regulation) and *icsA* (cell-to-cell spread) mutants of *Shigella flexneri*: vaccine candidates and probes to study the pathogenesis of Shigellosis. Vaccine *9*, 416-422.

Sansonetti, P.J., d'Hauteville, H., Formal, S.B., and Toucas, M. (1982a). Plasmid-mediated invasiveness of "*Shigella*-like" *Escherichia coli*. Ann Microbiol (Inst Pasteur) *132A*, 351-355.

Sansonetti, P.J., Kopecko, D.J., and Formal, S.B. (1981). *Shigella sonnei* plasmids: evidence that a large plasmid is necessary for virulence. Infect Immun *34*, 75-83.

Sansonetti, P.J., Kopecko, D.J., and Formal, S.B. (1982b). Involvement of a plasmid in the invasive ability of *Shigella flexneri*. Infect Immun *35*, 852-860.

Sansonetti, P.J., Mounier, J., Prevost, M.C., and Mege, R.-M. (1994). Cadherin expression is required for the spread of *Shigella flexneri* between epithelial cells. Cell *76*, 829-839.

Sansonetti, P.J., and Phalipon, A. (1999). M cells as ports of entry for enteroinvasive pathogens: mechanisms of interaction, consequences for the disease process. Semin Immunol *11*, 193-203.

Sansonetti, P.J., Ryter, A., Clerc, P., Maurelli, A.T., and Mounier, J. (1986). Multiplication of *Shigella flexneri* within HeLa cells: lysis of the phagocytic vacuole and plasmid-mediated contact hemolysis. Infect Immun *51*, 461-469.

Sasakawa, C. (2010). A new paradigm of bacteria-gut interplay brought through the study of *Shigella*. Proc Jpn Acad Ser B Phys Biol Sci *86*, 229-243.

Sasakawa, C., Makino, S., Kamata, K., and Yoshikawa, M. (1986). Isolation, characterization, and mapping of Tn*5* insertions into the 140-megadalton invasion plasmid defective in the mouse Sereny test in *Shigella flexneri* 2a. Infect Immun *54*, 32-36.

Scheffers, D.J., Robichon, C., Haan, G.J., den Blaauwen, T., Koningstein, G., van Bloois, E., Beckwith, J., and Luirink, J. (2007). Contribution of the FtsQ transmembrane segment to localization to the cell division site. J Bacteriol *189*, 7273-7280.

Shere, K.D., Sallustio, S., Manessis, A., D'Aversa, T.G., and Goldberg, M.B. (1997). Disruption of IcsP, the major *Shigella* protease that cleaves IcsA, accelerates actin-based motility. Mol Microbiol *25*, 451-462.

Shibata, T., Takeshima, F., Chen, F., Alt, F.W., and Snapper, S.B. (2002). Cdc42 facilitates invasion but not the actin-based motility of *Shigella*. Curr Biol *12*, 341-345.

Shim, D.H., Suzuki, T., Chang, S.Y., Park, S.M., Sansonetti, P.J., Sasakawa, C., and Kweon, M.N. (2007). New animal model of Shigellosis in the Guinea pig: its usefulness for protective efficacy studies. J Immunol *178*, 2476-2482.

Snapper, S.B., Takeshima, F., Anton, I., Liu, C.H., Thomas, S.M., Nguyen, D., Dudley, D., Fraser, H., Purich, D., Lopez-Ilasaca, M., *et al.* (2001). N-WASP deficiency reveals distinct pathways for cell surface projections and microbial actin-based motility. Nat Cell Biol *3*, 897-904.

Steinhauer, J., Agha, R., Pham, T., Varga, A.W., and Goldberg, M.B. (1999). The unipolar *Shigella* surface protein IcsA is directly targeted to the old pole; IcsP cleavage of IcsA occurs over the entire bacterial surface. Mol Microbiol *32*, 367-378.

Stevens, J.M., Galyov, E.E., and Stevens, M.P. (2006). Actin-dependent movement of bacterial pathogens. Nat Rev Microbiol *4*, 91-101.

Strahl, H., and Hamoen, L.W. (2010). Membrane potential is important for bacterial cell division. Proc Natl Acad Sci *107*, 12281-12286.

Suzuki, T., Miki, H., Takenawa, T., and Sasakawa, C. (1998). Neural Wiskott-Aldrich syndrome protein is implicated in the actin-based motility of *Shigella flexneri*. EMBO J *17*, 2767-2776.

Suzuki, T., Mimuro, H., Suetsugu, S., Miki, H., Takenawa, T., and Sasakawa, C. (2002). Neural Wiskott-Aldrich syndrome protein (N-WASP) is the specific ligand for *Shigella* VirG among the WASP family and determines the host cell type allowing actin-based spreading. Cell Microbiol *4*, 223-233.

Suzuki, T., Saga, S., and Sasakawa, C. (1996). Functional analysis of *Shigella* VirG domains essential for interaction with vinculin and actin-based motility. J Biol Chem *271*, 21878-21885.

Takeuchi, A., Formal, S.B., and Sprintz, H. (1968). Experimental acute colitis in the rhesus monkey following peroral infection with *Shigella flexneri*. Am J Pathol *52*, 503-529.

Teh, M., and Morona, R. (2013). Identification of *Shigella flexneri* IcsA Residues Affecting Interaction with N-WASP, and Evidence for IcsA-IcsA Co-Operative Interaction. PLoS ONE *8*, e55152.

Teh, M., Tran, E., and Morona, R. (2012). Absence of O antigen suppresses *Shigella flexneri* IcsA autochaperone region mutations. Microbiology *158*, 2835-2850.

Tran, E., Doyle, M., and Morona, R. (2013). LPS Unmasking of *Shigella flexneri* Reveals Preferential Localisation of Tagged Outer Membrane Protease IcsP to Septa and New Poles. PLoS ONE *8*, e70508.

Trofa, A.F., Ueno-Olsen, H., Oiwa, R., and Yoshikawa, M. (1999). Dr. Kiyoshi Shiga: discoverer of the dysentery bacillus. Clin Infect Dis *29*, 1303-1306.

Urbanus, M.L., Scotti, P.A., Froderberg, L., Saaf, A., de Gier, J.W., Brunner, J., Samuelson, J.C., Dalbey, R.E., Oudega, B., and Luirink, J. (2001). Sec-dependent membrane protein insertion: sequential interaction of nascent FtsQ with SecY and YidC. EMBO Rep 2, 524-529.

van den Ent, F., Vinkenvleugel, T.M., Ind, A., West, P., Veprintsev, D., Nanninga, N., den Blaauwen, T., and Lowe, J. (2008). Structural and mutational analysis of the cell division protein FtsQ. Mol Microbiol *68*, 110-123.

van der Laan, M., Houben, E.N., Nouwen, N., Luirink, J., and Driessen, A.J. (2001). Reconstitution of Sec-dependent membrane protein insertion: nascent FtsQ interacts with YidC in a SecYEG-dependent manner. EMBO Rep *2*, 519-523.

Vasselon, T., Mounier, J., Hellio, R., and Sansonetti, P.J. (1992). Movement along actin filaments of the perijunctional area and de novo polymerization of cellular actin are required for *Shigella flexneri* colonization of epithelial Caco-2 cell monolayers. Infect Immun *60*, 1031-1040.

Venkatesan, M.M., Goldberg, M.B., Rose, D.J., Grotbeck, E.J., Burland, V., and Blattner, F.R. (2001). Complete DNA sequence and analysis of the large virulence plasmid of *Shigella flexneri*. Infect Immun *69*, 3271-3285.

Wagner, J., Heindl, J., Gray, A., Jain, S., and Goldberg, M. (2009). Contribution of the Periplasmic Chaperone Skp to Efficient Presentation of the Autotransporter IcsA on the Surface of *Shigella flexneri*. J Bacteriol *191*, 815-821.

Wei, J., Goldberg, M.B., Burland, V., Venkatesan, M.M., Deng, W., Fournier, G., Mayhew, G.F., Plunkett, G., 3rd, Rose, D.J., Darling, A., *et al.* (2003). Complete genome sequence and comparative genomics of *Shigella flexneri* serotype 2a strain 2457T. Infect Immun *71*, 2775-2786.

Welch, M.D., and Way, M. (2013). Arp2/3-mediated actin-based motility: a tail of pathogen abuse. Cell Host Microbe *14*, 242-255.
Xie, K., and Dalbey, R.E. (2008). Inserting proteins into the bacterial cytoplasmic membrane using the Sec and YidC translocases. Nat Rev Microbiol *6*, 234-244.
Zeile, W.L., Purich, D.L., and Southwick, F.S. (1996). Recognition of two classes of oligoproline sequences in profilin-mediated acceleration of actin-based *Shigella* motility. J Cell Biol *133*, 49-59.
Zychlinsky, A., Prevost, M., and Sansonetti, P. (1992). *Shigella flexneri* induces apoptosis in infected macrophages. Nature *358*, 167-168.

9

Trans-epithelial Communication in the Context of *Shigella* Infection

Ana Maldonado and Beth A. McCormick

Microbiology and Physiological Systems Department, University of Massachusetts Medical School, 55 Lake Avenue North, Worcester, MA 01604, USA
beth.mccormick@umassmed.edu

Abstract

Shigella ssp. are the causative agents of bacillary dysentery, a disease characterized by a severe form of bloody diarrhea. Every year there are about two billion cases of diarrheal disease resulting in millions of deaths. Ninety-nine percent of these cases occur in developing countries, where diarrhea is the second-leading cause of death in children. *Shigella* is one of the most communicable bacterial agents of diarrheal disease and has evolved to be a highly efficient pathogen, as only humans (and certain higher primates) serve as the only natural host and reservoir. Shigellosis consists of painful abdominal cramps, nausea, fever, tenesmus, and frequently blood and mucus in the stools. These symptoms reflect the invasion of the colonic submucosa by *Shigella*. Pathogenesis of *Shigella* ssp. comprise an intricate cross-talk between bacterial factors and host components that leads to the destruction of the epithelium, and the characteristic signs of inflammation exhibited by massive infiltration and exudation of polymorphonuclear cells evidenced in histopathologic analysis of infected patients. At the same time the host activates a sophisticated defense mechanism aimed at clearing the infection. Here we summarize the bacterial-epithelial cell interactions during the course of shigellosis.

Intestinal epithelium architecture

The gastrointestinal tract is the largest mucosal surface of the human body. It has a surface area of about 200 m^2 and is a dynamic environment with continuous exposure to passaging food, containing dietary antigens, nutrients, and a diverse array of microorganisms. Due to the continuous presence of nutrients, the gastrointestinal tract is a highly colonized niche. In fact, about 70% of the total microbes co-existing in the body reside in the colon alone (Ley et al., 2006; Whitman et al., 1998).

The entire colonic surface is lined by a polarized single layer of epithelial cells that folds into a serpentine architecture to increase the absorption area, and in addition creates a tightly regulated barrier to limit the access of 100 million microbes to the internal milieu. The columnar epithelium sits on top of the lamina propria and is divided into regions known as the crypt and villus (Figure 9.1A). The crypts are populated by stem cells that differentiate into enterocytes, enteroendocrine cells, goblet cells or Paneth cells. As enterocytes migrate up the protruding villi they become mature absorptive cells. Globet cells produce mucins that form a layer of mucus, which covers the epithelium, preventing direct exposure of the cell layer with microbes and contents of the lumen. Paneth and enteroendrocrine cells secrete antimicrobial peptides and intestinal hormones, respectively, into the lumen. Additional surveillance is carried out by the gut-associated lymphoid tissue, which is located within the mucosal lining of the gastrointestinal tract. This is a unique niche that houses a collection of antigen-presenting cells. Such antigen-presenting cells include conventional cells from the innate and adaptive immune systems (i.e., T and B cells, macrophages, and dendritic cells), in addition to lamina propria and intraepithelial lymphocytes.

While serving a critical role as a barrier for foreign molecules/microbes, the intestinal epithelium also needs to be sufficiently permeable to allow for the universal fluid exchange of molecules and ions that functions at any mucosae surface. There are two transport routes in the intestinal epithelium:

• Paracellular: Exchange of molecules through the space between cells, mainly mediated by tight and adherent junctions (discussed in detail bellow).
• Transcellular: Across the plasma membrane of the epithelial cells.

At the most apical side of the polarized epithelium, the tight junction complex mediates cell-to-cell interaction maintaining cell polarity (Figure 9.1B). Tight junction proteins also

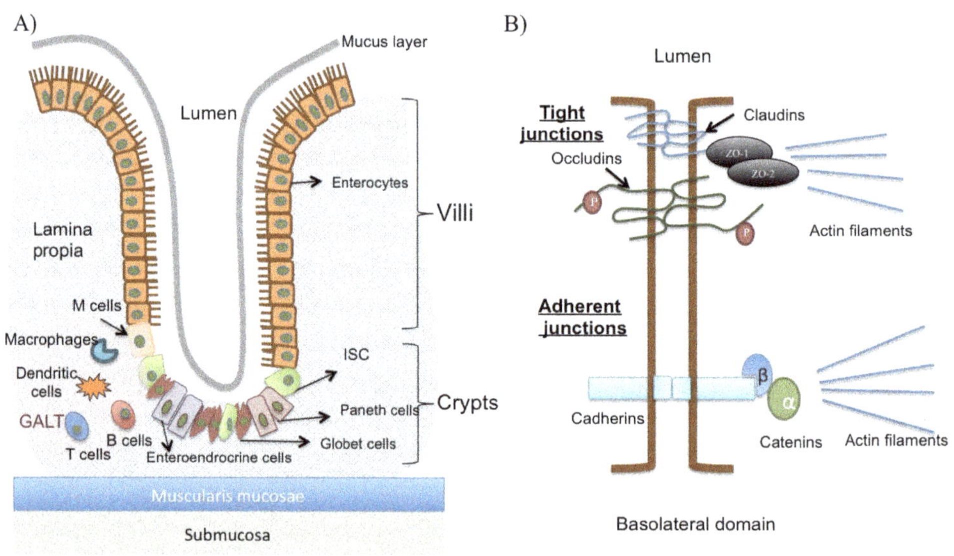

Figure 9.1. Architecture of the intestinal epithelium. A) Cellular organization on the epithelium: Villi contains mature enterocytes with absorptive function, while crypts are compose of intestinal stem cells (ISC) and additional set of cells dedicated to secretion of hormones (enteroendocine cells), mucus (globet cells) or antimicrobial peptides (Paneth cells) into the lumen. This serpentine structure is supported by the lamina propria where gut-associated lymphoid tissue (GALT) contains immune cells. B) A closer look at the inter-cell junctional proteins that maintains the structure, allow intercellular communication and also play a key role in permeability of the epithelium. Tight junctions are located in the most apical side of the cells while adherent junctions seal the intercellular space at the basolateral side.

control epithelial permeability by preventing paracellular transport of enteric bacteria, and by modulating the lateral diffusion of water and solutes. Proteins belonging to the occludin and claudin families (respectively: from the Latin *occludere* and *claudere* meaning 'to close') are the major structural components of the tight junction complex. Occludins are tetraspanning membrane proteins characterized by large cytoplasmic tails with numerous residues that can be phosphorylated. Phosphorylated occludin appears to be associated with apical tight junctions while lack of phosphorylation leads to accumulation of occludin both in the cytosol and at the basolateral membrane (Wong, 1997).

The role of occludins remains controversial. Occludins have been associated with the establishment and maintenance of the physiological properties of the tight junction complex (Furuse et al., 1996; McCarthy et al., 1996). While occludin-null mice exhibit physiologically and structurally normal tight junctions (Saitou et al., 2000), the null phenotype displays numerous abnormalities in many organs. It has also been suggested that occludins are required for signaling events that result in barrier remodeling, such as cytokine-induced changes to the epithelial barrier properties, linked to caveolin-occludin interactions (Nusrat et al., 2000; Van Itallie et al., 2010).

Following the discovery of occludin as a critical junctional protein, claudins were next identified as an additional group of proteins with an important role in maintaining barrier function. There are at least 24 isoforms of claudins, all of which contain a strand of four

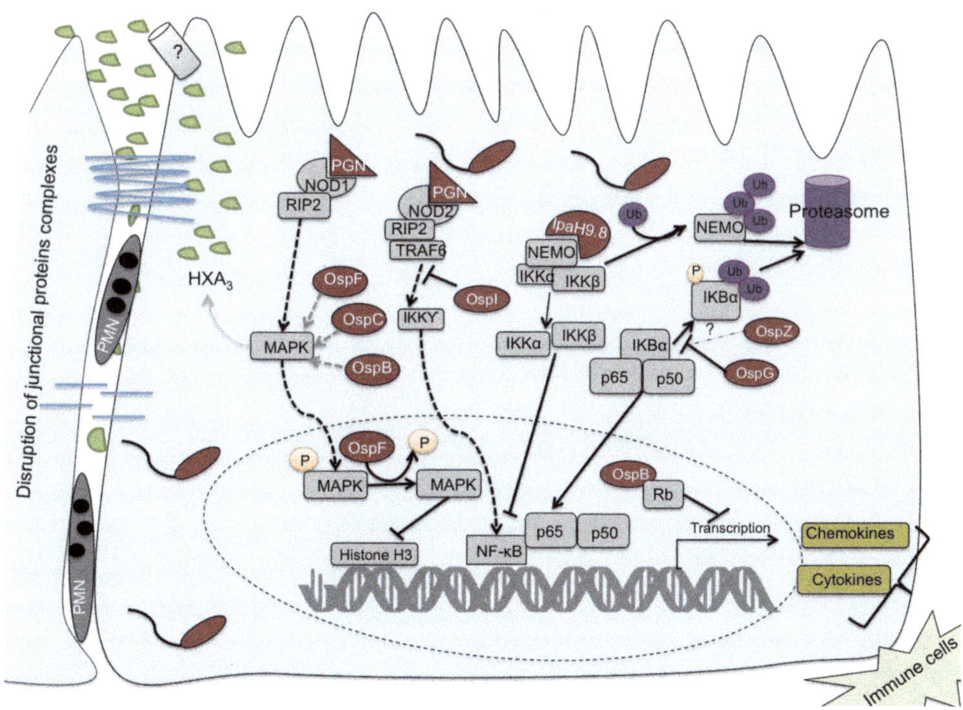

Figure 9.2. Host-pathogen communication. Schematically illustrated in the interaction between *Shigella*'s T3SS secreted effectors and host proteins. The *Shigella* secreted effectors coopt host signal transduction pathways during infection. Interrupted arrows means: activation; continued arrows refers to induction.

transmembrane domains: two extracellular domains, a short cytosolic N-terminus, and a longer cytosolic C-terminus that includes a PDZ-binding domain (Furuse et al., 1998). Inside the cells, the C-terminus of claudins interact via PDZ domains with zonula occludens proteins (ZO1 and ZO2), which in turn bind directly to actin filaments (Fanning et al., 1998; Itoh et al., 1999; Wittchen et al., 1999). ZO proteins are membrane-associated guanylate kinase-like homologues (Anderson, 1996) that cross-link tight junction strands to actin filaments. Electron microscopy studies uncovered that the tight junctions are continuous structures, forming complexes with tight junctions from the membranes of adjacent cells, limiting the intercellular space (Staehelin, 1974). Thus, the tight junction complex is the rate-limiting step for paracellular solute flux and is the principal determinant of epithelial mucosa permeability.

Subjacent to the tight junction complex, adherent junctions and desmosomes provide a strong adhesive tie that governs the proximity between cells and mediates cell-to-cell interactions. Adherens junctions are comprised of members of the cadherin family. Cadherins are transmembrane proteins that form a strong bond through homotypic interactions with molecules on adjacent cells. Epithelial (E) cadherin is found throughout the entire intestinal epithelium. The cytoplasmic domain of E-cadherin is linked to catenin δ1 (also referred as p120) and β-catenin. In turn, β-catenin binds to α-catenin and also interacts with actin and several actin-binding proteins (Kobielak and Fuchs, 2004), thereby mediating local actin assembly to form the peri-junctional actomyosin ring. Thus, adherent junctions play an essential role in cell-to-cell and cell-to-matrix interactions, effective polarization and differentiation of epithelial cells, regulation of apoptosis, and tissue morphogenesis (Halbleib and Nelson, 2006; Hermiston and Gordon, 1995).

Several studies have identified an association between intestinal barrier function and disease. Patients with inflammatory bowel disease, for instance exhibit a decrease in the number and complexity of tight junction strands due to changes in expression and distribution of claudin 1, 2, 5 and 8 (Schmitz et al., 1999; Weber et al., 2008; Zeissig et al., 2007). In addition, the integrity of the barrier function can be modulated in response to bacterial infection. Some bacterial products such as *Clostridium difficile* toxins (Nusrat et al., 1995; Nusrat et al., 2001), *Clostridium perfringens* enterotoxin (Sonoda et al., 1999), cytotoxic necrotizing factor 1 from *Escherichia coli* (Oswald et al., 1994), *Bordetella dermonecrotic* toxin (Horiguchi et al., 1995), and zonula occludens toxin from *Vibrio cholerae* (Fasano *et al.*, 2000) can alter actin polymerization and cytoskeletal organization, thus triggering a barrier function defect at the tight junction complex.

Likewise, Enteropathogenic *Escherichia coli* (EPEC) infection causes permeability disruption by increasing myosin light chain kinase (MLCK) phosphorylation (Yuhan et al., 1997). Phosphorylated MLCK promotes cytoskeletal contraction presumably by causing increased tension on the zonula occludens localized in the cell membrane. Moreover, EPEC can cause de-phosphorylation and dissociation of occludin from the tight junction complex, which also correlates with decreases in the transepithelial electrical resistance (TER) of cell monolayers in *in vitro* models (Simonovic et al., 2000). Infection with Enterohaemorrhagic *E. coli* and Enteroaggregative *E. coli* (EHEC and EAEC, respectively) is also associated with epithelial barrier dysfunction (Philpott et al., 1998; Strauman et al., 2010)

Intestinal barrier function and *Shigella* infection

Like the pathogenic organisms described above, *Shigella spp* infection, such as *S. flexneri*, also promotes deregulation of the barrier function. Once an individual gets infected with contaminated water or food containing *Shigella flexneri*, the bacteria survive the passage through the hostile acidic environment of the stomach to reach the colonic intestinal epithelium. Here, *Shigella* must trespass the epithelial barrier in order to gain access to the basolateral domain of the epithelium. Upon contacting the colonic epithelium, *Shigella* activates its type III secretion system (T3SS), deploying an armament of effector proteins that leads to the internalization of the bacterial into the epithelial cells.

It has long been known that *S. flexneri* invades the intestinal epithelium from the basolateral surface, but how *Shigella* gets there is incompletely understood. Infections carried out in rabbit ligated ileal loops put forward the notion that M cells located over mucosal lymphoid follicles are a major entry site at early stages of *Shigella* infection (Sansonetti et al., 1991). M cells are specialized epithelial cells localized in the mucosa-associated lymphoid tissue, and their main function is to sample microorganisms and any other macromolecules from the lumen and transport them, via transcytosis, into the subepithelial dome region where macrophages and other immune cells of the lymphoid tissue reside. Upon entering the dome region, *Shigella* is taken up into macrophages, inducing apoptosis of this immune cell (Zychlinsky et al., 1992; Zychlinsky et al., 1996), and is eventually released. This event not only allows the bacterial to avoid elimination but also promotes the secretion of inflammatory cytokines, such as IL-1β and IL-8 by apoptotic macrophages. However, given the vast epithelial surface area relative to the M cell surface area in the human large intestine (far less than 1%), it is likely that other mechanisms allowing access to the basolateral membrane domain exist (Perdomo et al., 1994).

Perdomo *et al.*, raised the possibility that transmigration of PMN across the intestinal epithelium may play an active role during the initial stages of invasion by opening the paracellular pathway for bacterial entry into differentiating colonocytes (Perdomo et al., 1994). As an alternative tenet, we have further proposed that *S. flexneri* alters the intestinal barrier function (i.e., the tight junctional complex), to facilitate paracellular passage of the organism across the intestinal epithelial barrier (Sakaguchi et al., 2002). In studying the regulation of tight junction proteins during the course of infection we have shown that infection with only wild type *Shigella* was capable of causing a decrease in expression of membrane-bound claudin-1 (Sakaguchi et al., 2002). Such a decrease in claudin-1 expression was observed as early as 10 min after infection, and partial recovery of expression was detected after 90 min of infection. We also observed other changes in the tight junction composition. After 60 min of *S. flexneri* infection, dephosphorylation of membrane-bound occludin occurs. This dephosphorylation event is also correlated with removal of occludin from the tight junction seal, since a parallel increase in cytosolic non-phosphorylated occludin was also observed. In addition, *S. flexneri* infection from the apical domain of epithelial cells had a profound effect on ZO-1 expression, causing the protein to completely disappear from the cell membrane fractions. Moreover, both ZO-2 and E-cadherin were recruited from the cytosol pool to the cell membrane fractions following *S. flexneri* infection. Thus, our studies put forth the concept that *S. flexneri* has evolved the ability to regulate functional components of tight junctions at the apical pole of model intestinal epithelia (Sakaguchi et al., 2002) to provide access to the paracellular space, hence promoting *Shigella* invasion.

At the basolateral domain

Once *Shigella* trespasses the epithelial barrier it gains access to the basolateral mileu from where invasion into the epithelium commences. S. *flexneri* possesses a 220-Kb. plasmid that harbors most of its virulence determinants (Buchrieser et al., 2000;

Sansonetti et al., 1982). A 31 Kb. region of this plasmid, termed the *mxi-spa* locus, encodes the proteins necessary to assemble a type III secretion system (T3SS) (Parsot et al., 1995; Sasakawa et al., 1993). The T3SS is a needle-like complex, which upon contact of *Shigella* with epithelial cells, is directed to secrete effector proteins. These effectors proteins govern subsequent pathogenesis by facilitating bacterial entry and expansion, suppressing the host immune response, and modulating host cell biology (Bruno et al., 2009; Phalipon and Sansonetti, 2007; Simpson et al., 2006).

The genes necessary for bacterial entry are also encoded on the *mxi-spa* locus: IpaA, IpaB, IpaC, IpgB1, IpgD and VirA are involved in triggering F-actin and microtubule cytoskeletons rearrangement, and promoting bacterial uptake by host cells. In concert, these bacterial effector proteins orchestrate (See review (Ogawa and Sasakawa, 2006)): Changes in actin dynamics through nucleation and depolymerization, in addition microtubule degradation around the bacterial entry site.

Membrane ruffles in epithelial cells via: i) stimulation of Rac1, Cdc42, RhoA, and phosphorylation of cortactin and Crk by Abl-family tyrosin kinases, and ii) accumulation of c-Src. Thus, the intricate interplay between *Shigella* effectors proteins, their target host proteins, and Rho-GTPases and tyrosine kinases, is fundamental in promoting bacterial phagocytosis. In a separate chapter, the role of each effector protein is discussed in detail.

Inside the cell

Avoiding autophagy

Autophagy is a catabolic mechanism that involves cell degradation of unnecessary or dysfunctional cellular components through the lysosomal machinery (from the Greek words, *auto* "self" and *phagein* "to eat"). The breakdown of cellular components can ensure cellular survival during starvation by maintaining cellular energy levels. Thus, autophagy, if regulated, ensures the synthesis, degradation and recycling of cellular components. Cells also can employ autophagy as means to eliminate foreign intruders, such as microbes.

Eradication of such unwanted intruders by autophagy is initiated by encapsulation of the bacterial cell by membrane-bound organelles called phagophores that elongate to form a double membrane vacuole called autophagosomes. These structures ultimately fuse with lysosomes (autolysosome), resulting in degradation and processing of the encapsulated bacteria. Autophagosome initiation, elongation, and maturation depend on the expression of more than thirty autophagy related (ATG) proteins. The mechanism underlying bacterial recognition by autophagy is still unknown. However, autophagy receptors recognize ubiquitinated substrates, and have also been shown to be involved in the recruitment of membranes to form autophagosomes through interaction with ATG family proteins (Guarner and Malagelada, 2003; Swidsinski et al., 2002).

In the case of *Shigella* infection, once the bacteria escape the vacuole compartment and reach the cytoplasm, polarized expression of IcsA (VirG) occurs. IcsA promotes actin nucleation and polymerization, thereby forming actin-tails that confers a propulsive force that enables the movement of the bacteria within the cytosol. Autophagy of *Shigella* starts with recognition of IcsA by ATG5 (Ogawa et al., 2005). However, *Shigella* is able to subvert autophagy given its ability to multiply freely in the cytoplasm and to disseminate to adjacent cells due to secretion of IcsB via the type III secretion system. IcsB competitively binds to IcsA, thus avoiding any further interaction between IcsA and ATG5

(Ogawa et al., 2005). Mutated *Shigella* strains that lack the *icsB* gene are trapped by autophagy within the epithelial cells (Ogawa et al., 2005). It is now known that the targeting of IcsA by ATG5 occurs at one pole of the bacterial surface where IcsA is being expressed and this initiates phagophore formation, suggesting that ATG5 acts as a cargo receptor protein in selective autophagy of *Shigella* (Ogawa et al., 2005).

More recently, Tectonin domain-containing protein (Tecpr1) and WIPI-2, both ubiquitously expressed in humans, have also been found to play a key role in *Shigella's* selective autophagy. In one study, in cells infected with *ΔiscB Shigella*, Tecpr1 and WIPI-2 were found to co-localize with LC-3 (a specific autophagy marker) and ATG5 of autophagosomes that had surrounded *Shigella.* Furthermore, in Tecpr1-deficient mice, embryonic fibroblasts were defective for selective autophagy, and intracellular multiplication of *ΔiscB Shigella* was increased compared to wild-type mice. In WIPI-2 knockdown cells, the association between Tecpr1 and phagophores was impaired (Ogawa et al., 2011). Thus, a complicated interplay between WIPI-2, Tecpr1, and ATG5 may be required for targeted autophagy of *Shigella* that lack IcsB.

Cell-to-cell spreading

Upon escaping autophagy, *Shigella* freely multiplies within the cytoplasm and spreads from cell to cell. The steps for bacterial spreading are comprised of: i) Actin nucleation and elongation to form the actin tail at one pole of the bacterial cell, mediated by IcsA; ii) Actin-based bacterial intracellular movement to approach and engage junctional proteins; iii) Bacterial impingement into the neighboring host-cell causing protrusion formation; iv) Uptake of the protrusion-containing bacteria by the adjacent cell via endocytosis; and v) Lysis of the double membrane surrounding the bacteria. The newly liberated bacteria will proliferate in the cytoplasm and spread to the neighboring, repeating this cycle over and over.

Shigella's ability to spread depends on IcsA, which play a crucial role in the recruitment of actin. Expression of IcsA is polarized to one pole of the organism and interacts with vinculin (Suzuki et al., 1996) and N-WASP (neural Wiskott-Aldrich syndrome Protein) (Egile et al., 1999). Vinculin is an actin-binding cytoskeletal protein, which during *Shigella* infection has been shown to accumulate around polar IcsA, specifically to the exposed domain of the bacterial effector. This observation suggests that accumulation of vinculin might aid recruitment of F-actin to IcsA (Suzuki et al., 1996). Moreover, IcsA also interacts with N-WASP and activates it in a Cdc42-like manner. N-WASP is a member of the WASP family required to mediate actin polymerization by interacting with the Arp2/3 complex. N-WASP interacts with the glycine-rich repeat domain of IcsA, an indispensable domain for *Shigella* recruitment of F-actin. Furthermore, IcsA-N-WASP recruits and activates the Arp2/3 complex, upon which the IcsA-N-WASP-Arp2/3 complex induces nucleation and elongation of the actin tail (Egile et al., 1999; Suzuki et al., 2000).

In addition to the host factors described above, other factors are found to be involved in the actin-based motility of *Shigella*. These include plastin (Prevost et al., 1992), filamin (Prevost et al., 1992), α-actinin (Gouin et al., 1999; Tran Van Nhieu et al., 1997; Zeile et al., 1996), vasodilator-simulated phosphoprotein (Goldberg, 1997), ezrin, cofilin, capping protein, profilin and CapZ (Gouin et al., 1999; Loisel et al., 1999). These proteins co-localize either with the actin tail or at the posterior end of intracellular *Shigella*. The underlying mechanism by which most of these host factors are activated by bacterial effector proteins remains to be determined. However, their activation promotes *Shigella's* propulsive force, harnessed in the form of actin recruitment at one pole of the bacterial cell to create a protrusion (or finger-like projection) into the adjacent cell.

A robust body of literature supports the hypothesis that spreading of *Shigella* between cells requires intact adherent junctional protein complexes. Rathman *et al.* (2000) showed that pharmacological inhibition of MLCK blocks dissemination of intracellular *Shigella* but neither its replication nor the actin-tail formation inside the epithelial cell. MLCK is responsible for phosphorylation of Myosin II, an ATP-dependent protein associated with membrane events such as endocytosis, phagocytosis, and formation of cellular protrusions. MLCK activation is required for the formation of *Shigella*-containing protrusions, and even when the bacterial cells were localized close to the adjacent cell, the protrusion was rarely initiated when MLCK activity was inhibited (Rathman et al., 2000). This implies a role for activated Myosin II in the formation of protrusions in *Shigella*-infected cells.

Additional host proteins have been found to play a role in bacterial dissemination. Sansonetti *et al.* (1994) showed that cadherins are essential for the spreading of *S. flexneri*. In this study, wild type- and cadherin-null epithelial cell lines were infected with *S. flexneri*. In the cell line expressing cadherin, *Shigella* was rapidly internalized and efficiently translocated from cell to cell. In contrast, cells not expressing cadherin did not allow cell-to-cell spread of the bacteria. As previously described, cadherins concentrate within the junctional space between cells, forming complexes that are tightly associated with actin filaments. It has been shown that a *S. flexneri* IcsA mutant strain (that neither recognizes nor polarizes actin) is unable to spread from one cell to another (Sansonetti et al., 1994), thus confirming that spreading of *S. flexneri* infection is based on actin engagement, which may be mediated by interaction with cadherins within the paracellular space.

In another study, connexin proteins were also associated with *Shigella* movement between adjacent epithelial cells (Tran Van Nhieu et al., 2003). Connexins are four-pass transmembrane proteins arranged in hexamers that form hemichannels. These channels change conformation to open or close allowing/preventing the passage of calcium, ATP, and other second messengers (Goodenough and Paul, 2009). Tran Van Nhieu *et al.* (2003) demonstrated that cell lines stably expressing connexin 26 exhibited significantly higher levels of bacterial dissemination foci than the non-expressing parental cell line. Consequently, blockade of this gap-junction protein has a severe adverse effect on bacterial dissemination (Tran Van Nhieu et al., 2003). This study also demonstrated that in adjacent cells, *Shigella* infection triggers the release of ATP throughout the opened connexin 26 hemichannels. This resulted in oscillatory levels of intracellular calcium in neighboring cells. The increase of released ATP was associated with membrane ruffling, which facilitates *Shigella* uptake. Extracellularly applied ATP resulted in a threefold increase in the level of *Shigella* dissemination (Tran Van Nhieu et al., 2003).

More recently, tricellulin, a protein localized at the junction between three neighboring intestinal epithelial cells, appears to play a more important role in *Shigella* cellular spreading than either E-cadherin or occludin (Fukumatsu et al., 2012a). The intestinal epithelium forms a honeycomb-like net where cell-cell contact can occur between: i) two adjacent cells (interaction modulated by bicellular tight junctions); and ii) three or four adjacent cells (governed by tricellular tight junctions). These tricellular tight junctions contain tricellulin, a recently discovered transmembrane protein (Ikenouchi et al., 2005). Tricellulin is involved in regulating the permeability of macromolecules at the epithelium

(Ikenouchi et al., 2005; Krug et al., 2009), and it also plays a role in maintaining the integrity of tight junction strands in the plasma membrane (Ikenouchi et al., 2008). To some extent, tricellulin can also be found at the bicellular junctions. Fukumatsu *et al.* (2012) reported that *Shigella* predominately navigates between adjacent cells at the tricellular contact points. Moreover, knockdown of tricellulin (and to a lesser extent of E-cadherin or occluding), all reduced bacterial dissemination (Fukumatsu et al., 2012a). Furthermore, the ability of *Shigella* to spread requires phosphoinositide (PI) 3-kinase, a protein involved in membrane surface remodeling and cytoskeleton dynamics (Fukumatsu et al., 2012b).

Interplay of Shigella *effector proteins and inflammatory host signals*
The ability of the host to defend against bacterial infection is crucial for limiting colonization and eventual cell death. Thus, the host responds by mounting an immune response aimed at eliminating the invading organisms. However, all pathogens have to successfully overcome these broad immune mechanisms, by having evolved distinct strategies to avoid or even exploit the host responses for their own benefit.

In the case of *S. flexneri*, the invasion of the colonic epithelium causes an intense inflammatory response and the acute recto-colitis that follows is a paradigm of the disease process. Histopathological analysis of infected patients reveals an apparent demise of the epithelium, along with the characteristic signs of inflammation exhibited most profoundly by massive infiltration and exudation of PMN (Mathan and Mathan, 1986). Inflammatory effects of *S. flexneri* infection can occur when the bacteria is becomes engulfed by macrophages where release of various pathogen-associated molecular patterns (PAMPs) and T3SS effector proteins occur.

These components, such as lipopolysaccharides (LPS), peptidoglycans (PGN) and many T3SS-delivered effectors proteins, are then recognized by Nod-like receptors that form part of the inflammasome protein complex (namely, NLRC4 and NALP3), provoking the production of interleukins, specifically IL-1β and IL-18, and pyroptosis (Ashida et al., 2011a). Pyroptosis refers to a form of necrotic cell death related to inflammation dependent on caspase-1 expression, nuclear condensation, and damage to the plasma membrane. At early time points following *S. flexneri* infection, NLRC4 recognizes the basal body component of the T3SS apparatus, in particular MxiI, causing macrophages to undergo caspase-1-dependent pyroptosis (Miao et al., 2010). However, necrosis-like caspase-1-independent cell death, termed pyronecrosis, is also induced during *Shigella* infection and involves recognition of PAMPs by NLRP3 (Willingham et al., 2007).

The interplay between bacterial effectors, PAMPs, and host receptors leads to macrophage cell death and release of *Shigella* into the basolateral space. Upon internalization of *Shigella* into the epithelium, PAMPs and danger-associated molecular patterns (DAMPs) are detected by Toll-like receptors (TLRs) and Nod-like receptors (NLRs), along with components of the autophagic machinery (as described above). Sensing of *Shigella* PAMPs and DAMPs triggers inflammatory signaling cascades involving mitogen activated protein kinases (MAPK) and nuclear factor κB (NF-κB) that result in the expression of pro-inflammatory cytokines, chemokines (e.g. IL-8 and TNF-α), and antimicrobial peptides (e.g. β-defensins and LL37) (Ashida et al., 2011b; Takeuchi and Akira, 2010). An overview of the inflammatory network provoked by *Shigella* infection is presented in Figure 9.1. Here, we illustrate that Nod 1 and Nod 2 receptors recognize *Shigella*-released PGN, which leads to the downstream activation of MAPK and NF-κB, and with the help of additional adapters/chaperones fosters an initial immune response against *Shigella*. As a consequence of this network of interactions, phagocytic cells including neutrophils, macrophages, and dendritic cells are recruited to the site of infection. It should be emphasized that at the initial stages of *Shigella* infection, paracellular migration of neutrophils actually contributes to bacterial invasion due to disruption of the cell-to-cell junctional complex (Perdomo et al., 1994).

Despite the occurrence of exacerbated inflammation and immune signaling, *S. flexneri* is able to avoid elimination and instead manipulates the expression of host proteins for its own benefit by injecting effector proteins into the cytoplasm that targets host cell signal transduction pathways, and host cell transcription. Thus, both the pathogen and the host are in combat to establish the appropriate balance between pro and anti-inflammatory signals to favor invasion/pathogenesis or neutralization/clearance, respectively. Regulation of inflammatory pathways, including MAPKs and NF-κB activation, is mainly achieved by causing post-transcriptional modifications (e.g., phosphorylation and ubiquitination). In addition to the effectors involved in facilitating invasion, *Shigella* relies on another set of effectors, namely members of the multigene *ospB-G* (outer membrane proteins) and *ipaH* families, to carry out post-transcriptional modifications to components of the host immune response.

MAPKs and ERK are components of cell signaling pathways by which host cells respond to extracellular stimuli through targeting of transcription factors, resulting in the modulation of gene expression. During *Shigella* multiplication within the intestinal epithelium, NOD1 senses bacteria-released PGN. This, in turn activates the NOD–RIP2 pathway and the downstream MAPK and NF-κB signaling network that up-regulates the transcription of IL-8. However, *Shigella* effector proteins can also be translocated to the nucleus causing gene repression of inflammatory cytokines via histone re-modeling (Reiterer et al., 2011; Zurawski et al., 2008). These disparate effects of *Shigella* signals do not occur simultaneously but rather in different stages of the infection (early vs. later stages of infection).

Pro-inflammation signaling

The Osp effectors (OspF, OspC1, OspB, and OspZ) are injected into the host cell to activate MAPK/ERK signaling pathways, thereby evoking a massive transepithelial infiltration of PMN, a hallmark of shigellosis (Zurawski et al., 2006; Zurawski et al., 2008). Such extensive inflammation causes epithelial destruction that further facilitates bacterial pathogenesis at the initial stages of infection.

Activation of the MAPK/ERK pathway is also required for $cPLA_2$ activation and the production of hepoxilin A_3 (HXA_3), a potent chemoattractant that ultimately promotes PMN transepithelial migration (Mumy et al., 2008). Cells infected with *Shigella* strains lacking *ospB*, *ospC1*, *ospF* or *ospZ* genes exhibit a severe decrease in PMN transpithelial migration compared with cells infected with wild type *Shigella* (Zurawski et al., 2006; Zurawski et al., 2008). As wild type infection progresses, increased tight junction disruption caused by the infiltration of PMN into the lumen leads to acute inflammation and epithelial destruction which is associated with diarrhea.

Anti-inflammation signaling

At late stages of infection, OspF also appears to facilitate anti-inflammatory responses by translocating to the nucleus to irreversibly dephosphorylate MAPKs, specifically p38 and ERK, thereby inhibiting histone H3-phosphorylation, leading to a repression in a subset of NF-κB regulated genes, including IL-8 (Arbibe et al., 2007; Li et al., 2007; Zurawski et al.,

2009). Likewise, OspB has been observed to localize in the nucleus where it mediates chromatin remodeling, resulting in reduced transcription of inflammatory cytokines. By reducing inflammation, it is inferred that *Shigella* may prevent premature clearance from the host in an attempt to prolong its proliferation.

Another set of effector proteins known to target molecules and pathways that inhibit the NF-κB pathway are OspG, OspI, and IpaH. OspG is a protein kinase that binds UbcH5, thereby promoting ubiquitination of the NF-κB inhibitor type alpha (phospho-IκBα) by E3 ubiquiting ligase $SCF^{\beta\text{-}TrCP}$. NF-κB activation is tightly regulated by the IKK (IκB kinase) complex and by IκBα degradation. Degradation of IκBα by OspG impedes NF-κB activation (Figure 9.1) (Kim et al., 2005). As an example, *in vivo* experiments have shown that infection with an *ospG Shigella* mutant induces destruction of the mucosa to a greater extent than infection with the wild-type strain, exhibited by more necrotic areas, complete disappearance of villi, extensive hemorrhages, and massive infiltrates of PMN (Kim et al., 2005).

OspI is also involved in the inhibition of NF-κB (Sanada et al., 2012), which overall leads to a reduced inflammatory response. This effector is crucial for down-regulation of chemokine and chemokine receptors, cytokines, IκBα phosphorylation, and NF-κB nuclear translocation factor (p65). Furthermore, OspI is a glutamine deamidase, which through deamidation of UBC13 inhibits TRAF6 activation. TRAF6 is an E3 ubiquitin ligase capable of promoting NF-kB activation. Thus, OspI through inactivation of TRAF6 also restricts NF-kB activation (Figure 9.1) (Sanada et al., 2012).

The IpaH family of proteins comprises yet another subset of effectors that regulate inflammation in response to *Shigella* infection. Structurally, all IpaH proteins exhibit an N-terminus leucine rich repeat and a highly conserved C-terminal region. Functionally, the IpaH family has been shown to possess a conserved E3 ubiquitin ligase activity within the C-terminal region (Rohde et al., 2007). Inside the host cell, *Shigella* secretes IpaH9.8, which is subsequently translocated to the nucleus. While in the cytoplasm, IpaH9.8 avoids NF-kB activation via its E3 ubiquitin-ligase activity. As previously described, IKK is a key regulator of NF-kB signaling. IKK activation depends on polyubiquitin post-transcriptional modification of NEMO. IpaH9.8 targets the IKK/NEMO complex, thereby eliciting aberrant NEMO polyubiquitination causing the complex to be recognized for proteosomal degradation, thus promoting a decrease in NF-kB activity (Ashida et al., 2010). More recently, the IpaH family member IpaH0722 has been associated with impairing the severe inflammatory response against *Shigella.* IpaH0722 inhibits protein kinase C-mediated activation of NF-κB by promoting proteasome-dependent degradation TRAF2 through ubiquitination targeting; TRAF2 is molecule downstream PKC-dependent activation of NF-κB (Ashida *et al.*, 2013).

Location, location, location

In an effort to reconcile the pro- and anti-inflammatory properties of OspB and OsF, we found that these two effectors possess distinct activities in different compartments of the host cell (cytoplasm vs. nucleus) (Zurawski et al., 2009). When OspF enters the cell, it first encounters the cytoplasm and phosphorylates ERK1/2. This event triggers a pro-inflammatory program that leads to the release of HXA_3, which is secreted apically by intestinal epithelial cells and subsequently forms a paracellular gradient through the tight junctional complex that prompts the guidance of PMNs across the epithelium (McCormick, 2007). Later on, phosphorylated ERK and the OspF and OspB effectors enter the nucleus. In the case of OspB, the N-terminus is required for nuclear transport whereas OspF associates with microtubules, ERK1/2, and retinoblastoma protein (Zurawski et al., 2009). OspB associates directly with Rb via an unknown binding site or intermediary protein. Once there, OspB/F recruit factors such as histone deacetylases, which decreases IL-8 production as well as expression of other pro-inflammatory mediators via chromatin remodeling. Support for this model is further substantiated by the observation that OspF binding to retinoblastoma protein is not a prerequisite for induction of PMN migration.

Similarly, OspZ has been found to have both pro- and anti-inflammatory effects. The mechanism utilized by OspZ to up-regulate inflammation remains unclear but it has been reported that infection with *Shigella flexneri* 2a strains lacking the *ospZ* gene exhibit reduced PMN transepithelial migration and NF-κB activation compared to infection with the wild type (Zurawski et al., 2008). In contrast, OspZ from *Shigella flexneri 6* strains have been associated with restriction of nuclear translocation of p65, while allowing nuclear import of a transcriptional repressor that suppresses of IL-8 expression (Newton et al., 2010). The NF-κB subunit p65 is required for the activation of transcription from NF-κB-binding sites in target genes. Thus, by preventing p65 nuclear translocation, OspZ causes NF-κB inhibition. This apparent discrepancy is explained by the fact that the two serotypes used in the different studies - *S. flexneri 2a* and *S. flexneri 6* - have different versions of OspZ. *S. flexneri 2a* possesses a truncated OspZ protein lacking the functional domain for NF-κB suppression. Establishing the prevalence of these different OspZ isotypes among *Shigella flexneri* isolates and their resulting phenotypes may aid in determining the overall function of OspZ in *Shigella* pathogenesis.

References

Anderson, J.M. (1996). Cell signalling: MAGUK magic. Curr Biol *6*, 382-384.

Arbibe, L., Kim, D.W., Batsche, E., Pedron, T., Mateescu, B., Muchardt, C., Parsot, C., and Sansonetti, P.J. (2007). An injected bacterial effector targets chromatin access for transcription factor NF-kappaB to alter transcription of host genes involved in immune responses. Nat Immunol *8*, 47-56.

Ashida, H., Kim, M., Schmidt-Supprian, M., Ma, A., Ogawa, M., and Sasakawa, C. (2010). A bacterial E3 ubiquitin ligase IpaH9.8 targets NEMO/IKKgamma to dampen the host NF-kappaB-mediated inflammatory response. Nat Cell Biol *12*, 66-73; sup pp 61-69.

Ashida, H., Nakano, H., and Sasakawa, C. (2013). Shigella IpaH0722 E3 ubiquitin ligase effector targets TRAF2 to inhibit PKC-NF-kappaB activity in invaded epithelial cells. PLoS Pathog *9*, e1003409.

Ashida, H., Ogawa, M., Kim, M., Suzuki, S., Sanada, T., Punginelli, C., Mimuro, H., and Sasakawa, C. (2011a). Shigella deploy multiple countermeasures against host innate immune responses. Curr Opin Microbiol *14*, 16-23.

Ashida, H., Ogawa, M., Mimuro, H., Kobayashi, T., Sanada, T., and Sasakawa, C. (2011b). Shigella are versatile mucosal pathogens that circumvent the host innate immune system. Curr Opin Immunol *23*, 448-455.

Bruno, V.M., Hannemann, S., Lara-Tejero, M., Flavell, R.A., Kleinstein, S.H., and Galan, J.E. (2009). Salmonella Typhimurium type III secretion effectors stimulate innate immune responses in cultured epithelial cells. PLoS pathogens *5*, e1000538.

Buchrieser, C., Glaser, P., Rusniok, C., Nedjari, H., D'Hauteville, H., Kunst, F., Sansonetti, P., and Parsot, C. (2000). The virulence plasmid pWR100 and the repertoire of proteins secreted by the type III secretion apparatus of Shigella flexneri. Mol Microbiol *38*, 760-771.

Egile, C., Loisel, T.P., Laurent, V., Li, R., Pantaloni, D., Sansonetti, P.J., and Carlier, M.F. (1999). Activation of the CDC42 effector N-WASP by the Shigella flexneri IcsA protein promotes actin nucleation by Arp2/3 complex and bacterial actin-based motility. J Cell Biol *146*, 1319-1332.

Fanning, A.S., Jameson, B.J., Jesaitis, L.A., and Anderson, J.M. (1998). The tight junction protein ZO-1 establishes a link between the transmembrane protein occludin and the actin cytoskeleton. J Biol Chem *273*, 29745-29753.

Fasano, A., Not, T., Wang, W., Uzzau, S., Berti, I., Tommasini, A., and Goldblum, S.E. (2000). Zonulin, a newly discovered modulator of intestinal permeability, and its expression in coeliac disease. Lancet *355*, 1518-1519.

Fukumatsu, M., Ogawa, M., Arakawa, S., Suzuki, M., Nakayama, K., Shimizu, S., Kim, M., Mimuro, H., and Sasakawa, C. (2012a). Shigella targets epithelial tricellular junctions and uses a noncanonical clathrin-dependent endocytic pathway to spread between cells. Cell Host Microbe *11*, 325-336.

Fukumatsu, M., Ogawa, M., Kim, M., Mimuro, H., and Sasakawa, C. (2012b). Uptake of Shigella-containing pseudopodia by neighboring epithelial cells at tricellular junctions via non-canonical clathrin-dependent trafficking pathway. Virulence *3*, 515-518.

Furuse, M., Fujimoto, K., Sato, N., Hirase, T., and Tsukita, S. (1996). Overexpression of occludin, a tight junction-associated integral membrane protein, induces the formation of intracellular multilamellar bodies bearing tight junction-like structures. J Cell Sci *109 (Pt 2)*, 429-435.

Furuse, M., Fujita, K., Hiiragi, T., Fujimoto, K., and Tsukita, S. (1998). Claudin-1 and -2: novel integral membrane proteins localizing at tight junctions with no sequence similarity to occludin. J Cell Biol *141*, 1539-1550.

Goldberg, M.B. (1997). Shigella actin-based motility in the absence of vinculin. Cell Motil Cytoskeleton *37*, 44-53.

Goodenough, D.A., and Paul, D.L. (2009). Gap junctions. Cold Spring Harb Perspect Biol *1*, a002576.

Gouin, E., Gantelet, H., Egile, C., Lasa, I., Ohayon, H., Villiers, V., Gounon, P., Sansonetti, P.J., and Cossart, P. (1999). A comparative study of the actin-based motilities of the pathogenic bacteria Listeria monocytogenes, Shigella flexneri and Rickettsia conorii. J Cell Sci *112 (Pt 11)*, 1697-1708.

Guarner, F., and Malagelada, J.R. (2003). Gut flora in health and disease. Lancet *361*, 512-519.

Halbleib, J.M., and Nelson, W.J. (2006). Cadherins in development: cell adhesion, sorting, and tissue morphogenesis. Genes Dev *20*, 3199-3214.

Hermiston, M.L., and Gordon, J.I. (1995). In vivo analysis of cadherin function in the mouse intestinal epithelium: essential roles in adhesion, maintenance of differentiation, and regulation of programmed cell death. J Cell Biol *129*, 489-506.

Horiguchi, Y., Senda, T., Sugimoto, N., Katahira, J., and Matsuda, M. (1995). Bordetella bronchiseptica dermonecrotizing toxin stimulates assembly of actin stress fibers and focal adhesions by modifying the small GTP-binding protein rho. J Cell Sci *108 (Pt 10)*, 3243-3251.

Ikenouchi, J., Furuse, M., Furuse, K., Sasaki, H., and Tsukita, S. (2005). Tricellulin constitutes a novel barrier at tricellular contacts of epithelial cells. J Cell Biol *171*, 939-945.

Ikenouchi, J., Sasaki, H., Tsukita, S., and Furuse, M. (2008). Loss of occludin affects tricellular localization of tricellulin. Mol Biol Cell *19*, 4687-4693.

Itoh, M., Furuse, M., Morita, K., Kubota, K., Saitou, M., and Tsukita, S. (1999). Direct binding of three tight junction-associated MAGUKs, ZO-1, ZO-2, and ZO-3, with the COOH termini of claudins. J Cell Biol *147*, 1351-1363.

Kim, D.W., Lenzen, G., Page, A.L., Legrain, P., Sansonetti, P.J., and Parsot, C. (2005). The Shigella flexneri effector OspG interferes with innate immune responses by targeting ubiquitin-conjugating enzymes. Proceedings of the National Academy of Sciences of the United States of America *102*, 14046-14051.

Kobielak, A., and Fuchs, E. (2004). Alpha-catenin: at the junction of intercellular adhesion and actin dynamics. Nat Rev Mol Cell Biol *5*, 614-625.

Krug, S.M., Amasheh, S., Richter, J.F., Milatz, S., Gunzel, D., Westphal, J.K., Huber, O., Schulzke, J.D., and Fromm, M. (2009). Tricellulin forms a barrier to macromolecules in tricellular tight junctions without affecting ion permeability. Mol Biol Cell *20*, 3713-3724.

Ley, R.E., Peterson, D.A., and Gordon, J.I. (2006). Ecological and evolutionary forces shaping microbial diversity in the human intestine. Cell *124*, 837-848.

Li, H., Xu, H., Zhou, Y., Zhang, J., Long, C., Li, S., Chen, S., Zhou, J.M., and Shao, F. (2007). The phosphothreonine lyase activity of a bacterial type III effector family. Science *315*, 1000-1003.

Loisel, T.P., Boujemaa, R., Pantaloni, D., and Carlier, M.F. (1999). Reconstitution of actin-based motility of Listeria and Shigella using pure proteins. Nature *401*, 613-616.

Mathan, M.M., and Mathan, V.I. (1986). Ultrastructural pathology of the rectal mucosa in *Shigella* dysentery. Am J Pathol *123*, 25-38.

McCarthy, K.M., Skare, I.B., Stankewich, M.C., Furuse, M., Tsukita, S., Rogers, R.A., Lynch, R.D., and Schneeberger, E.E. (1996). Occludin is a functional component of the tight junction. J Cell Sci *109 (Pt 9)*, 2287-2298.

McCormick, B.A. (2007). Bacterial-induced hepoxilin A3 secretion as a pro-inflammatory mediator. Febs J *274*, 3513-3518.

Miao, E.A., Mao, D.P., Yudkovsky, N., Bonneau, R., Lorang, C.G., Warren, S.E., Leaf, I.A., and Aderem, A. (2010). Innate immune detection of the type III secretion apparatus through the NLRC4 inflammasome. Proc Natl Acad Sci U S A *107*, 3076-3080.

Mumy, K.L., Bien, J.D., Pazos, M.A., Gronert, K., Hurley, B.P., and McCormick, B.A. (2008). Distinct isoforms of phospholipase A2 mediate the ability of *Salmonella enterica* serotype typhimurium and *Shigella flexneri* to induce the transepithelial migration of neutrophils. Infect Immun *76*, 3614-3627.

Newton, H.J., Pearson, J.S., Badea, L., Kelly, M., Lucas, M., Holloway, G., Wagstaff, K.M., Dunstone, M.A., Sloan, J., Whisstock, J.C., *et al.* (2010). The type III effectors NleE and NleB from enteropathogenic E. coli and OspZ from Shigella block nuclear translocation of NF-kappaB p65. PLoS Pathog *6*, e1000898.

Nusrat, A., Giry, M., Turner, J.R., Colgan, S.P., Parkos, C.A., Carnes, D., Lemichez, E., Boquet, P., and Madara, J.L. (1995). Rho protein regulates tight junctions and perijunctional actin organization in polarized epithelia. Proc Natl Acad Sci U S A *92*, 10629-10633.

Nusrat, A., Parkos, C.A., Verkade, P., Foley, C.S., Liang, T.W., Innis-Whitehouse, W., Eastburn, K.K., and Madara, J.L. (2000). Tight junctions are membrane microdomains. J Cell Sci *113 (Pt 10)*, 1771-1781.

Nusrat, A., von Eichel-Streiber, C., Turner, J.R., Verkade, P., Madara, J.L., and Parkos, C.A. (2001). Clostridium difficile toxins disrupt epithelial barrier function by altering membrane microdomain localization of tight junction proteins. Infect Immun *69*, 1329-1336.

Ogawa, M., and Sasakawa, C. (2006). Intracellular survival of Shigella. Cell Microbiol *8*, 177-184.

Ogawa, M., Yoshikawa, Y., Kobayashi, T., Mimuro, H., Fukumatsu, M., Kiga, K., Piao, Z., Ashida, H., Yoshida, M., Kakuta, S., *et al.* (2011). A Tecpr1-dependent selective autophagy pathway targets bacterial pathogens. Cell Host Microbe *9*, 376-389.

Ogawa, M., Yoshimori, T., Suzuki, T., Sagara, H., Mizushima, N., and Sasakawa, C. (2005). Escape of intracellular Shigella from autophagy. Science *307*, 727-731.

Oswald, E., Sugai, M., Labigne, A., Wu, H.C., Fiorentini, C., Boquet, P., and O'Brien, A.D. (1994). Cytotoxic necrotizing factor type 2 produced by virulent *Escherichia coli* modifies the small GTP-binding proteins Rho involved in assembly of actin stress fibers. Proc Natl Acad Sci U S A *91*, 3814-3818.
Parsot, C., Menard, R., Gounon, P., and Sansonetti, P.J. (1995). Enhanced secretion through the Shigella flexneri Mxi-Spa translocon leads to assembly of extracellular proteins into macromolecular structures. Mol Microbiol *16*, 291-300.
Perdomo, J.J., Gounon, P., and Sansonetti, P.J. (1994). Polymorphonuclear leukocyte transmigration promotes invasion of colonic epithelial monolayer by *Shigella flexneri*. J Clin Invest *93*, 633-643.
Phalipon, A., and Sansonetti, P.J. (2007). Shigella's ways of manipulating the host intestinal innate and adaptive immune system: a tool box for survival? Immunol Cell Biol *85*, 119-129.
Philpott, D.J., McKay, D.M., Mak, W., Perdue, M.H., and Sherman, P.M. (1998). Signal transduction pathways involved in enterohemorrhagic Escherichia coli-induced alterations in T84 epithelial permeability. Infect Immun *66*, 1680-1687.
Prevost, M.C., Lesourd, M., Arpin, M., Vernel, F., Mounier, J., Hellio, R., and Sansonetti, P.J. (1992). Unipolar reorganization of F-actin layer at bacterial division and bundling of actin filaments by plastin correlate with movement of Shigella flexneri within HeLa cells. Infect Immun *60*, 4088-4099.
Rathman, M., de Lanerolle, P., Ohayon, H., Gounon, P., and Sansonetti, P. (2000). Myosin light chain kinase plays an essential role in S. flexneri dissemination. J Cell Sci *113 Pt 19*, 3375-3386.
Reiterer, V., Grossniklaus, L., Tschon, T., Kasper, C.A., Sorg, I., and Arrieumerlou, C. (2011). Shigella flexneri type III secreted effector OspF reveals new crosstalks of proinflammatory signaling pathways during bacterial infection. Cell Signal *23*, 1188-1196.
Rohde, J.R., Breitkreutz, A., Chenal, A., Sansonetti, P.J., and Parsot, C. (2007). Type III secretion effectors of the IpaH family are E3 ubiquitin ligases. Cell Host Microbe *1*, 77-83.
Saitou, M., Furuse, M., Sasaki, H., Schulzke, J.D., Fromm, M., Takano, H., Noda, T., and Tsukita, S. (2000). Complex phenotype of mice lacking occludin, a component of tight junction strands. Mol Biol Cell *11*, 4131-4142.
Sakaguchi, T., Kohler, H., Gu, X., McCormick, B.A., and Reinecker, H.C. (2002). *Shigella flexneri* regulates tight junction-associated proteins in human intestinal epithelial cells. Cell Microbiol *4*, 367-381.
Sanada, T., Kim, M., Mimuro, H., Suzuki, M., Ogawa, M., Oyama, A., Ashida, H., Kobayashi, T., Koyama, T., Nagai, S., *et al.* (2012). The Shigella flexneri effector OspI deamidates UBC13 to dampen the inflammatory response. Nature *483*, 623-626.
Sansonetti, P.J., Arondel, J., Fontaine, A., d'Hauteville, H., and Bernardini, M.L. (1991). OmpB (osmo-regulation) and icsA (cell-to-cell spread) mutants of *Shigella flexneri:* vaccine candidates and probes to study the pathogenesis of shigellosis. Vaccine *9*, 416-422.
Sansonetti, P.J., Kopecko, D.J., and Formal, S.B. (1982). Involvement of a plasmid in the invasive ability of *Shigella flexneri*. Infect Immun *35*, 852-860.
Sansonetti, P.J., Mounier, J., Prevost, M.C., and Mege, R.M. (1994). Cadherin expression is required for the spread of Shigella flexneri between epithelial cells. Cell *76*, 829-839.
Sasakawa, C., Komatsu, K., Tobe, T., Suzuki, T., and Yoshikawa, M. (1993). Eight genes in region 5 that form an operon are essential for invasion of epithelial cells by *Shigella flexneri* 2a. J Bacteriol *175*, 2334-2346.

Schmitz, H., Barmeyer, C., Fromm, M., Runkel, N., Foss, H.D., Bentzel, C.J., Riecken, E.O., and Schulzke, J.D. (1999). Altered tight junction structure contributes to the impaired epithelial barrier function in ulcerative colitis. Gastroenterology *116*, 301-309.

Simonovic, I., Rosenberg, J., Koutsouris, A., and Hecht, G. (2000). Enteropathogenic *Escherichia coli* dephosphorylates and dissociates occludin from intestinal epithelial tight junctions. Cell Microbiol 2, 305-315.

Simpson, N., Shaw, R., Crepin, V.F., Mundy, R., FitzGerald, A.J., Cummings, N., Straatman-Iwanowska, A., Connerton, I., Knutton, S., and Frankel, G. (2006). The enteropathogenic Escherichia coli type III secretion system effector Map binds EBP50/NHERF1: implication for cell signalling and diarrhoea. Mol Microbiol *60*, 349-363.

Sonoda, N., Furuse, M., Sasaki, H., Yonemura, S., Katahira, J., Horiguchi, Y., and Tsukita, S. (1999). *Clostridium perfringens* enterotoxin fragment removes specific claudins from tight junction strands: Evidence for direct involvement of claudins in tight junction barrier. J Cell Biol *147*, 195-204.

Staehelin, L.A. (1974). Structure and function of intercellular junctions. Int Rev Cytol *39*, 191-283.

Strauman, M.C., Harper, J.M., Harrington, S.M., Boll, E.J., and Nataro, J.P. (2010). Enteroaggregative Escherichia coli disrupts epithelial cell tight junctions. Infect Immun *78*, 4958-4964.

Suzuki, T., Mimuro, H., Miki, H., Takenawa, T., Sasaki, T., Nakanishi, H., Takai, Y., and Sasakawa, C. (2000). Rho family GTPase Cdc42 is essential for the actin-based motility of Shigella in mammalian cells. J Exp Med *191*, 1905-1920.

Suzuki, T., Saga, S., and Sasakawa, C. (1996). Functional analysis of Shigella VirG domains essential for interaction with vinculin and actin-based motility. J Biol Chem *271*, 21878-21885.

Swidsinski, A., Ladhoff, A., Pernthaler, A., Swidsinski, S., Loening-Baucke, V., Ortner, M., Weber, J., Hoffmann, U., Schreiber, S., Dietel, M., *et al.* (2002). Mucosal flora in inflammatory bowel disease. Gastroenterology *122*, 44-54.

Takeuchi, O., and Akira, S. (2010). Pattern recognition receptors and inflammation. Cell *140*, 805-820.

Tran Van Nhieu, G., Ben-Ze'ev, A., and Sansonetti, P.J. (1997). Modulation of bacterial entry into epithelial cells by association between vinculin and the Shigella IpaA invasin. Embo J *16*, 2717-2729.

Tran Van Nhieu, G., Clair, C., Bruzzone, R., Mesnil, M., Sansonetti, P., and Combettes, L. (2003). Connexin-dependent inter-cellular communication increases invasion and dissemination of Shigella in epithelial cells. Nat Cell Biol *5*, 720-726.

Van Itallie, C.M., Fanning, A.S., Holmes, J., and Anderson, J.M. (2010). Occludin is required for cytokine-induced regulation of tight junction barriers. J Cell Sci *123*, 2844-2852.

Weber, C.R., Nalle, S.C., Tretiakova, M., Rubin, D.T., and Turner, J.R. (2008). Claudin-1 and claudin-2 expression is elevated in inflammatory bowel disease and may contribute to early neoplastic transformation. Lab Invest *88*, 1110-1120.

Whitman, W.B., Coleman, D.C., and Wiebe, W.J. (1998). Prokaryotes: the unseen majority. Proc Natl Acad Sci U S A *95*, 6578-6583.

Willingham, S.B., Bergstralh, D.T., O'Connor, W., Morrison, A.C., Taxman, D.J., Duncan, J.A., Barnoy, S., Venkatesan, M.M., Flavell, R.A., Deshmukh, M., *et al.* (2007). Microbial pathogen-induced necrotic cell death mediated by the inflammasome components CIAS1/cryopyrin/NLRP3 and ASC. Cell Host Microbe 2, 147-159.

Wittchen, E.S., Haskins, J., and Stevenson, B.R. (1999). Protein interactions at the tight junction. Actin has multiple binding partners, and ZO-1 forms independent complexes with ZO-2 and ZO-3. J Biol Chem *274*, 35179-35185.

Wong, V. (1997). Phosphorylation of occludin correlates with occludin localization and function at the tight junction. Am J Physiol *273*, C1859-1867.

Yuhan, R., Koutsouris, A., Savkovic, S.D., and Hecht, G. (1997). Enteropathogenic Escherichia coli-induced myosin light chain phosphorylation alters intestinal epithelial permeability. Gastroenterology *113*, 1873-1882.

Zeile, W.L., Purich, D.L., and Southwick, F.S. (1996). Recognition of two classes of oligoproline sequences in profilin-mediated acceleration of actin-based Shigella motility. J Cell Biol *133*, 49-59.

Zeissig, S., Burgel, N., Gunzel, D., Richter, J., Mankertz, J., Wahnschaffe, U., Kroesen, A.J., Zeitz, M., Fromm, M., and Schulzke, J.D. (2007). Changes in expression and distribution of claudin 2, 5 and 8 lead to discontinuous tight junctions and barrier dysfunction in active Crohn's disease. Gut *56*, 61-72.

Zurawski, D.V., Mitsuhata, C., Mumy, K.L., McCormick, B.A., and Maurelli, A.T. (2006). OspF and OspC1 are *Shigella flexneri* type III secretion system effectors that are required for postinvasion aspects of virulence. Infect Immun *74*, 5964-5976.

Zurawski, D.V., Mumy, K.L., Badea, L., Prentice, J.A., Hartland, E.L., McCormick, B.A., and Maurelli, A.T. (2008). The NleE/OspZ family of effector proteins is required for polymorphonuclear transepithelial migration, a characteristic shared by enteropathogenic *Escherichia coli* and *Shigella flexneri* infections. Infect Immun *76*, 369-379.

Zurawski, D.V., Mumy, K.L., Faherty, C.S., McCormick, B.A., and Maurelli, A.T. (2009). Shigella flexneri type III secretion system effectors OspB and OspF target the nucleus to downregulate the host inflammatory response via interactions with retinoblastoma protein. Mol Microbiol *71*, 350-368.

Zychlinsky, A., Prevost, M.C., and Sansonetti, P.J. (1992). Shigella flexneri induces apoptosis in infected macrophages. Nature *358*, 167-169.

Zychlinsky, A., Thirumalai, K., Arondel, J., Cantey, J.R., Aliprantis, A.O., and Sansonetti, P.J. (1996). In vivo apoptosis in Shigella flexneri infections. Infection and immunity *64*, 5357-5365.

10

Imaging Techniques to Track *Shigella* Infection at the Cellular Level

Nora Mellouk and Jost Enninga

Institut Pasteur, Research Unit "Dynamics of host pathogen interactions", 25 rue du Dr. Roux, 75724 Paris, France, Jost Enninga: jostenn@pasteur.fr

Abstract

Cellular microbiology has benefitted tremendously from the usage, adaptation, and development of novel imaging tools. This is particularly true for the study of intracellular bacterial pathogens, such as *Listeria* or *Shigella,* due to their complex interactions with the host at the tissue and cellular level. In the case of *Shigella*, microscopy has been instrumental in uncovering a large number of its infection paradigms, for example how it interacts with different host cell types during the infection of the epithelial barrier, how it employs its type III secretion system (T3SS), how it is internalized into host cells and reaches its cytoplasmic niche, and how it moves both intra- and inter-cellularly. More recently, imaging has also been used to monitor the response of individual host cells to the pathogen during infection in a dynamic fashion. Now, dimensions reaching from only a few nanometers to hundreds of micrometers can be bridged correlating different light and electron microscopy approaches. This is important to obtain functional insights into the infection strategy of *Shigella*. Imaging probes are being continuously improved and developed to uncover functional paradigms of the *Shigella* infection process. Furthermore, image processing of the experimental data is increasingly done in an automated way to avoid a potential bias of its interpretation. Improving the links between different imaging modalities and their integration for use in dynamic imaging will provide comprehensive information on the *Shigella* infection process *in situ* and *in vivo*, and for the study of other infection processes as well.

Why is imaging increasingly important for our understanding of *Shigella* infection?

Introduction

Over the last years, fluorescence microscopic techniques have significantly altered our capacities to monitor the *Shigella* infection of different host cells types (Bobard et al., 2011), and so we will focus on methods that use this imaging modality. Progress has been in three distinct technology domains: (1) the microscope hardware for tracking host-pathogen interactions has been changing, (2) a large number of functional reporter assays and fluorescent probes have become available, and (3) microscopy has become tightly linked with bioinformatics analysis. The revolution in cellular imaging is still ongoing, and every year brings even more accurate and exciting technologies.

Improvements in the imaging hardware

Among the fluorescence microscopy techniques that have been established and improved over the last decade, a number stand out in the context of *Shigella* infection studies: First, time-lapse studies of *Shigella* internalization into host cells using inverted fluorescence microscopes with autofocus systems within heated chambers that allow the acquisition of numerous movies in multi-well formats simultaneously improving the throughput (numbers of acquired situations) and the content (dynamic imaging versus snap-shots) (Bobard et al., 2011; Ehsani et al., 2012) . Secondly, confocal microscopy, either through laser scanning or spinning disc approaches, yields data-sets of *Shigella* infection in three dimensions (3D) at high spatial resolution, for example during T3SS effector secretion, formation of actin comet tails, or cell-to-cell spread (Ehsani et al., 2012; Enninga et al., 2005; Fukumatsu et al., 2012). Especially, the spinning disc approach can be combined with rapid time-lapse movies for the measurements of dynamic events in 3D, which cannot be achieved through laser scanning microscopes. Combining spinning disc microscopy with highly sensitive back illuminated electron multiplying charged coupled device (EMCCD) cameras yields about 100 to 1000 images per second. (Frischknecht et al., 2006). Third, non-linear multiphoton microscopes allow the imaging of *Shigella* infection in deeper tissue up to 400 µm, and it has been used to measure its interaction with immune cells (Salgado-Pabon et al., 2013). These relatively novel fluorescent methodologies have become important for the dynamic studies of *Shigella* entry into host cells, nevertheless two additional approaches will potentially bring a second revolution to the field: First, over the past eight years novel light microscopes have been built that are capable to break the resolution limit of light (about 200 nm), for example via photoactivated localization microscopy (PALM), optical reconstruction microscopy (STORM), stimulated emission depletion (STED) or structured illumination microscopy (SIM) (Coltharp and Xiao, 2012). Secondly, correlative light electron microscopy (CLEM) provides specific information on molecules in the *Shigella*-host interaction process (through fluorescence labeling) and yields simultaneous ultrastructural information on the infection process via the ultrastructure in identical specimen (Campbell-Valois et al., 2014; Mironov and Beznoussenko, 2012).

Novel probes for Shigella studies

Fluorescent probes for the labeling of specific molecular and cellular structures have been at the heart of the progress made by fluorescence microscopy (Giepmans et al., 2006). These probes can be used to localize low numbers of labeled molecules or even individual molecules inside fixed or living cells. They have been mainly applied to proteins, however also nucleic acids, lipids or sugars can be imaged via fluorescence, for example through a high affinity between the specific molecule and a small organic fluorophore fused to a sensor, such as fluorescently labeled phalloidin intercalating into F-actin structures. Additionally, genetically encoded fluorophores allow the expression of fluorescent protein chimeras of molecules of interest or genetically fluorescent sensors. These include the engineered green fluorescent protein and its derivates spanning most of the visual spectrum of light, such as mCherry (Shaner et al., 2004), plant derived fluorescent proteins such as miniSOG or iLov, or others such as dsRed or the firefly luciferase (Giepmans et al., 2006). Each of these proteins has specific spectral properties and characteristics that need to be evaluated for a specific experimental design. In some cases, the fluorescent proteins impede the function of the tagged proteins, hence alternative approaches have been developed to overcome issues arising through the bulkiness of GFP derivates. One approach that has been relevant for *Shigella* T3SS injection is based on a two-compound labeling technique. It takes advantage of the high affinity of the metallo-organic FlAsH compound to an engineered tetracysteine tag added to any protein of interest (Gaietta et al., 2002). The FlAsH compound only starts to fluoresce upon binding to the tetracysteine sequence, and represents a smaller alternative to GFP. It can also be used in pulse chase experiments and works in conjunction with CLEM (Gaietta et al., 2002). Other properties of fluorophores have also been used to track host pathogen interactions. One approach irreversibly destroys fluorophores for the measurements of mobility within living cells via fluorescence recovery after photobleaching (FRAP), the other approach takes advantage of fluorescence transfer from a donor fluorophore to an acceptor molecule during fluorescence resonance energy transfer (FRET) (Frischknecht et al., 2006). We will give examples of their use later in the text.

Processing of complex imaging data

Automated image processing and analysis has been very important for the appropriate interpretation of imaging data. Even though a number of commercial programs exist and microscopes are piloted by software developed by the hardware providers (for example in the case of Leica, Zeiss, Nikon or Olympus), most tasks during fluorescence imaging can be achieved through freely available programs. The advantage of the freeware is the possibility to trace each step of the processing and allows the implementation of custom-tailored protocols. Four examples of widely spread image analysis software include ImageJ (http://rsbweb.nih.gov/ij/), FIJI (http://fiji.sc/Fiji), CellProfiler (http://www.cellprofiler.org), and ICY (http://icy.bioimageanalysis.org). Recently, an in-depth comparison of their capacities has been published (Chenouard et al., 2014). Freeware is also available to drive microscopes (e.g. "micromanager", https://micro-manager.org), which is useful for the development of novel imaging application, for example in the context of super-resolution microscopy.

Some commercial programs, such as "Metamorph" (Molecular Devices) or "Volocity" (Improvision/Perkin Elmer) drive microscopes, and are also equipped with in-depth analytical tools. Yet, other commercial solutions, like "Imaris" (Bitplane), or "Huygens" (Scientific Volume Imaging, SVI) perform specific tasks, for example 3D reconstruction or deconvolution and are often considered to be more user friendly than the freeware.

Researchers should consider commercial and freeware options depending on their own expertise. It is of note that some free solutions, for example ICY, allow the integration of other programs like ImageJ. Online discussion groups exist for the different programs for the exchange about problems and the development of novel tools.

Imaging the regulation and the activity of the *Shigella* weaponry: the type III secretion apparatus

Visualization of the T3SS Regulation

The expression and activity of the T3SS is tightly regulated for all bacteria that express these secretion machineries. In the case of *Shigella*, it is encoded by the mxi-spa pathogenicity island on a large virulence plasmid (Buchrieser et al., 2000). Expression of the structural genes of the needle apparatus and a first wave of effectors is induced at 37°C through the transcription factor VirF, however the translated effectors remain trapped within the bacteria since the tip of the needles is sealed by the translocon/needle-tip complex that includes the proteins IpaD and IpaB (Espina et al., 2006; Olive et al., 2007; Veenendaal et al., 2007). Upon T3SS activation and effector injection a second wave of effectors is induced through the transcriptional activator *mxiE*. It involves the *Shigella* chaperone *ipgC* after translocation of the translocator/effectors IpaB/IpaC

(Mavris et al., 2002). Time-lapse epifluorescence microscopy has identified the polar storage of the effector protein IpaC before the induction of the T3SS (Jaumouille et al., 2008). The dynamic imaging of this work has also revealed that the T3SS is activated preferentially at one of the bacterial poles, however no correlation could be made whether this could be linked with the old or new pole after bacterial scission. Ultra-structural studies combined with fluorescence imaging have also been crucial for the investigation of the activity status and sensing qualities of the T3SS (Blocker et al., 1999; Dickenson and Picking, 2012; Veenendaal et al., 2007). For example, IpaD could be visualized at the needle tips through indirect immunofluorescence, and through immunogold labeling of transmission micrographs (Espina et al., 2006; Sani et al., 2007).

A major question has been whether the T3SS is down-regulated and reactivated during the course of cellular infection and cell-to-cell spread. This has been measured through a recently established reporter approach, named transcription-based secretion reporters (TSAR) that are based on the tight regulation of the second wave of T3S effectors (Campbell-Valois et al., 2014). In the TSAR assays T3SS activity is assessed in single bacteria using a fast-folding destabilized variant of GFP driven by a modified IpaH7.8 promoter. IpaH7.8 belongs to the second wave effectors only expressed upon activation of the T3SS. Performing dynamic imaging and CLEM, the TSAR reporters demonstrated that T3SS is activated upon contact with the host plasma membrane and is rapidly turned off after internalization. Then, T3SS requires actin-polymerization on one of the bacterial poles to become reactivated during the formation of membrane protrusions that are formed before cell-to cell spread of *Shigella*. Even though TSAR can demonstrate already correlations between the step of bacterial invasion and the activity of the T3SS, its temporal resolution is still limited to about 10 to 15 minutes. Another opportunity for further TSAR development is the adaptation of the employed fluorophores for two-photon microscopy for *in vivo* studies.

Imaging the effector secretion process

The induction of the T3S apparatus and the injection of effector proteins into host cells by *Shigella* are crucial for cell invasion and subverting the host immune response. Imaging this process has remained challenging for a long time due to the limitations of the genetically encoded GFP and its variants, chimeric fusions of the bulky GFP and bacterial T3SS effectors simply impede their own secretion and clog the T3SS. To overcome these limitations, another fluorescence labeling approach has been established to directly visualize T3S effectors via the tetracysteine-FlAsH labeling procedure (see Figure 10.1) (Gaietta et al., 2002). The principle of this approach is the high affinity of the metallo-organic FlAsH component to a small tetracysteine peptide motive of ten amino acids containing that form highly stable complexes. Upon complex formation, the FlAsH molecule emits in green, and it can readily be used on fluorescence microscopes equipped with filters for green fluorophores, such as GFP. Another aspect of this method is that FlAsH can be replaced by another component, ReAsH that fluoresces in red and can precipitate DAB (diaminobenzidine) that can then be used for CLEM.

The tetracysteine-FlAsH labeling method has been used to measure the secretion of the *Shigella* T3S effectors IpaB and IpaC from individual bacteria into host cells by spinning-disc confocal microscopy (Enninga et al., 2005). Tracking the secretion of these effectors revealed that their injection takes place concomitantly after contact with host cells, and indicated that about 50% of each effector, stored within the bacteria before host cellular contact, was released within four minutes after the induction of the T3SS. In another study, the tetracysteine-FlAsH method was further improved to monitor T3S of *Salmonella*

SPI-1 effectors. This showed differential secretion kinetics of two *Salmonella* effectors, SopE2 and SptP (Van Engelenburg and Palmer, 2008). Another interesting feature of the tetracysteine-FlAsH labeling approach is the blinking behavior of the fluorophore. This physical property has been exploited for super-resolution microscopy via PALM imaging. FlAsH-PALM has been used to track the transport of HIV virions through the nuclear pore complex of infected host cells (Lelek et al., 2012).

A few drawbacks limit the versatility of the tetracysteine-FlAsH labeling approach as reviewed in Simpson and colleagues (Simpson et al., 2010). First, the FlAsH compound is an arsenic derivate requiring a tight control on its potential toxicity when non-chelated. Secondly, for time-lapse studies, only effectors synthesized before activation of the T3S apparatus can be measured due to the need to perform the labeling procedure before the contact with challenged host cells. Third, the visualized effector proteins have to be expressed at rather high quantities, which is not the case for all T3SS effectors. Therefore, other visualization approaches of T3SS activity upon host contact have been

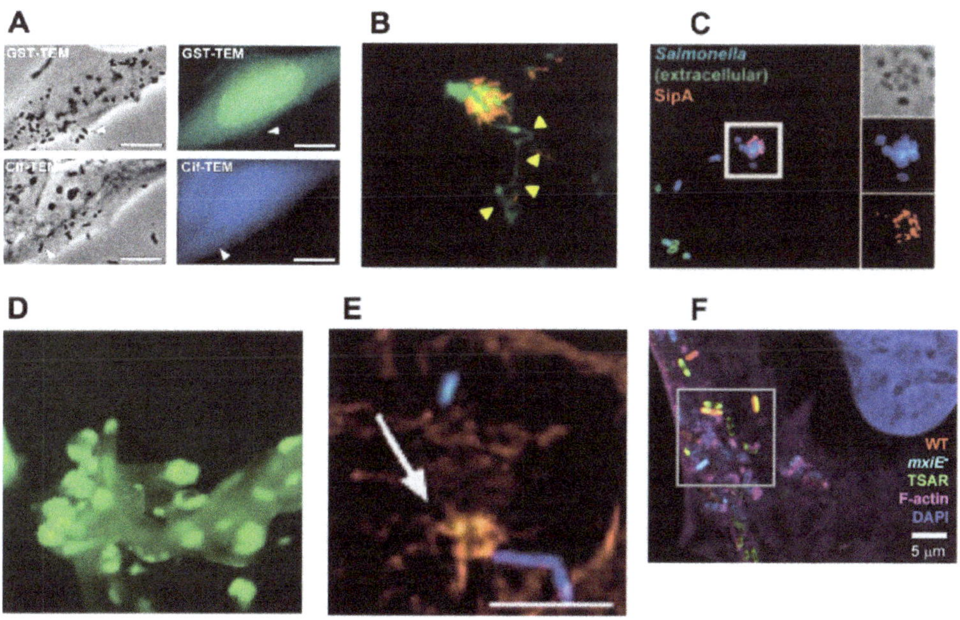

Figure 10.1. Imaging the activity of the T3SS of bacterial pathogens. Fluorescent imaging has been used to measure injection of T3SS effectors into host cells, or to monitor the activity of the T3SS. A) The CCF4 FRET reporter yields information on cells that are injected with T3SS effectors through a color switch from green to blue. Here, the translocation of IpgD into T-cells was studied (Konradt et al., 2011). B) The Split GFP approach allows the tracking of T3SS after their translocation. This showed the association of the Salmonella effector with Sif structures around the SCV (Van Engelenburg and Palmer, 2010). C) Fluorescently labeled chaperones identify the accumulation of T3SS effectors. In the case of *Salmonella*, this allowed the quantification of the effector translocation kinetics of SipA into host cells (Schlumberger et al., 2005). D) The CreLox system has been initially used to track injection of type 4 secretion system effectors into host cells, however it could potentially also applied to measure T3SS effectors (Vergunst et al., 2005). The reporter system is entirely genetically encoded and is suitable for *in vivo* studies. E) The tetracysteine-FlAsH approaches measures directly the translocation from the bacteria into host cells. It works in live cell imaging and fixed cell analysis (Enninga et al., 2005). F) The TSAR reporter measures the activity status of the Shigella T3SS fusing a rapidly folding GFP downstream of the IpaH7.8 promoter (Campbell-Valois et al., 2014).

developed for a number of T3SS containing pathogens. All of them can be easily adapted for *Shigella* studies.

A widely used approach relies on a FRET reporter consisting of a cephalosporin-derived molecule called CCF4-AM with a FRET donor linked to one extremity and a FRET acceptor linked to its other extremity (Charpentier and Oswald, 2004). Cells can be loaded with this FRET probe, and the AM-ester moieties are subsequently cleaved off. In its intact form the CCF4 molecule fluoresces via FRET in the green channel. This FRET activity is lost after cleavage by β-lactamase, and the probe fluoresces in blue allowing simple ratiometric measurements. Host cells are labeled with the FRET probe before bacterial challenge, and the bacteria require to express T3S effectors that are fused to β-lactamase. After translocation of the effectors, the enzyme cleaves the FRET probe within the host cytosol (Charpentier and Oswald, 2004). This has been used to measure effector translocation into host cells for a number of different bacteria for example *EPEC*, *Salmonella*, or *Legionella*. Its versatility is underlined by the potential to use this method for a range of different secretion machineries not being limited to the T3SS. In the context of EPEC infection, it has revealed the existence of specific secretion patterns among the different effectors EspF, EspG, EspH, EspZ, Map, and Tir (Mills et al., 2008). Another approach takes advantage of the fact that numerous T3SS effectors require bacterial chaperones within the bacterial cytoplasm before the translocation process as is the case for the SPI-1 effector SipA (Schlumberger et al., 2005). The authors used the *Salmonella* effector SipA chaperone InvB fused to GFP. Chaperone-GFP fusion proteins were stably expressed within host cells before bacterial challenge. Then, the injected effectors were tracked at the bacterial contact site by the tethering of InvB-GFP. Quantification of the fluorescent signals revealed that SipA secretion started about 90 seconds after host cellular contact and lasted up to 10 minutes after secretion initiation. An advantage of this approach is that the pathogen does not need to be treated (e.g. stained with a fluorophore) before the infection process, however it is limited for studies on effectors with chaperones and requires their localization at specific sites within the host cell upon translocation to allow quantification of the chaperone-GFP accumulation. A recently developed method exploits another genetically encoded fluorophore, iLov that can be fused to any bacterial effector of interest. iLov is a small 100 amino acid long flavin-derived fluorescent domain that fluoresces without the need of ambient oxygen. Fusions between iLov and bacterial effectors fluoresce brightly, and in the case of EHEC it has been shown that the effector-iLov fusions translocate and do not clog the T3S ((Gawthorne et al., 2012) and our own data). This needs to be confirmed in the case of *Shigella*, however it is already clear that this alternative labeling approach bears a great potential to simplify studies on T3S effector kinetics and their subsequent fate within the host cell.

The "life" of the T3SS effectors after translocation

Imaging of T3SS effectors upon translocation is challenging due to dilution effects within the host cells and the turnover of the proteins themselves. Classic approaches, such as indirect immunofluorescence have been applied to a number of *Shigella* effectors, however this can only be done in fixed samples, and it requires highly specific antibodies. These limitations were partially overcome via the tetracysteine-FlAsH approach demonstrating an accumulation of IpaB at the proximal ends of the entry foci and IpaC at the bottom of it (Enninga et al., 2005). As mentioned above, combining the FlAsH labeling with PALM has the potential to reveal unprecedented details of their localization and role within the host cell. Alternatively, fluorescence complementation has been used for the dynamic imaging of *Salmonella* SPI2 effectors upon translocation. For these studies, one β-barrel strand of GFP is fused to a bacterial T3SS effector, and the remaining part expressed within the host cell challenged by the pathogen. Formation of a fluorescent GFP complex of the two split halves takes about 30 minutes allowing the imaging of slow events. For example, the recruitment of T3SS effectors to the filamentous structures around the *Salmonella* containing vacuoles called Sifs could be imaged using this method (Van Engelenburg and Palmer, 2010). Since the events during *Shigella* entry and intracellular niche formation are very rapid, we predict that the split GFP approach is only of limited use for this pathogen.

Novel approaches to monitor entry into host cells

Tracking the capture of Shigella at the host surface and attachment of Shigella

Details of the earliest contacts between *Shigella* and host cells have been identified through novel imaging and micromanipulation technologies. Phase-contrast time-lapse imaging showed the capture of *Shigella* at the surface of epithelial cells through filopodia-like structures that were coined "nanometer-thin micropodial extensions (NMEs)" (Romero et al., 2011). Capture is dependent on proteins of the T3SS needle tip, together with ATP- and connexin-mediated signaling. In turn, ATP stimulates ERK1/2 and alters the actin retrograde flow leading to the bacterial tethering onto the host cell surface. During these studies, the actin retrograde flow could be directly tracked through FRAP measurements of the NMEs and quantitative analysis of the resulting kymographs. In a subsequent work, the forces of the retracting filopodia were determined by optical tweezers that captured beads coated with *Shigella* surface membranes as model for the pathogen (Romero et al., 2012). This indicated that individual filopodia exert retraction forces between 42 and 80 pN that could be related with the density of ligands on the bead surface. Further, the authors showed that very few bacterial T3SS are sufficient to result in the retraction of the T3SS.

Adhesion to the host cell surface has been imaged by fluorescence microscopy and scanning electron microscopy (Brotcke Zumsteg et al., 2014; Skoudy et al., 2000). On the bacterial side this implicated IpaB and more recently IcsA in the adhesion process. On the host side, it was shown that IpaB interacts with CD44, and its recruitment to the bacterial entry site could be imaged by indirect immunofluorescence (Skoudy et al., 2000). Interestingly, similar imaging approaches have also found the recruitment of cytoplasmic sensors of pathogens, such as Nod1, to the bacterial entry site before the internalization process (Kufer et al., 2008).

Imaging approaches for bacterial entry

It appears that the internalization process of *Shigella* into host cells has profited the most from dynamic fluorescence imaging. The entry dynamics were imaged by time-lapse microscopy implicating numerous bacterial T3SS effectors in the modulation of the host, for example a role of the C-terminus of IpaC in recruitment of the Src tyrosine kinase to the entry site (Mounier et al., 2009). It also revealed the rapid tethering of other actin-modulating factors, including RhoA, Rac, Abl, just to name a few (Burton et al., 2003; Ehsani et al., 2012). Quantification of parameters associated with entry foci has been attempted, however due to their large morphological variety there is no standardized procedure for their automatic counting and shape analysis (Dumenil et al., 1998; Handa et al., 2007). Nevertheless, analyses of foci revealed an impact of interferon-α on the size of the entry foci, and it also revealed a role of the T3SS effector IpgB1 in the re-localization of the ELMO-Dock180 complex to the plasma membrane surface.

The regulators of actin reorganization represent molecular switches, for example small GTPases or kinases that exist in an active form and an inactive form. Functional fluorescent probes have been developed to measure the activity status of these switches through fluorescence energy transfer (FRET). Numerous of such probes are available now, and a few of them have been used for bacterial entry. For example, it was shown via FRET image analysis that Rac1 is activated at the site of *Yersinia* entry in typically non-phagocytic cells using the interaction of the FRET pair Rac1-CFP with a YFP-fusion of a domain of its downstream adaptor PAK1 (Wong and Isberg, 2005). In another study on *Listeria* entry, the same principle was used to monitor Rac1 activation and the implication of small signaling lipids of the inositol-phosphatidyl family (Seveau et al., 2007). Future studies could profit from two developments. First, novel FRET-based biosensors for intracellular membrane trafficking would be relevant, for example Raichu-Rab5 to monitor the spatio-temporal activity of small GTPases during bacterial invasion (Kitano et al., 2008). Secondly, FRET measurements via fluorescence lifetime imaging (FLIM) allows measurements in real time which are needed to trace the highly dynamic events during the rapid entry of *Shigella*.

Imaging of the intracellular localization

Dynamic imaging of the vacuolar state

Upon entry, bacterial pathogens can remain within the vacuole and avoid fusion with lysosomes in order to create a unique replicative niche. Alternatively, some pathogens rupture their vacuole to access the nutrient-rich host cytosol. During vacuolar maturation, diverse bacterial effectors and host factors modulate the lipid and protein composition of the vacuolar membrane. Small GTPases such as Rabs, ARFs and Rhos are key regulators of host vesicular trafficking and the cytoskeleton. Tracking their localization by fluorescence video microscopy has provided important information on the host pathways subverted by each pathogen during invasion. A remarkable example is *Legionella*, which secretes numerous bacterial effectors that control the recruitment of Rab1-positive ER vesicles to the *Legionella*-containing vacuole (Brumell and Scidmore, 2007) (Alix et al., 2011). Recently, Becken et al. have developed a fluorescence microscopy-based *in vitro* assay to study fusion between purified bacteria-containing phagosomes and different host endocytic compartments. This assay robustly confirmed that *Salmonella*-containing phagosomes barely fuse with lysosomes (Becken et al., 2010). Separately, numerous studies take advantage of lipid-binding domains (e.g. PH, PX, FYVE, etc.) fused with a fluorescent protein probes to analyze the distribution of phosphoinositides during bacteria-containing vacuoles maturation (Pizarro-Cerda and Cossart, 2004). Fluorescence microscopy-based approaches can be used to unravel the interplay between pathogens and their host. *Shigella* remains only briefly within a vacuole, therefore these approaches are particularly fruitful with time-lapse studies capable to capturing rapid and transient events.

Visualizing vacuolar rupture, electron microscopy and other novel functional approaches

Over the last decades, electron microscopy (EM) has been extensively used to investigate the intracellular localization of bacterial pathogens (Figure 10.2A). This high-resolution microscopy technique has provided the first evidence that *Shigella*, *Listeria*, *Rickettsia* and *Francisella* species access the host cytoplasm (Fredlund and Enninga, 2014). Alternatively, upon permeabilization of infected cells, immunofluorescent staining of host marker of the vacuolar membrane such as LAMP1 or pathogen-specific antibodies to detect only the cytosolic bacteria have been used to assess vacuolar integrity (Beuzon et al., 2000). However, the main drawback of these techniques include potential artifacts caused by harsh sample preparation and the inability to track vacuolar rupture in real time. Novel approaches based on fluorescent reporters have been developed to overcome these limitations. For instances, Galectin-3 is a host lectin that can be used as a dynamic marker for vacuolar escape simply by expressing a fluorescently tagged version of the protein within the host cell (Figure 10.2B). As long as the vacuolar integrity is maintained, the Galectin-3 fluorescent signal is uniformly distributed within the entire host cell. Upon vacuolar rupture, it accumulates in the vicinity of the damaged membrane of bacteria-containing vacuole, forming a readily visible "Galectin-3 ghost". This has first revealed that *Shigella* escapes into the cytoplasm in about 10 minutes upon bacterial entry (Ray et al., 2010) (Paz et al., 2010). Besides, the Galectin-3 marker or other galectins have been successfully used to follow vacuolar rupture in the context of various bacterial pathogens (e.g. *Listeria*, *Salmonella*, *Legionella*, etc.) (Paz et al., 2010) (Creasey and Isberg, 2012) (Thurston et al., 2012). Recently, Ray et al. have established a robust assay to track vacuolar rupture in both fixed and living cell conditions. This method is based on a cephalosporin-derived FRET probe, CCF4-AM that can be cleaved by β-lactamase (Figure 10.2C). This assay has been portrayed above, and it can be used to measure vacuolar rupture due to the enzymatic activity of β-lactamases at the surface of *Shigella* containing plasmids

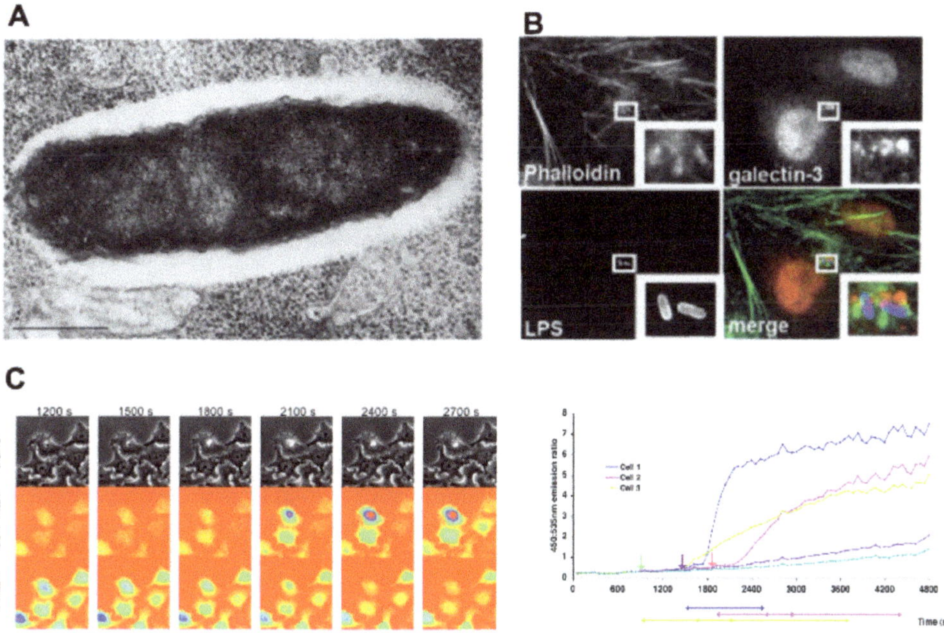

Figure 10.2. Visualization of vacuolar rupture. Different imaging modalities have been used to track vacuolar rupture by *Shigella* either in a direct or indirect fashion. A) Transmission electron microscopy has been used to reveal the cytoplasmic localization of a large number of intracellular pathogens through the absence of a vacuolar membrane. This has also identified *Shigella* within the cytoplasm (Sansonetti et al., 1986). B) Galectin-3 binds to the inner leaflet of the ruptured membranes after *Shigella* accesses the host cytoplasm. Transfecting cells with fluorescent chimeras of galectin-3 (here galectin-3-eGFP) temporarily identifies the escape into the cytoplasm (Paz et al., 2010). C) As described in the text, the CCF4 FRET probe tracks cytoplasmic access of *Shigella* in fixed cells and in real time. Upon vacuolar rupture the FRET-probe is cleaved leading to a switch of signal from green to blue. This showed that *Shigella* access the host cytoplasm within 10 minutes upon internalization into host cells as indicated by the quantification on the right side (Ray et al., 2010).

encoding bLa genes. Using this assay in real time, this revealed the rapid escape of *Shigella* from the vacuole within less than 10 minutes after cellular entry, consistent with the Galectin-3 results (Ray et al., 2010). This technique has been successfully adapted for other pathogens such as *M. tuberculosis*, *Francisella* and *Listeria* ((Juruj et al., 2013) (Simeone et al., 2012) and unpublished data) and can be used for high-throughput screening (Keller et al., 2013). Another fluorescence imaging approach takes advantage of the observation that fluorescent dextran, a fluid-phase marker for endocytic compartments accumulates within the *Salmonella*-containing vacuole (SCV) (Drecktrah et al., 2006). Cells are pre-loaded with Alexa488-Dextran overnight and challenged with mCherry expressing *Salmonella*. After 2-3h post-infection, SCV integrity was surveyed by time-lapse imaging assessing the co-localization of mCherry-*Salmonella* with Alexa488-Dextran. This showed that the *Salmonella* T3SS SPI-2 is required for bacterial replication within the vacuole but not for cytosolic replication (Malik-Kale et al., 2012). This technique can differentiate between vacuolar and cytosolic bacteria but unlike the Galectin-3 and FRET probe methods, it does not yield precise quantitative information on the timing of vacuolar rupture.

Quantifying actin comet tails

Many cytosolic bacteria (e.g. *Shigella*, *Listeria*, *Rickettsia*, etc.) subvert the host cytoskeleton machinery to move intra- and inter-cellularly. Remarkably, these bacterial pathogens promote polymerization of actin at the surface of one pole of the bacterium forming a so-called actin tail, which propel them within the cytosol (Goldberg, 2001). Gouin et al. performed a careful analysis to compare the actin-based motility mechanisms of *Shigella*, *Listeria* and *Rickettsia* species using three imaging techniques. First, electron microscopy provided ultrastructural details of the actin tail highlighting that *Shigella* and *Listeria* display short and highly branched filaments whereas the one from *Rickettsia* are long and contain mainly parallel F-actin stretches. Furthermore, phase-contrast video microscopy measured the rate of movements in infected cells, with *Rickettsia* moving threefold slower than *Shigella* and *Listeria*. Besides, immunofluorescence staining of actin-binding proteins allowed identifying which one are co-localize with the actin tail depending on the pathogen. Collectively, this study underlines the advantages of multi-modular imaging approach to reveal molecular, ultrastructural and dynamic characteristics of actin-based motility of intracellular bacteria (Gouin et al., 1999).

Outstanding progress in our understanding of the molecular mechanism of bacterial actin tail motility was accomplished by establishing two distinct *in vitro* assays. One assay used purified proteins of the actin cytoskeleton to reconstitute *Shigella* and *Listeria* movement, thereby determining the essential proteins required for these processes (Loisel et al., 1999). The second assay took advantage of the *Listeria* effector ActA that function as a nucleation-promoting factor. Polystyrene beads were coated with purified ActA to generate actin-based motility of the inert beads in cell-free system without other bacterial factors (Cameron et al., 1999). Interestingly, Soo and Theriot have developed a high-throughput quantitative analysis of fluorescence video microscopy data to automatically track the movement of a large number of bacteria in reconstituted *Xenopus* egg extract. By this mean, they addressed the cause of variation in the rate of bacterial actin-based movement (Rafelski and Theriot, 2005) (Soo and Theriot, 2005). Taken together, these methods shed light on the mechanisms of bacterial actin-based motility as well as fundamental actin assembly in eukaryotic cells. Since most of the studies were performed *in vitro*, it remains to be determined if the same mechanisms occur *in vivo* in the highly packed and complex host cytosol. Future siRNA or chemical compound screening for selective interference with actin tail movement of pathogens could then lead to the design of novel therapeutic treatments.

Septin cages, dynamic imaging, super-resolution

Septins are GTP-binding proteins known to assemble as filaments and interact with membrane, F-actin and microtubules (Mostowy and Cossart, 2012). Recently, Mostowy et al. has revealed that a subset of intracytosolic *Shigella* was confined in septin cage-like structure using immunofluorescence confocal microscopy. In addition, super-resolution microscopy was performed via STORM and PALM to provide high-resolution images of fluorescently labeled *Shigella*-septin cages. These single-molecule imaging techniques are based on the sequential activation and time-resolved localization of photoswitchable fluorophores to reconstitute a super-resolution image. The only technical difference between STORM and PALM is the nature of the photoswitchable fluorophore respectively using synthetic dyes (e.g. Cy5) or genetically encoded fluorescent proteins (e.g. mEOS2). Further ultrastructural details of *Shigella* entrapped within a septin cage were obtained using CLEM via scanning EM. Briefly, cells were plated on glass-bottom dishes with locator grid that allow imaging of the exact same position at both microscopes. The fluorescently labeled *Shigella*-septin cage was first localized and imaged by light microscopy and subsequently processed for SEM. Important steps include removal of the cell membrane using 0.5% Triton in a cytoskeleton stabilization buffer and immunogold labeling of endogenous septins. The same *Shigella*-septin cage location was recovered at the SEM and imaged using a secondary electron detector (SEI) and a backscattered electron detector (YAG) respectively showing the morphology of the septin cage and the immunogold labeling for septin. Since *Shigella*-septin cages were observed to last several hours, FRAP was performed to investigate the dynamic of septins recruitment. After photobleaching using 100% laser power, the fluorescence signal of GFP fused septin was monitored over time and did not recover demonstrating that *Shigella*-septin cages are stable. Next, using living cell video microscopy, Mostowy et al. showed that actomyosin activity is required for *Shigella*-septin caging. Infected cells co-expressing fluorescently tagged actin and septin were imaged and after 5 min cells were treated with blebbistatin (i.e. chemical inhibitor of myosin II). Time-lapse imaging showed that upon myosin II inhibition septin cages disassembled and subsequently the bacteria could initiate actin-based motility (Mostowy et al., 2010). Collectively, numerous state-of-the-art imaging techniques including FRAP, super-resolution microscopy, CLEM and video microscopy have been applied to provide insight on the compartmentalization of *Shigella* within septin cage.

Shigella and autophagy

Autophagy is a host cell defense mechanism against bacterial invasion by delivering intracellular pathogens to the lysosome for degradation. This process is characterized by the sequestration of intracellular bacteria within double-membrane compartment named autophagosome (AP). One key component of the autophagy pathway is LC3 (microtubule-associated protein 1 Light Chain 3), which is widely used as a marker of AP (Shahnazari and Brumell, 2011). Pioneering work from Ogawa et al. revealed that 50% of *Shigella ΔicsB* were associated with GFP fused LC3 using fluorescence microscopy. In addition, cell treatment with chemical inhibitors for autophagy (e.g. wortmannin, bafilomycin-A1 and 3-methyl-adenine) prevented the association of *Shigella ΔicsB* with LC3-GFP, further suggesting that *ΔicsB* bacteria were targeted by autophagy. Because it is challenging to observe the classical double-membrane structure of AP by fluorescence microscopy, high-resolution EM is commonly used to identify AP. By this means, *Shigella ΔicsB* was often observed to be surrounded by multi-lamellar membrane structure. In contrast, most *Shigella* WT lacked the lamellar membrane and instead had actin tails. Moreover, LC3-GFP was colocalized with the multi-lamellar structures by immunogold labeling with antibody against GFP (Ogawa et al., 2005). Similar imaging techniques have been used to demonstrate that vacuolar membrane remnants labeled with Galectin-3 or Galectin-8 can be targeted for autophagy (Thurston et al., 2012) (Dupont et al., 2009). Besides, *Shigella*-septin cages co-localize with many autophagy markers and knockdown of septins or autophagy components lead in loss of both septin caging and autophagy of *Shigella* (Mostowy et al., 2010). More recently, a novel *in vivo* model for the study of *Shigella* invasion has been established within zebrafish as host (Mostowy et al., 2013). The early stage larva of zebrafish is naturally transparent, which makes it readily suitable for real time or fixed fluorescence microscopy imaging at single cell resolution. While zebrafish do not have adaptive immune system until reaching the adult stage, zebrafish larva already possesses an innate immune system including macrophage-like cells that can phagocytize bacterial pathogens (Coombes and Robey, 2010). By using confocal video microscopy and EM approaches, Mostowy et al. found that upon engulfment within zebrafish larva macrophages, *Shigella* escape into the cytosol and can induce septin-caging and autophagy *in vivo*. Another advantage of the zebrafish model is the ability to specifically knockdown a host target by injecting morpholino oligonucleotides. Using this knockdown tool to deplete p62, an adaptor protein involved in autophagy and septin caging, showed that p62-mediated autophagy is important for bacterial clearance and subsequent host survival *in vivo*. Further *in vivo* perturbation of autophagy via rapamycin stimulation (i.e. inhibitor of the autophagy suppressor mTOR) surprisingly leads to the same phenotype than p62 depletion instead of promoting bacterial clearance (Mostowy et al., 2013). There has been outstanding progress in our understanding of how bacteria interact with the autophagy pathway mainly by using EM and fluorescence microscopy in combination with autophagy markers, siRNA knockdown and chemical inhibitors. Further studies will be required to identify specific inhibitor and activator of the different autophagy pathway to better characterize them and eventually be used as anti-bacterial molecules.

Cell-to-cell spread and multicellular context of *Shigella* infection

Different approaches to assess Shigella cell-to-cell spread

A simple way to measure cell-to-cell spread during *Shigella* invasion is to infect with very low bacterial numbers in cell-culture dishes, and then measure the size of the plaques arising from *Shigella*-infected cells (for example (Purdy et al., 2002)). During the cell-to-cell spread, *Shigella* forms protrusions emanating from the infected cell containing the pathogen on top of an actin tail. Importantly, the spread into neighboring cells requires the rupture of two membranes, one from the plasma membrane of the previously infected cell, and one from the newly infected cell. These double membranes could be visualized by electron microscopy in *Listeria* and *Shigella* (eg. (Mounier et al., 1990)). More recently, the size of the plaques due to cell-to-cell spread could be visualized through fluorescently labeled bacteria. Together, this has implicated the necessity of actin-polymerization through the formin Dia, the tyrosine kinases, and the T3SS effectors in the process of cell-to-cell spread. More sophisticated time-lapse imaging of bacterial spread revealed further details, intriguingly cell-to-cell spread appeared preferentially at tri-cellular junctions through a non-canonical clathrin pathway (Fukumatsu et al., 2012). This study also revealed the involvement of numerous host factors in the spreading process including tricellulin, phosphor-inositide-3-kinase, clathrin, Epsin-1 and Dynamin-2. Another host factor that has been implicated through imaging into cell-to-cell spread is myosin-X playing a role in *Shigella* and *Listeria* infection (Bishai et al., 2013).

Imaging infection in the multicellular context and infection of different cell types
Shigella cannot only be imaged during its spread from cell-to-cell, however, the cells surrounding infected *Shigella* have also been analyzed by microscopy. Using host cells expressing a fluorescent NF-κB reporter and low multiplicity of infection with only few cells infected within a cell culture plate, it became clear that the NF-κB reporter preferentially translocated into the nucleus of the uninfected bystander cells. Furthermore, one bacterial effector OspF was capable for the dampening of the inflammatory pathway within the infected cells (Kasper et al., 2010). Imaging in the multicellular context of the intestine was also done in isolated tissue such as in ileal loops from rabbits. For example, this model was used to investigate the dependence of the T3SS at low oxygen levels that could be directly tracked using GFP and its oxygen dependent fluorescence as reporter (Marteyn et al., 2010). During the infection process, *Shigella* also interacts with lymphocytes, and 3D confocal imaging demonstrated the T3SS dependent internalization into T-cells (Konradt et al., 2011). Further investigation of T-cell mobility during *Shigella* infection within lymph nodes by two-photon imaging underlined that the pathogen-lymphocyte interaction stalls T-cell migration (Salgado-Pabon et al., 2013). Quantification of the T-cell trajectories showed a decrease of a mean velocity from 5.4 µm/min to 2.9 µm/min. Analysis of fluorescent bacteria with fluorescent lymphocytes also indicated that a direct bacterial contact was not always required for their deceleration, nevertheless the molecular components that control this process still need to be discovered. It is evident from these descriptive studies that a combination of two-photon imaging with fluorescent probes that monitor cellular functions has a strong potential for the discovery of key steps during *Shigella* infections that interfere with the adaptive immune response.

Conclusion and outlook
Our understanding of *Shigella* infection has benefitted enormously from the implementation and adaptation of novel imaging approaches. Especially, this has brought insights into the intimate contacts between the pathogen with individual host cells, their invasion, and the motility of *Shigella* within the cytoplasm. Delineating these processes with even more detail will require the implementation of novel technologies, and more importantly will need still improved data analysis. First, the functional imaging probes presented in this chapter have important limitations. This is apparent for those that are not genetically encoded, such as the tetracysteine-FlAsH labeling of T3SS effectors or the CCF4 probe for the vacuolar rupture, because they cannot readily be used within an *in vivo* context. On the other hand, genetically encoded sensors lack spatiotemporal sensitivity, for example the galectin-3 labels ruptured membranes only during short periods of time, and FRET probes for GTPase activity measurements do not necessarily localize to their endogenous sites of action. Some probes, like the PH-domain sensors for phosphoinositides content assessment impede the function of the endogenous biomolecules and interfere with the infection process. Therefore, a focus needs to be put on probes that are specific, sensitive and exhibit only little interference with the pathophysiology of *Shigella* infection. Upon improvement of these probes, they can be readily used in commercially available high-content/high-throughput imaging systems to screen loss-of-function phenotypes (through siRNA or microRNA libraries) or chemical compound libraries. A second field of technology improvement will be integrative imaging using multimodular systems that allow the bridging of the different dimensions at which an infection occurs. This requires technologies that span the behavior of biomolecules at the nanoscale level to the morphological rearrangements at the organ level. Such integration appears in sight due to the relatively new super-resolution approaches that allow optical imaging down to about 10 nm. It is anticipated that merging super-resolution microscopy

with ultra-structural methods will bridge cellular morphology with molecular specificity. One limitation is that most available methodologies in those fields are relatively "slow" and difficult to adapt for the rapid internalization of *Shigella* into host cells. A third important direction of technology development in the context of imaging is the establishment of improved and automated data analysis. As described in this chapter different freely available solutions can be downloaded from the web, and they are improved continuously. Further, online databases, for example through the OMERO network render data more accessible and will allow improved comparability between data generated by different research groups. Finally, *Shigella* research becomes increasingly interdisciplinary. In the context of cellular imaging, microengineering has become relevant, for example microfluidic devices allow the miniaturization of experimental set-ups, functionalized surfaces through micro-patterning increase the precision of measured phenotypes, and intelligent microscopy allows feedback between microscopic acquisition and a specific imaging mode. Together, all these improvements of our imaging capabilities will make an important impact on a comprehensive understanding of *Shigella* infection that can also be applied to the study of other intracellular pathogens.

References

Alix, E., S. Mukherjee, and C.R. Roy. 2011. Subversion of membrane transport pathways by vacuolar pathogens. *J Cell Biol.* 195:943-952.

Becken, U., A. Jeschke, K. Veltman, and A. Haas. 2010. Cell-free fusion of bacteria-containing phagosomes with endocytic compartments. *Proc Natl Acad Sci U S A.* 107:20726-20731.

Beuzon, C.R., S. Meresse, K.E. Unsworth, J. Ruiz-Albert, S. Garvis, S.R. Waterman, T.A. Ryder, E. Boucrot, and D.W. Holden. 2000. Salmonella maintains the integrity of its intracellular vacuole through the action of SifA. *Embo J.* 19:3235-3249.

Bishai, E.A., G.S. Sidhu, W. Li, J. Dhillon, A.B. Bohil, R.E. Cheney, J.H. Hartwig, and F.S. Southwick. 2013. Myosin-X facilitates Shigella-induced membrane protrusions and cell-to-cell spread. *Cell Microbiol.* 15:353-367.

Blocker, A., P. Gounon, E. Larquet, K. Niebuhr, V. Cabiaux, C. Parsot, and P. Sansonetti. 1999. The tripartite type III secreton of Shigella flexneri inserts IpaB and IpaC into host membranes. *J Cell Biol.* 147:683-693.

Bobard, A., N. Mellouk, and J. Enninga. 2011. Spotting the right location- imaging approaches to resolve the intracellular localization of invasive pathogens. *Biochim Biophys Acta.* 1810:297-307.

Brotcke Zumsteg, A., C. Goosmann, V. Brinkmann, R. Morona, and A. Zychlinsky. 2014. IcsA is a Shigella flexneri adhesin regulated by the type III secretion system and required for pathogenesis. *Cell Host Microbe.* 15:435-445.

Brumell, J.H., and M.A. Scidmore. 2007. Manipulation of rab GTPase function by intracellular bacterial pathogens. *Microbiol Mol Biol Rev.* 71:636-652.

Buchrieser, C., P. Glaser, C. Rusniok, H. Nedjari, H. D'Hauteville, F. Kunst, P. Sansonetti, and C. Parsot. 2000. The virulence plasmid pWR100 and the repertoire of proteins secreted by the type III secretion apparatus of Shigella flexneri. *Mol Microbiol.* 38:760-771.

Burton, E.A., R. Plattner, and A.M. Pendergast. 2003. Abl tyrosine kinases are required for infection by Shigella flexneri. *Embo J.* 22:5471-5479.

Cameron, L.A., M.J. Footer, A. van Oudenaarden, and J.A. Theriot. 1999. Motility of ActA protein-coated microspheres driven by actin polymerization. *Proc Natl Acad Sci U S A.* 96:4908-4913.

Campbell-Valois, F.X., P. Schnupf, G. Nigro, M. Sachse, P.J. Sansonetti, and C. Parsot. 2014. A fluorescent reporter reveals on/off regulation of the Shigella type III secretion apparatus during entry and cell-to-cell spread. *Cell Host Microbe*. 15:177-189.

Charpentier, X., and E. Oswald. 2004. Identification of the secretion and translocation domain of the enteropathogenic and enterohemorrhagic Escherichia coli effector Cif, using TEM-1 beta-lactamase as a new fluorescence-based reporter. *J Bacteriol*. 186:5486-5495.

Chenouard, N., I. Smal, F. de Chaumont, M. Maska, I.F. Sbalzarini, Y. Gong, J. Cardinale, C. Carthel, S. Coraluppi, M. Winter, A.R. Cohen, W.J. Godinez, K. Rohr, Y. Kalaidzidis, L. Liang, J. Duncan, H. Shen, Y. Xu, K.E. Magnusson, J. Jalden, H.M. Blau, P. Paul-Gilloteaux, P. Roudot, C. Kervrann, F. Waharte, J.Y. Tinevez, S.L. Shorte, J. Willemse, K. Celler, G.P. van Wezel, H.W. Dan, Y.S. Tsai, C. Ortiz de Solorzano, J.C. Olivo-Marin, and E. Meijering. 2014. Objective comparison of particle tracking methods. *Nature methods*. 11:281-289.

Coltharp, C., and J. Xiao. 2012. Superresolution microscopy for microbiology. *Cell Microbiol*. 14:1808-1818.

Coombes, J.L., and E.A. Robey. 2010. Dynamic imaging of host-pathogen interactions in vivo. *Nat Rev Immunol*. 10:353-364.

Creasey, E.A., and R.R. Isberg. 2012. The protein SdhA maintains the integrity of the Legionella-containing vacuole. *Proc Natl Acad Sci U S A*. 109:3481-3486.

Dickenson, N.E., and W.D. Picking. 2012. Forster resonance energy transfer (FRET) as a tool for dissecting the molecular mechanisms for maturation of the Shigella type III secretion needle tip complex. *Int J Mol Sci*. 13:15137-15161.

Drecktrah, D., L.A. Knodler, R. Ireland, and O. Steele-Mortimer. 2006. The mechanism of Salmonella entry determines the vacuolar environment and intracellular gene expression. *Traffic*. 7:39-51.

Dumenil, G., J.C. Olivo, S. Pellegrini, M. Fellous, P.J. Sansonetti, and G.T. Nhieu. 1998. Interferon alpha inhibits a Src-mediated pathway necessary for Shigella-induced cytoskeletal rearrangements in epithelial cells. *J Cell Biol*. 143:1003-1012.

Dupont, N., S. Lacas-Gervais, J. Bertout, I. Paz, B. Freche, G.T. Van Nhieu, F.G. van der Goot, P.J. Sansonetti, and F. Lafont. 2009. Shigella phagocytic vacuolar membrane remnants participate in the cellular response to pathogen invasion and are regulated by autophagy. *Cell Host Microbe*. 6:137-149.

Ehsani, S., J.C. Santos, C.D. Rodrigues, R. Henriques, L. Audry, C. Zimmer, P. Sansonetti, G. Tran Van Nhieu, and J. Enninga. 2012. Hierarchies of host factor dynamics at the entry site of Shigella flexneri during host cell invasion. *Infect Immun*. 80:2548-2557.

Enninga, J., J. Mounier, P. Sansonetti, and G. Tran Van Nhieu. 2005. Secretion of type III effectors into host cells in real time. *Nat Methods*. 2:959-965.

Espina, M., A.J. Olive, R. Kenjale, D.S. Moore, S.F. Ausar, R.W. Kaminski, E.V. Oaks, C.R. Middaugh, W.D. Picking, and W.L. Picking. 2006. IpaD localizes to the tip of the type III secretion system needle of Shigella flexneri. *Infect Immun*. 74:4391-4400.

Fredlund, J., and J. Enninga. 2014. Cytoplasmic access by intracellular bacterial pathogens. *Trends Microbiol*. 22:128-137.

Frischknecht, F., O. Renaud, and S.L. Shorte. 2006. Imaging today's infectious animalcules. *Curr Opin Microbiol*. 9:297-306.

Fukumatsu, M., M. Ogawa, S. Arakawa, M. Suzuki, K. Nakayama, S. Shimizu, M. Kim, H. Mimuro, and C. Sasakawa. 2012. Shigella targets epithelial tricellular junctions and uses a noncanonical clathrin-dependent endocytic pathway to spread between cells. *Cell Host Microbe*. 11:325-336.

Gaietta, G., T.J. Deerinck, S.R. Adams, J. Bouwer, O. Tour, D.W. Laird, G.E. Sosinsky, R.Y. Tsien, and M.H. Ellisman. 2002. Multicolor and electron microscopic imaging of connexin trafficking. *Science*. 296:503-507.

Gawthorne, J.A., L.E. Reddick, S.N. Akpunarlieva, K.S. Beckham, J.M. Christie, N.M. Alto, M. Gabrielsen, and A.J. Roe. 2012. Express your LOV: an engineered flavoprotein as a reporter for protein expression and purification. *PLoS One*. 7:e52962.

Giepmans, B.N., S.R. Adams, M.H. Ellisman, and R.Y. Tsien. 2006. The fluorescent toolbox for assessing protein location and function. *Science*. 312:217-224.

Goldberg, M.B. 2001. Actin-based motility of intracellular microbial pathogens. *Microbiol Mol Biol Rev*. 65:595-626, table of contents.

Gouin, E., H. Gantelet, C. Egile, I. Lasa, H. Ohayon, V. Villiers, P. Gounon, P.J. Sansonetti, and P. Cossart. 1999. A comparative study of the actin-based motilities of the pathogenic bacteria Listeria monocytogenes, Shigella flexneri and Rickettsia conorii. *J Cell Sci*. 112 (Pt 11):1697-1708.

Handa, Y., M. Suzuki, K. Ohya, H. Iwai, N. Ishijima, A.J. Koleske, Y. Fukui, and C. Sasakawa. 2007. Shigella IpgB1 promotes bacterial entry through the ELMO-Dock180 machinery. *Nat Cell Biol*. 9:121-128.

Jaumouille, V., O. Francetic, P.J. Sansonetti, and G. Tran Van Nhieu. 2008. Cytoplasmic targeting of IpaC to the bacterial pole directs polar type III secretion in Shigella. *Embo J*. 27:447-457.

Juruj, C., V. Lelogeais, R. Pierini, M. Perret, B.F. Py, Y. Jamilloux, P. Broz, F. Ader, M. Faure, and T. Henry. 2013. Caspase-1 activity affects AIM2 speck formation/stability through a negative feedback loop. *Front Cell Infect Microbiol*. 3:14.

Kasper, C.A., I. Sorg, C. Schmutz, T. Tschon, H. Wischnewski, M.L. Kim, and C. Arrieumerlou. 2010. Cell-cell propagation of NF-kappaB transcription factor and MAP kinase activation amplifies innate immunity against bacterial infection. *Immunity*. 33:804-816.

Keller, C., N. Mellouk, A. Danckaert, R. Simeone, R. Brosch, J. Enninga, and A. Bobard. 2013. Single cell measurements of vacuolar rupture caused by intracellular pathogens. *J Vis Exp*:e50116.

Kitano, M., M. Nakaya, T. Nakamura, S. Nagata, and M. Matsuda. 2008. Imaging of Rab5 activity identifies essential regulators for phagosome maturation. *Nature*. 453:241-245.

Konradt, C., E. Frigimelica, K. Nothelfer, A. Puhar, W. Salgado-Pabon, V. di Bartolo, D. Scott-Algara, C.D. Rodrigues, P.J. Sansonetti, and A. Phalipon. 2011. The Shigella flexneri type three secretion system effector IpgD inhibits T cell migration by manipulating host phosphoinositide metabolism. *Cell Host Microbe*. 9:263-272.

Kufer, T.A., E. Kremmer, A.C. Adam, D.J. Philpott, and P.J. Sansonetti. 2008. The pattern-recognition molecule Nod1 is localized at the plasma membrane at sites of bacterial interaction. *Cell Microbiol*. 10:477-486.

Lelek, M., F. Di Nunzio, R. Henriques, P. Charneau, N. Arhel, and C. Zimmer. 2012. Superresolution imaging of HIV in infected cells with FlAsH-PALM. *Proc Natl Acad Sci U S A*. 109:8564-8569.

Loisel, T.P., R. Boujemaa, D. Pantaloni, and M.F. Carlier. 1999. Reconstitution of actin-based motility of Listeria and Shigella using pure proteins. *Nature*. 401:613-616.

Malik-Kale, P., S. Winfree, and O. Steele-Mortimer. 2012. The bimodal lifestyle of intracellular Salmonella in epithelial cells: replication in the cytosol obscures defects in vacuolar replication. *PLoS One*. 7:e38732.

Marteyn, B., N.P. West, D.F. Browning, J.A. Cole, J.G. Shaw, F. Palm, J. Mounier, M.C. Prevost, P. Sansonetti, and C.M. Tang. 2010. Modulation of Shigella virulence in response to available oxygen in vivo. *Nature*. 465:355-358.

Mavris, M., A.L. Page, R. Tournebize, B. Demers, P. Sansonetti, and C. Parsot. 2002. Regulation of transcription by the activity of the Shigella flexneri type III secretion apparatus. *Mol Microbiol*. 43:1543-1553.

Mills, E., K. Baruch, X. Charpentier, S. Kobi, and I. Rosenshine. 2008. Real-time analysis

of effector translocation by the type III secretion system of enteropathogenic Escherichia coli. *Cell Host Microbe*. 3:104-113.
Mironov, A.A., and G.V. Beznoussenko. 2012. Correlative light-electron microscopy a potent tool for the imaging of rare or unique cellular and tissue events and structures. *Methods Enzymol*. 504:201-219.
Mostowy, S., M. Bonazzi, M.A. Hamon, T.N. Tham, A. Mallet, M. Lelek, E. Gouin, C. Demangel, R. Brosch, C. Zimmer, A. Sartori, M. Kinoshita, M. Lecuit, and P. Cossart. 2010. Entrapment of intracytosolic bacteria by septin cage-like structures. *Cell Host Microbe*. 8:433-444.
Mostowy, S., L. Boucontet, M.J. Mazon Moya, A. Sirianni, P. Boudinot, M. Hollinshead, P. Cossart, P. Herbomel, J.P. Levraud, and E. Colucci-Guyon. 2013. The zebrafish as a new model for the in vivo study of Shigella flexneri interaction with phagocytes and bacterial autophagy. *PLoS Pathog*. 9:e1003588.
Mostowy, S., and P. Cossart. 2012. Septins: the fourth component of the cytoskeleton. *Nat Rev Mol Cell Biol*. 13:183-194.
Mounier, J., M.R. Popoff, J. Enninga, M.C. Frame, P.J. Sansonetti, and G.T. Van Nhieu. 2009. The IpaC carboxyterminal effector domain mediates Src-dependent actin polymerization during Shigella invasion of epithelial cells. *PLoS Pathog*. 5:e1000271.
Mounier, J., A. Ryter, M. Coquis-Rondon, and P.J. Sansonetti. 1990. Intracellular and cell-to-cell spread of Listeria monocytogenes involves interaction with F-actin in the enterocytelike cell line Caco-2. *Infect Immun*. 58:1048-1058.
Ogawa, M., T. Yoshimori, T. Suzuki, H. Sagara, N. Mizushima, and C. Sasakawa. 2005. Escape of intracellular Shigella from autophagy. *Science*. 307:727-731.
Olive, A.J., R. Kenjale, M. Espina, D.S. Moore, W.L. Picking, and W.D. Picking. 2007. Bile salts stimulate recruitment of IpaB to the Shigella flexneri surface, where it colocalizes with IpaD at the tip of the type III secretion needle. *Infect Immun*. 75:2626-2629.
Paz, I., M. Sachse, N. Dupont, J. Mounier, C. Cederfur, J. Enninga, H. Leffler, F. Poirier, M.C. Prevost, F. Lafont, and P. Sansonetti. 2010. Galectin-3, a marker for vacuole lysis by invasive pathogens. *Cell Microbiol*. 12:530-544.
Pizarro-Cerda, J., and P. Cossart. 2004. Subversion of phosphoinositide metabolism by intracellular bacterial pathogens. *Nat Cell Biol*. 6:1026-1033.
Purdy, G.E., M. Hong, and S.M. Payne. 2002. Shigella flexneri DegP facilitates IcsA surface expression and is required for efficient intercellular spread. *Infect Immun*. 70:6355-6364.
Rafelski, S.M., and J.A. Theriot. 2005. Bacterial shape and ActA distribution affect initiation of Listeria monocytogenes actin-based motility. *Biophys J*. 89:2146-2158.
Ray, K., A. Bobard, A. Danckaert, I. Paz-Haftel, C. Clair, S. Ehsani, C. Tang, P. Sansonetti, G.V. Tran, and J. Enninga. 2010. Tracking the dynamic interplay between bacterial and host factors during pathogen-induced vacuole rupture in real time. *Cell Microbiol*. 12:545-556.
Romero, S., G. Grompone, N. Carayol, J. Mounier, S. Guadagnini, M.C. Prevost, P.J. Sansonetti, and G.T. Van Nhieu. 2011. ATP-mediated Erk1/2 activation stimulates bacterial capture by filopodia, which precedes Shigella invasion of epithelial cells. *Cell Host Microbe*. 9:508-519.
Romero, S., A. Quatela, T. Bornschlogl, S. Guadagnini, P. Bassereau, and G. Tran Van Nhieu. 2012. Filopodium retraction is controlled by adhesion to its tip. *J Cell Sci*. 125:4999-5004.
Salgado-Pabon, W., S. Celli, E.T. Arena, K. Nothelfer, P. Roux, G. Sellge, E. Frigimelica, P. Bousso, P.J. Sansonetti, and A. Phalipon. 2013. Shigella impairs T lymphocyte dynamics in vivo. *Proc Natl Acad Sci U S A*. 110:4458-4463.
Sani, M., A. Botteaux, C. Parsot, P. Sansonetti, E.J. Boekema, and A. Allaoui. 2007. IpaD

is localized at the tip of the Shigella flexneri type III secretion apparatus. *Biochim Biophys Acta*. 1770:307-311.

Sansonetti, P.J., A. Ryter, P. Clerc, A.T. Maurelli, and J. Mounier. 1986. Multiplication of Shigella flexneri within HeLa cells: lysis of the phagocytic vacuole and plasmid-mediated contact hemolysis. *Infect Immun*. 51:461-469.

Schlumberger, M.C., A.J. Muller, K. Ehrbar, B. Winnen, I. Duss, B. Stecher, and W.D. Hardt. 2005. Real-time imaging of type III secretion: Salmonella SipA injection into host cells. *Proc Natl Acad Sci U S A*. 102:12548-12553.

Seveau, S., T.N. Tham, B. Payrastre, A.D. Hoppe, J.A. Swanson, and P. Cossart. 2007. A FRET analysis to unravel the role of cholesterol in Rac1 and PI 3-kinase activation in the InlB/Met signalling pathway. *Cell Microbiol*. 9:790-803.

Shahnazari, S., and J.H. Brumell. 2011. Mechanisms and consequences of bacterial targeting by the autophagy pathway. *Curr Opin Microbiol*. 14:68-75.

Shaner, N.C., R.E. Campbell, P.A. Steinbach, B.N. Giepmans, A.E. Palmer, and R.Y. Tsien. 2004. Improved monomeric red, orange and yellow fluorescent proteins derived from Discosoma sp. red fluorescent protein. *Nat Biotechnol*. 22:1567-1572.

Simeone, R., A. Bobard, J. Lippmann, W. Bitter, L. Majlessi, R. Brosch, and J. Enninga. 2012. Phagosomal rupture by Mycobacterium tuberculosis results in toxicity and host cell death. *PLoS Pathog*. 8:e1002507.

Simpson, N., L. Audry, and J. Enninga. 2010. Tracking the secretion of fluorescently labeled type III effectors from single bacteria in real time. *Methods Mol Biol*. 619:241-256.

Skoudy, A., J. Mounier, A. Aruffo, H. Ohayon, P. Gounon, P. Sansonetti, and G. Tran Van Nhieu. 2000. CD44 binds to the Shigella IpaB protein and participates in bacterial invasion of epithelial cells. *Cell Microbiol*. 2:19-33.

Soo, F.S., and J.A. Theriot. 2005. Large-scale quantitative analysis of sources of variation in the actin polymerization-based movement of Listeria monocytogenes. *Biophys J*. 89:703-723.

Thurston, T.L., M.P. Wandel, N. von Muhlinen, A. Foeglein, and F. Randow. 2012. Galectin 8 targets damaged vesicles for autophagy to defend cells against bacterial invasion. *Nature*. 482:414-418.

Van Engelenburg, S.B., and A.E. Palmer. 2008. Quantification of real-time Salmonella effector type III secretion kinetics reveals differential secretion rates for SopE2 and SptP. *Chem Biol*. 15:619-628.

Van Engelenburg, S.B., and A.E. Palmer. 2010. Imaging type-III secretion reveals dynamics and spatial segregation of Salmonella effectors. *Nature methods*. 7:325-330.

Veenendaal, A.K., J.L. Hodgkinson, L. Schwarzer, D. Stabat, S.F. Zenk, and A.J. Blocker. 2007. The type III secretion system needle tip complex mediates host cell sensing and translocon insertion. *Mol Microbiol*. 63:1719-1730.

Vergunst, A.C., M.C. van Lier, A. den Dulk-Ras, T.A. Stuve, A. Ouwehand, and P.J. Hooykaas. 2005. Positive charge is an important feature of the C-terminal transport signal of the VirB/D4-translocated proteins of Agrobacterium. *Proc Natl Acad Sci U S A*. 102:832-837.

Wong, K.W., and R.R. Isberg. 2005. Yersinia pseudotuberculosis spatially controls activation and misregulation of host cell Rac1. *PLoS Pathog*. 1:e16.

11

Immune Responses Against *Shigella* During Natural Infection and Induced by Vaccination

Francisco Martinez-Becerra

Department of Microbiology and Molecular Genetics, Oklahoma State University, Stillwater, OK 74078 USA, francisco.martinez@ku.edu

Abstract

Shigella is an important gastrointestinal pathogen that affects primarily small children (0-5 years of age) in developing countries. This Gram negative bacterium is a facultative intracellular pathogen that is capable of invading numerous cell types in its human host. Microfold or M cells, macrophages and epithelial cells are the 'main' targets for *Shigella* invasion during infection, however, dendritic cells and T lymphocytes are also susceptible to bacterial invasion with direct consequences on the resulting the immune response. Many host factors contribute to mounting an immune response against *Shigella*, but these are not enough to prevent the onset of infection. Furthermore, the host's adaptive immune response tends to be skewed toward LPS recognition which leads to the development of a serotype specific immunity. Therefore, the host is usually not protected against subsequent infections by *Shigella* bearing different O antigen serotypes. By understanding the immune responses elicited by *Shigella*, we will more fully appreciate the ways in which *Shigella* evades the host response and identify the immune mechanisms that provide the most efficient bacterial clearance. This will allow us to identify adequate correlates of protection that could ultimately lead to effective vaccine candidates to prevent shigellosis.

Introduction

Shigella infection and pathogenicity are closely associated with the immune responses they elicit in host tissues. These immune responses are aimed at preventing infection and eliminating the pathogen, however, *Shigella* is able to bypass them and ultimately induce severe gut pathology. More importantly, the adaptive immune response that results from *Shigella* infection has strong serotype specificity, thus opening the possibility of subsequent attack by different serotypes. This serotype-dependent immunity can play an important role in the context of vaccination. There is currently no approved *Shigella* vaccine.

Shigella interactions with the host innate and adaptive immune responses have been the subject of several reviews (Ashida et al., 2011a; Ashida et al., 2011b; Phalipon and Sansonetti, 2007). In this chapter, immune responses against *Shigella* infection will be examined with an eye toward the future development of an efficacious vaccine. The caveat of this context lies in the possibility that natural infection might follow a different pattern of immune recognition than what would occur in a protective vaccine. This may especially be the case for subunit vaccines, where the antigenic composition and complexity can vary greatly relative to an invading pathogen. Given this possibility, a characterization of immune mechanisms involved in protection by both subunit and live-attenuated vaccines will help to optimize vaccine formulations and ultimately lead to improved vaccine candidates (Pasetti et al., 2011).

Innate barriers and *Shigella*

Shigella infects through the oral route and exposure typically occurs due to ingestion of contaminated food or water. Outbreaks are usually associated to poor hygiene during food handling or the consumption of water contaminated with fecal matter (Black et al., 1978). Houseflies have been also implicated in the transmission of *Shigella*, and control of these insects has been proposed as an effective method of disease prevention (Farag et al., 2013). *Shigella* is able to survive the highly acidic environment of the stomach for periods of at least 2 hours (Gorden and Small, 1993). This high resistance to the acidic environments is not shared by other enteric pathogens such as *Salmonella*, which is thought to account for its relatively low infectious dose of about 100 colony forming units (CFUs) in humans (DuPont et al., 1989).

After surviving passage through the stomach, *Shigella* arrives in the small and then large intestines. There, the first barriers against infection involve a mucoid layer that provides a physical separation between bacteria present in the lumen and the epithelial cells underneath (Duerkop et al., 2009). This barrier not only limits the access of these bacterial populations to the epithelial layer but provides a scaffold for antimicrobial peptides and immunoglobulin A (IgA). Mucins are one of the main protein components of this barrier. These glycoproteins are secreted by Goblet cells present in the intestine and form a gel matrix (Gum et al., 1994). *Shigella* shows a preference for binding to human colonic mucin, thus providing an initial binding step (Sudha et al., 2001). It has been recently proposed that *Shigella* also alters the process of mucin production to enhance its ability to access and invade the underlying epithelial cells (Sperandio et al., 2013). This process is dependent on the presence of the type three secretion system (T3SS) and involves the upregulation of MUC5AC, a mucin which generates a gel structure that favors cell surface accessibility.

Shigella is not able to invade the polarized intestinal epithelial cells through their apical face, so in order to transverse the epithelial barrier this pathogen initially targets Microfold (M) cells. M cells are specialized enterocytes present in the epithelial layer that transport antigens and bacteria from the apical side of the intestine to the lymphoid tissues that are associated underneath (Mabbott et al., 2013). This allows for an efficient method of transport dedicated to antigen and microbiota sampling that promotes immune responses in the gut-associated lymphoid tissues or GALT. During infection, M cells internalize *Shigella* (Sansonetti and Phalipon, 1999; Wassef et al., 1989) and transport the bacterium to the basal side of the epithelial barrier. This transport is independent of bacterial viability, as dead bacteria were also shown to be internalized in a rabbit intestinal loop model (Wassef et al., 1989), therefore targeting the bacterial/M cell interaction may prove to be difficult, however, advances in the analysis of this interaction may prove beneficial to our understanding of this process.

Epithelial cells as activators of the immune response

A primary function of intestinal epithelial cells is absorption of nutrients and water, however, they are also capable of initiating or enhancing the immune response by responding to signals elicited by bacterial products and invasion. One example of these processes is LPS recognition by epithelial cells, which leads to IL-8 expression mediated by NF-κB activation (Philpott et al., 2000). This process recognizes intracellular LPS through the CARD4/Nod1 sensor present in the cytoplasm of epithelial cells (Girardin et al., 2001). IL-8 secretion is important for recruitment of cellular effectors to the site of infection, but it also represents a molecule involved in inflammation mediated damage (Sansonetti et al., 1999). The natural lack of IL-8 in mice seems to play an important role in several animal models of *Shigella* infection, as the observed inflammatory profile is different from the one observed in humans. Studies using Caco2 cells were performed by infecting monolayers with both wild-type and attenuated *Shigella* strains. Secretion of IL-8 was observed with both the wild-type strain and the attenuated vaccine strains (Fiorentino et al., 2014).

The role of antimicrobial peptides during *Shigella* infection

Antimicrobial peptides form a natural barrier against bacterial infections, especially at mucosal sites in the gut (Muniz et al., 2012). *Shigella* interacts with the host antimicrobial peptide systems in many ways (Gudmundsson et al., 2010). The levels of antimicrobial peptides in biopsies obtained from patients infected with *Shigella* suggest there is a down-regulation in both transcript and protein levels for LL37 and HBD-1. Both of these antimicrobial peptides have shown to be able to kill all four species of *Shigella* in *in vitro* studies. This down-regulation effect is also observed when epithelial and macrophage lines are incubated with bacterial extracts (Islam et al., 2001). The relevance of this expression is highlighted when oral infection of newborn mice is analyzed. While 4 day old animals are susceptible to intragastric infection, 7 day old animals are not. One of the differences in gene expression is in antimicrobial peptides (*i.e.* α-5 defensin) linked to Paneth cell maturation (Fernandez et al., 2008). Other antimicrobial peptides important for this differential susceptibility in mice include cryptidins and lysozyme, whose expression levels follow a pattern of resistance/susceptibility (Shim et al., 2010). *Shigella* is also able to directly down regulate epithelial cell expression of several other antimicrobial peptides, including hBD-3, the antimicrobial chemokine CCL20 and LL37. These peptides are known to be able to directly kill *Shigella* in antimicrobial assays (Sperandio et al., 2008). This down regulation of antimicrobial peptides is reversible when epithelial cells are stimulated with phenylbutyrate in rabbit infection models (Sarker et al., 2011). In agreement with this, butyrate induction of the antimicrobial peptide CAP-18 reduced severity of disease in experimentally infected rabbits (Raqib et al., 2006). Interestingly, TH17 responses are associated with antimicrobial responses via IL-17 and IL-22 (Liang et al., 2006). The fact that these responses are modulated by components of the adaptive immune response components opens the possibility that vaccines can modify or enhance antimicrobial responses via upregulation of specific T cell populations that enhance these responses in order to eliminate *Shigella* during infection. This link will be further discussed below in section on TH17 cells.

Neutrophils and NETS

During infection by *Shigella*, one of the most common histological findings is the presence of a large infiltrate to the infected tissues that contains large numbers of neutrophils (Anand et al., 1986). Among the mechanisms that neutrophils use to eliminate bacterial infections is the secretion of granules containing a variety of enzymes and antibacterial molecules, which can have a direct effect on *Shigella* survival (Francois et al., 2000; Mandic-Mulec et al., 1997). Interestingly, neutrophil-derived elastase is able to directly cleave bacterial invasion factors such as IpaB and C, which are necessary for virulence (Weinrauch et al., 2002). This may show a specific targeting of virulence associated proteins by the protease. Another recently described mechanism for neutrophil activity is the secretion of Neutrophil extracellular Traps, or NETS. NETs are composed of granules and chromatin that are released in order to bind and kill bacteria (Brinkmann et al., 2004). *S. flexneri* is susceptible to killing by this mechanism in a PAD-4 mediated manner, since cells lacking this molecule fail to reduce bacterial viability (Li et al., 2010). Interestingly, *Shigella* is able to exploit this response, as it has been shown that exposure to neutrophil NETs result in increased bacterial invasion, possibly due to binding of cationic molecules to the surface of bacteria that ultimately lead to enhanced adherence by the pathogen (Eilers et al., 2010). The NET mediated killing of bacteria can be activated by a pro-inflammatory environment (Brinkmann and Zychlinsky, 2007; Clark et al., 2007) caused by the CXCL-8 (IL-8) chemokine and/or LPS. In addition, opsonization by antibodies can also induce the formation of NETs (Fuchs et al., 2007). Therefore, it is possible that vaccines against *Shigella* can augment this response by inducing antibody production. An immune response that includes chemokine secretion and antibody mediated opsonization could provide a microenvironment that promotes bacterial clearance by such a mechanism.

Autophagy

Autophagy is the process by which cells eliminate damaged organelles and insoluble protein complexes. A related process called xenophagy may also be important during intracellular bacterial infections (Deretic, 2011). In the latter, host cells may be able to recognize and eliminate a variety of pathogens (Jo et al., 2013), including *Shigella* (Ogawa and Sasakawa, 2006). In the case of *Shigella*, an amino acid starvation program has been described as an initiator of autophagy in response to intracellular infection (Tattoli et al., 2012). This stress signal results in mTOR down-regulation which, in turn, leads to activation of autophagy mechanisms. Recognition of VirG during *Shigella* infection of epithelial cells is also a trigger for autophagy following an interaction with Atg5 (Ogawa et al., 2005). This recognition was demonstrated by immunoprecipitation experiments showing direct binding between these proteins. Other inductors of autophagy in *Shigella* include Shiga toxins, which are able to induce autophagy in THP-1 cells through caspase mediated cleavage of Atg5 and Beclin-1(Lee et al., 2011).

As would be required for intracellular survival, *Shigella* is able to escape autophagy mechanisms by blocking VirG recognition via IcsB activity (Kayath et al., 2010; Ogawa et al., 2005). IcsB activity seems to be based on masking of VirG recognition by Atg5. Mutants of IcsB are deficient in long term multiplication and survival in epithelial cells, probably due to increased susceptibility to autophagy. This mechanism of bacterial clearance seems to be highly relevant during *Shigella* infections. Thus, new models for studying these interaction, including a new zebrafish infection model (Mostowy et al., 2013), will provide tools for providing a full description of the role of autophagy in *Shigella* pathogenesis. The role of IcsB highlights the importance of this clearance mechanism since mutant bacteria that induce autophagy fail to invade and spread to adjacent cells. Furthermore, TNF-α and CXCL8 enhance autophagy against *Shigella* during infection (Dong et al., 2012; Mostowy et al., 2010; Mostowy et al., 2011), highlighting the role of cytokines in triggering autophagy. Once again, the latter mechanism suggests a host process that could be enhanced by vaccination regimens that generate the appropriate cytokine secretion profiles.

Macrophages

Macrophages are innate immune cells derived from the myeloid lineage with phagocytic activities. These cells are involved in many immune functions, including direct killing by internalization and phagosome maturation. These cells also have antigen presenting activities, which are responsible for activation of the early steps in some adaptive immune responses. When *Shigella* crosses intestinal M cells as described above, it encounters resident macrophages. One of the central interactions during *Shigella* infection is the induction of apoptosis in these first line macrophage cells. This killing was first demonstrated *in vitro* using J774 macrophages (Clerc et al., 1987) and it was later demonstrated *in vivo* during infection of rabbit Peyer's patches (Zychlinsky et al., 1996). Apoptosis is induced in macrophages by a caspase-1 mediated mechanism, which leads to IL-1β activation. This activity is induced by the *Shigella* invasion/translocator protein IpaB and macrophages incubated with this protein have been shown to undergo apoptosis *in vitro*. IpaB was shown by Zychlinsky's group to directly bind caspase-1 in macrophages (Hilbi et al., 1998). This interaction between bacterium and phagocyte is ultimately necessary for escape of the pathogen from these macrophages, which is an essential step in the pathogenesis of *Shigella* infection.

Recognition of *Shigella* by macrophages is achieved by interaction of NAIP proteins with the T3SS needle protein MxiH. This interaction results in caspase-1 activation and nod-like receptor NLRC4 inflammasome induction through NAIP1 (Yang et al., 2013). This activation was also found for T3SS needle proteins from other pathogens such as *Salmonella* and *E. coli*, suggesting a conserved pattern recognition mechanism. LPS from *Shigella* is also able to trigger TLR4 mediated recognition which leads to cytokine secretion and pathogen clearance activities. However, *Shigella* is able to modify the structure of LPS while inside epithelial cells to create a less immunogenic form (Paciello et al., 2013). This provides a mechanism for reducing the activation of inflammasome pathways in macrophages.

Macrophage activity against pathogens is induced by IFN-γ, and this activation is essential for a proper antibacterial activity by these cells in response to *Shigella* infection, as studies using mouse macrophages stimulated with IFN-γ showed that activated cells were able to control infection, while unstimulated macrophages failed to control infection (Way et al., 1998). This activation allows the upregulation of Nitric oxide (NO) activity in macrophages, however, this activity is not required for clearance *in vivo* (Way and Goldberg, 1998).

The studies described above imply that macrophages have an important role during infection, however, the nature of this role depends upon the activation state of these cells by the local cytokine environment. Furthermore, *Shigella* employs multiple effector mechanisms that alter the normal activities of the cells it encounters in order to escape clearance by macrophages. The resulting inflammation and cell death (Hathaway et al., 2002) can facilitate bacterial survival and infection by tissue disruption and release of nutrients (Guichon and Zychlinsky, 1996).

Natural killer cells

The role of the cytotoxic lymphocytes called natural killer (NK) cells has been studied in the context of *Shigella* infection. Experiments with mice deficient in NK cells showed that these cells are important for controlling infection, possibly due to their IFN-γ secretion activity, as mice deficient in IFN-γ demonstrated more exacerbated disease (Way et al., 1998). Furthermore, the presence of NK cells in immunodeficient animals lacking T and B cells was enough to provide resistance against a pulmonary infection in mice, which is in contrast to Rag°γc° mice that lack T, B and NK cells and fail to control infection (Le-Barillec et al., 2005). The observation that NK cells are able to secrete IFN-γ in response to macrophage death caused by *Shigella* infection expands the role of these cells during infection and its possible role in protection as effectors or modulators during vaccination(Kubota, 2006).

Linking innate and adaptive immunity: Dendritic cells

Dendritic cells are phagocytic cells that provide an important link between innate and adaptive immune responses. These cells are highly efficient in activating naïve T cells, thus initiating a pathway that leads to cell mediated immunity. Dendritic cells are also able to convey specific lineage programs upon lymphocytes, thereby modulating T helper (Th) polarization into Th1, 2 and 17 populations, in addition to others (Banchereau et al., 2000). Moreover, dendritic cell/T lymphocyte interaction instructs T cells to acquire specific migration patterns, which is controlled by the expression of homing molecules such as CCR9 and α4β7 integrin (Johansson-Lindbom and Agace, 2007). Meanwhile, dendritic cells located in the gut associated lymphoid tissue (GALT) are able to induce IgA secretion and increased levels of CCR9 and α4β7 (Mora et al., 2006) in B cells independently of T cells. Therefore, dendritic cells play an important role directing the proper localization of immune responses by instructing cells to migrate to specific organs and tissues.

Dendritic cells are susceptible to invasion by *Shigella* (Edgeworth et al., 2002; Kim et al., 2008) and this invasion results in caspase-mediated cell death in a process that requires an active *Shigella* T3SS and the secretion of IpaB. Another T3SS effector protein, OspF directly binds Erk 1/2 and modulates its phosphorylation state, thus down-regulating the immune response by reducing the activation state of dendritic cells (Kim et al., 2008). *Shigella* infection also results in an increase of expression of TLR4 but not TLR5 in dendritic cells, which in contrast to the up-regulation of both molecules by enteroinvasive *E. coli*. (Moreno et al., 2012). Dendritic cells are also able to detect needle proteins using NAIP1, which is similar to what was described above for macrophages (Yang et al., 2013).

In addition to directly influencing the properties of infected dendritic cells, *Shigella* is also able to alter the recruitment of dendritic cells to the site of epithelial cell infection. Epithelial cells infected with *Shigella* express lower levels of CCL20, a dendritic cell recruiting chemokine (Sperandio et al., 2008). This alteration could modulate the recruitment of large numbers of dendritic cells and other CCR6 expressing cells, such as certain T lymphocyte subpopulations, thereby reducing the overall number of immune cells that are able to respond to the infection.

B cells and antibodies

Several studies have explored the role of antibodies in *Shigella* infection. Administration of IgA antibodies directed against LPS to naïve mice has been shown to be sufficient to protect them using the *Shigella* lethal pulmonary challenge model (Phalipon et al., 1994). The presence of serotype specific mucosal antibodies is an important occurrence in humans (Rasolofo-Razanamparany et al., 2001), and it is thought by many to be the basis for the serotype dependent immunity observed in natural infections (Rasolofo-

Razanamparany et al., 2001). Antibody neutralization of LPS leads to a diminution of the inflammatory response, thereby reducing the tissue damage that accompanies *Shigella* infection of the intestinal epithelial barrier (Boullier et al., 2009). Another activity derived from antibody binding to LPS components is the suppression of *Shigella* T3SS activity, which leads to a reduction of the virulence and invasion activities of the pathogen. This reduction in virulence activity has been proposed to be related to a loss of membrane potential and concomitant decrease in intracellular ATP due to disruption of the proton motive force (Forbes et al., 2011). A mathematical analysis exploring the relationship of infection dynamics with antibody binding showed that the role of antibodies during vaccination relies heavily on the amount of immunoglobulins present in the site of infection, thus suggesting that the targeting of LPS alone is not be enough to provide protection (Davis et al., 2013). Other proteins such as Ipa proteins could provide additional targets for antibody activities that reduce pathogenesis (Simon et al., 2009).

T cell responses

While most of the described adaptive immune responses described for *Shigella* infection explore the secretion and activity of antibodies against a variety of components of the bacterium, the presence of antigen specific T cells was described in patients infected with *Shigella* (Islam and Christensson, 2000). In particular, populations of CD8+ cells and γδ T cells were increased in rectal biopsies obtained from patients infected with *S. dysenteriae* and *S. flexneri*. Interestingly, differences between the responses elicited by the two infecting species were observed. A role for T cells during adoptive transfer of immunity in mice has also been observed, most notably when antibodies against the TCR were used when transferring spleen cells from infected mice to protect naïve animals (Kowalewska and Wieczorek, 1980). Furthermore, Rag°γc° mice (which lack B, T and NK cells) are highly susceptible to *S. flexneri* pulmonary infection. IFN-γ secreting αβ T cells (which include CD4 and CD8 T cells) are able to rescue this phenotype, when transferred to Rag °γc° mice (Le-Barillec et al., 2005).

CD8 T cells are cells specialized for the clearance of intracellular pathogens via MHC-I recognition of host cells with subsequent lysis by granzyme secretion. These cells are generated in response to *Shigella* infection in patients infected with both *S. flexneri* and *S. dysenteriae* (Islam et al., 1995). However, the specific roles of these cells in protection are not clear because these cells are not required for the immunity conferred by sublethal *S. flexneri* 2a intranasal administration in mice (Way et al., 1999). Interestingly, the priming of these cells during infection of murine cells is compromised by the fact that infected cells exhibit low levels of peptide-presenting MHC class I molecules on their surface (Jehl et al., 2011). This reduction is dependent upon bacterial viability and active cellular invasion processes. It is worth noting, however, that an attenuated *Shigella* platform has been proposed as a vaccine vector for antiviral responses mediated by CD8 specific responses (Shata and Hone, 2001; Sizemore et al., 1995). This suggests that under certain attenuation conditions CD8 mediated responses can be promoted (Noriega et al., 1994). The role of CD8 cells seems to be limited in the case of sublethal challenges, but the use of attenuated strains as vectors for CD8 responses against viral antigens show that these cells might play a role in protection when these strains are used as vaccines against *Shigella*. Additional information is needed to evaluate the role of CD8 cells in the protection afforded by subunit vaccines, however, recognition of bacterial components presented in MHC-I molecules seems feasible.

Several T cell subpopulations are generated in response to a variety of bacterial infections. In the case of a mouse lethal pulmonary model, *Shigella* infection leads to the generation of Th17 cells. These cell populations secrete IL-22 and IL-17A and they have been described as being crucial for bacterial responses at mucosal sites. Interestingly, these cells are important for controlling bacterial growth during secondary infections. This observation is important especially when considering vaccine development because some of the immune responses elicited by vaccines seem to resemble a secondary infection (Sellge et al., 2010). During infection of human volunteers with non-Shiga toxin producing *S. dysenteria*, peripheral blood mononuclear cells (PBMCs) showed increased secretion levels of IFN-γ and IL-10 in response to *in vitro* stimulation with bacterial homogenates, as well as to the purified invasion plasmid antigens IpaC and IpaD, highlighting the recognition of these antigens by T cell populations during natural infection (Samandari et al., 2000).

Inhibition of T cell migration by *Shigella* has also been described as a means by which this pathogen can evade adaptive immunity in the host. One potential mechanism for this is hydrolysis of phosphatidylinositol biphosphate (PIPP) by the T3SS effector IpgD, which then interferes with signaling pathoways such as those involving G protein coupled receptors. *Shigella* is capable of invading activated $CD4^{+}$ T cells after which hydrolysis of PIPP leads to inhibition of shape change and ultimately inhibits chemokine mediated migration (Konradt et al., 2011). Injection of IpgD can also occur for T cells in the absence of cellular invasion (a bystander effect) where IpgD contributes to the depletion of PIPP in the cytoplasmic membrane and reduction of the signaling pathways governed by this phospholipid. These inhibitions can directly impair T cell mediated clearance of bacterial infections. Other alterations induced by *Shigella* include alteration of movement kinetics in lymph nodes, where infected lymphocytes fail to show normal movement and migration behavior (Salgado-Pabon et al., 2013). This process is mediated by T3SS, and invasion of CD4 T cells by *Shigella* can be detected.

Non-conventional T cells can also play a role in *Shigella* infection. Mucosa-associated invariant T (MAIT) cells (Treiner et al., 2003) are cells that recognize metabolites of vitamin B that are presented in an MHC related molecule that is recognized by their invariant TCR. MAIT cells are able to lyse *Shigella*-infected epithelial cells by secretion of granzyme after recognition of MR1 (Le Bourhis et al., 2013), in addition, this recognition leads to secretion of cytokines such as IFN-γ and TNF-α via CD161 (NKR-P1A, or killer cell lectin-like receptor). This activity could be of special relevance in mucosal sites, where these cells are enriched.

Discussion and future considerations

Full characterization of the immune responses involved in *Shigella* infection and pathogen clearance remains an important goal in the context of vaccine development. The role of antibodies against O-antigens in protection is a well characterized mechanism that allows control of invasion and bacterial clearance, however, with the development of potentially efficacious vaccines that do not contain O-antigens it is necessary to look for additional effector cells and molecules that participate in an alternative protective immune response. The important roles of multiple innate immune response mechanisms during *Shigella* infection and the potential activation of parallel adaptive immune responses opens the possibility that vaccine responses can target different aspects of the host response to infection. Different innate mechanisms that need to be considered include activation of activities related to the expression of antimicrobial peptides, autophagy and/ or neutrophil mediated clearance. The role of cells such as NK cells, neutrophils and macrophages can be evaluated in the context of vaccination to provide a clearer picture of the interactions that contribute to bacterial clearance.

Dendritic cells are an important cell lineage that is a target for many of *Shigella*'s immunomodulatory effector proteins. Because of the central role that dendritic cells have in antigen presentation and their potential value in directing the pathways of protective immunity, LPS-free vaccines against *Shigella* infection may well need target these cell populations at sites where they best give rise to protective immunity and in the context of appropriate adjuvants. T cells seem to also play an important role in both control of cytokine secretion in response to *Shigella* antigens and in modulation of the polarizing environments that will lead to cell differentiation and recruitment of specific effector populations. Preferential activation of those T cell populations most likely to lead to protection must also be the goal of any future successful *Shigella* vaccines. Emphasis should be put in analyzing unconventional T cell populations due to their enrichment in mucosal sites and their role in immunity against bacterial pathogens. All of these factors can play important roles during protective responses mediated by vaccines or during recurrent infections. Finally, due to the varying nature of the vaccine candidates against *Shigella*, multiple effector mechanisms will have different roles in protection against infection.

References

Anand, B.S., Malhotra, V., Bhattacharya, S.K., Datta, P., Datta, D., Sen, D., Bhattacharya, M.K., Mukherjee, P.P., and Pal, S.C. (1986). Rectal histology in acute bacillary dysentery. Gastroenterology *90*, 654-660.

Ashida, H., Ogawa, M., Kim, M., Suzuki, S., Sanada, T., Punginelli, C., Mimuro, H., and Sasakawa, C. (2011a). Shigella deploy multiple countermeasures against host innate immune responses. Current opinion in microbiology *14*, 16-23.

Ashida, H., Ogawa, M., Mimuro, H., Kobayashi, T., Sanada, T., and Sasakawa, C. (2011b). Shigella are versatile mucosal pathogens that circumvent the host innate immune system. Current opinion in immunology *23*, 448-455.

Banchereau, J., Briere, F., Caux, C., Davoust, J., Lebecque, S., Liu, Y.J., Pulendran, B., and Palucka, K. (2000). Immunobiology of dendritic cells. Annual review of immunology *18*, 767-811.

Black, R.E., Craun, G.F., and Blake, P.A. (1978). Epidemiology of common-source outbreaks of shigellosis in the United States, 1961-1975. American journal of epidemiology *108*, 47-52.

Boullier, S., Tanguy, M., Kadaoui, K.A., Caubet, C., Sansonetti, P., Corthesy, B., and Phalipon, A. (2009). Secretory IgA-mediated neutralization of Shigella flexneri prevents intestinal tissue destruction by down-regulating inflammatory circuits. J Immunol *183*, 5879-5885.

Brinkmann, V., Reichard, U., Goosmann, C., Fauler, B., Uhlemann, Y., Weiss, D.S., Weinrauch, Y., and Zychlinsky, A. (2004). Neutrophil extracellular traps kill bacteria. Science *303*, 1532-1535.

Brinkmann, V., and Zychlinsky, A. (2007). Beneficial suicide: why neutrophils die to make NETs. Nature reviews Microbiology *5*, 577-582.

Clark, S.R., Ma, A.C., Tavener, S.A., McDonald, B., Goodarzi, Z., Kelly, M.M., Patel, K.D., Chakrabarti, S., McAvoy, E., Sinclair, G.D., *et al.* (2007). Platelet TLR4 activates neutrophil extracellular traps to ensnare bacteria in septic blood. Nature medicine *13*, 463-469.

Clerc, P.L., Ryter, A., Mounier, J., and Sansonetti, P.J. (1987). Plasmid-mediated early killing of eucaryotic cells by Shigella flexneri as studied by infection of J774 macrophages. Infection and immunity *55*, 521-527.

Davis, C.L., Wahid, R., Toapanta, F.R., Simon, J.K., Sztein, M.B., and Levy, D. (2013). Applying mathematical tools to accelerate vaccine development: modeling Shigella immune dynamics. PloS one *8*, e59465.

Deretic, V. (2011). Autophagy in immunity and cell-autonomous defense against intracellular microbes. Immunological reviews *240*, 92-104.

Dong, N., Zhu, Y., Lu, Q., Hu, L., Zheng, Y., and Shao, F. (2012). Structurally distinct bacterial TBC-like GAPs link Arf GTPase to Rab1 inactivation to counteract host defenses. Cell *150*, 1029-1041.

Duerkop, B.A., Vaishnava, S., and Hooper, L.V. (2009). Immune responses to the microbiota at the intestinal mucosal surface. Immunity *31*, 368-376.

DuPont, H.L., Levine, M.M., Hornick, R.B., and Formal, S.B. (1989). Inoculum size in shigellosis and implications for expected mode of transmission. The Journal of infectious diseases *159*, 1126-1128.

Edgeworth, J.D., Spencer, J., Phalipon, A., Griffin, G.E., and Sansonetti, P.J. (2002). Cytotoxicity and interleukin-1beta processing following Shigella flexneri infection of human monocyte-derived dendritic cells. European journal of immunology *32*, 1464-1471.

Eilers, B., Mayer-Scholl, A., Walker, T., Tang, C., Weinrauch, Y., and Zychlinsky, A. (2010). Neutrophil antimicrobial proteins enhance Shigella flexneri adhesion and invasion. Cellular microbiology *12*, 1134-1143.

Farag, T.H., Faruque, A.S., Wu, Y., Das, S.K., Hossain, A., Ahmed, S., Ahmed, D., Nasrin, D., Kotloff, K.L., Panchilangam, S., *et al.* (2013). Housefly Population Density Correlates with Shigellosis among Children in Mirzapur, Bangladesh: A Time Series Analysis. PLoS neglected tropical diseases *7*, e2280.

Fernandez, M.I., Regnault, B., Mulet, C., Tanguy, M., Jay, P., Sansonetti, P.J., and Pedron, T. (2008). Maturation of paneth cells induces the refractory state of newborn mice to Shigella infection. J Immunol *180*, 4924-4930.

Fiorentino, M., Levine, M.M., Sztein, M.B., and Fasano, A. (2014). Effect of wild-type Shigella species and attenuated Shigella vaccine candidates on small intestinal barrier function, antigen trafficking, and cytokine release. PloS one *9*, e85211.

Forbes, S.J., Bumpus, T., McCarthy, E.A., Corthesy, B., and Mantis, N.J. (2011). Transient suppression of Shigella flexneri type 3 secretion by a protective O-antigen-specific monoclonal IgA. mBio *2*, e00042-00011.

Francois, M., Le Cabec, V., Dupont, M.A., Sansonetti, P.J., and Maridonneau-Parini, I. (2000). Induction of necrosis in human neutrophils by Shigella flexneri requires type III secretion, IpaB and IpaC invasins, and actin polymerization. Infection and immunity *68*, 1289-1296.

Fuchs, T.A., Abed, U., Goosmann, C., Hurwitz, R., Schulze, I., Wahn, V., Weinrauch, Y., Brinkmann, V., and Zychlinsky, A. (2007). Novel cell death program leads to neutrophil extracellular traps. The Journal of cell biology *176*, 231-241.

Girardin, S.E., Tournebize, R., Mavris, M., Page, A.L., Li, X., Stark, G.R., Bertin, J., DiStefano, P.S., Yaniv, M., Sansonetti, P.J., *et al.* (2001). CARD4/Nod1 mediates NF-kappaB and JNK activation by invasive Shigella flexneri. EMBO reports *2*, 736-742.

Gorden, J., and Small, P.L. (1993). Acid resistance in enteric bacteria. Infection and immunity *61*, 364-367.

Gudmundsson, G.H., Bergman, P., Andersson, J., Raqib, R., and Agerberth, B. (2010). Battle and balance at mucosal surfaces--the story of Shigella and antimicrobial peptides. Biochemical and biophysical research communications *396*, 116-119.

Guichon, A., and Zychlinsky, A. (1996). Apoptosis as a trigger of inflammation in Shigella-induced cell death. Biochemical Society transactions *24*, 1051-1054.

Gum, J.R., Jr., Hicks, J.W., Toribara, N.W., Siddiki, B., and Kim, Y.S. (1994). Molecular cloning of human intestinal mucin (MUC2) cDNA. Identification of the amino terminus and overall sequence similarity to prepro-von Willebrand factor. The Journal of biological chemistry *269*, 2440-2446.

Hathaway, L.J., Griffin, G.E., Sansonetti, P.J., and Edgeworth, J.D. (2002). Human monocytes kill Shigella flexneri but then die by apoptosis associated with suppression of proinflammatory cytokine production. Infection and immunity *70*, 3833-3842.

Hilbi, H., Moss, J.E., Hersh, D., Chen, Y., Arondel, J., Banerjee, S., Flavell, R.A., Yuan, J., Sansonetti, P.J., and Zychlinsky, A. (1998). Shigella-induced apoptosis is dependent on caspase-1 which binds to IpaB. The Journal of biological chemistry *273*, 32895-32900.

Islam, D., Bandholtz, L., Nilsson, J., Wigzell, H., Christensson, B., Agerberth, B., and Gudmundsson, G. (2001). Downregulation of bactericidal peptides in enteric infections: a novel immune escape mechanism with bacterial DNA as a potential regulator. Nature medicine *7*, 180-185.

Islam, D., Bardhan, P.K., Lindberg, A.A., and Christensson, B. (1995). Shigella infection induces cellular activation of T and B cells and distinct species-related changes in peripheral blood lymphocyte subsets during the course of the disease. Infection and immunity *63*, 2941-2949.

Islam, D., and Christensson, B. (2000). Disease-dependent changes in T-cell populations in patients with shigellosis. APMIS : acta pathologica, microbiologica, et immunologica Scandinavica *108*, 251-260.

Jehl, S.P., Doling, A.M., Giddings, K.S., Phalipon, A., Sansonetti, P.J., Goldberg, M.B., and Starnbach, M.N. (2011). Antigen-specific CD8(+) T cells fail to respond to Shigella flexneri. Infection and immunity *79*, 2021-2030.

Jo, E.K., Yuk, J.M., Shin, D.M., and Sasakawa, C. (2013). Roles of autophagy in elimination of intracellular bacterial pathogens. Frontiers in immunology *4*, 97.

Johansson-Lindbom, B., and Agace, W.W. (2007). Generation of gut-homing T cells and their localization to the small intestinal mucosa. Immunological reviews *215*, 226-242.

Kayath, C.A., Hussey, S., El hajjami, N., Nagra, K., Philpott, D., and Allaoui, A. (2010). Escape of intracellular Shigella from autophagy requires binding to cholesterol through the type III effector, IcsB. Microbes and infection / Institut Pasteur *12*, 956-966.

Kim, D.W., Chu, H., Joo, D.H., Jang, M.S., Choi, J.H., Park, S.M., Choi, Y.J., Han, S.H., and Yun, C.H. (2008). OspF directly attenuates the activity of extracellular signal-regulated kinase during invasion by Shigella flexneri in human dendritic cells. Molecular immunology *45*, 3295-3301.

Konradt, C., Frigimelica, E., Nothelfer, K., Puhar, A., Salgado-Pabon, W., di Bartolo, V., Scott-Algara, D., Rodrigues, C.D., Sansonetti, P.J., and Phalipon, A. (2011). The Shigella flexneri type three secretion system effector IpgD inhibits T cell migration by manipulating host phosphoinositide metabolism. Cell host and microbe *9*, 263-272.

Kowalewska, D., and Wieczorek, Z. (1980). Adoptive transfer of immunity to dysentery bacilli with immune T lymphocytes from mice orally immunized with Shigella sonnei bacilli. Archivum immunologiae et therapiae experimentalis *28*, 593-601.

Kubota, K. (2006). A novel functional T cell hybridoma recognizes macrophage cell death induced by bacteria: a possible role for innate lymphocytes in bacterial infection. J Immunol *176*, 7576-7588.

Le-Barillec, K., Magalhaes, J.G., Corcuff, E., Thuizat, A., Sansonetti, P.J., Phalipon, A., and Di Santo, J.P. (2005). Roles for T and NK cells in the innate immune response to Shigella flexneri. J Immunol *175*, 1735-1740.

Le Bourhis, L., Dusseaux, M., Bohineust, A., Bessoles, S., Martin, E., Premel, V., Core, M., Sleurs, D., Serriari, N.E., Treiner, E., *et al.* (2013). MAIT Cells Detect and Efficiently Lyse Bacterially-Infected Epithelial Cells. PLoS pathogens *9*, e1003681.

Lee, M.S., Cherla, R.P., Jenson, M.H., Leyva-Illades, D., Martinez-Moczygemba, M., and Tesh, V.L. (2011). Shiga toxins induce autophagy leading to differential signalling pathways in toxin-sensitive and toxin-resistant human cells. Cellular microbiology *13*, 1479-1496.

Li, P., Li, M., Lindberg, M.R., Kennett, M.J., Xiong, N., and Wang, Y. (2010). PAD4 is essential for antibacterial innate immunity mediated by neutrophil extracellular traps. The Journal of experimental medicine *207*, 1853-1862.

Liang, S.C., Tan, X.Y., Luxenberg, D.P., Karim, R., Dunussi-Joannopoulos, K., Collins, M., and Fouser, L.A. (2006). Interleukin (IL)-22 and IL-17 are coexpressed by Th17 cells and cooperatively enhance expression of antimicrobial peptides. The Journal of experimental medicine *203*, 2271-2279.

Mabbott, N.A., Donaldson, D.S., Ohno, H., Williams, I.R., and Mahajan, A. (2013). Microfold (M) cells: important immunosurveillance posts in the intestinal epithelium. Mucosal immunology *6*, 666-677.

Mandic-Mulec, I., Weiss, J., and Zychlinsky, A. (1997). Shigella flexneri is trapped in polymorphonuclear leukocyte vacuoles and efficiently killed. Infection and immunity *65*, 110-115.

Mora, J.R., Iwata, M., Eksteen, B., Song, S.Y., Junt, T., Senman, B., Otipoby, K.L., Yokota, A., Takeuchi, H., Ricciardi-Castagnoli, P., *et al.* (2006). Generation of gut-homing IgA-secreting B cells by intestinal dendritic cells. Science *314*, 1157-1160.

Moreno, A.C., Ferreira, K.S., Ferreira, L.G., Almeida, S.R., and Martinez, M.B. (2012). Recognition of enteroinvasive Escherichia coli and Shigella flexneri by dendritic cells: distinct dendritic cell activation states. Memorias do Instituto Oswaldo Cruz *107*, 138-141.

Mostowy, S., Bonazzi, M., Hamon, M.A., Tham, T.N., Mallet, A., Lelek, M., Gouin, E., Demangel, C., Brosch, R., Zimmer, C., *et al.* (2010). Entrapment of intracytosolic bacteria by septin cage-like structures. Cell host and microbe *8*, 433-444.

Mostowy, S., Boucontet, L., Mazon Moya, M.J., Sirianni, A., Boudinot, P., Hollinshead, M., Cossart, P., Herbomel, P., Levraud, J.P., and Colucci-Guyon, E. (2013). The zebrafish as a new model for the in vivo study of Shigella flexneri interaction with phagocytes and bacterial autophagy. PLoS pathogens *9*, e1003588.

Mostowy, S., Sancho-Shimizu, V., Hamon, M.A., Simeone, R., Brosch, R., Johansen, T., and Cossart, P. (2011). p62 and NDP52 proteins target intracytosolic Shigella and Listeria to different autophagy pathways. The Journal of biological chemistry *286*, 26987-26995.

Muniz, L.R., Knosp, C., and Yeretssian, G. (2012). Intestinal antimicrobial peptides during homeostasis, infection, and disease. Frontiers in immunology *3*, 310.

Noriega, F.R., Wang, J.Y., Losonsky, G., Maneval, D.R., Hone, D.M., and Levine, M.M. (1994). Construction and characterization of attenuated delta aroA delta virG Shigella flexneri 2a strain CVD 1203, a prototype live oral vaccine. Infection and immunity *62*, 5168-5172.

Ogawa, M., and Sasakawa, C. (2006). Shigella and autophagy. Autophagy *2*, 171-174.

Ogawa, M., Yoshimori, T., Suzuki, T., Sagara, H., Mizushima, N., and Sasakawa, C. (2005). Escape of intracellular Shigella from autophagy. Science *307*, 727-731.

Paciello, I., Silipo, A., Lembo-Fazio, L., Curcuru, L., Zumsteg, A., Noel, G., Ciancarella, V., Sturiale, L., Molinaro, A., and Bernardini, M.L. (2013). Intracellular Shigella remodels its LPS to dampen the innate immune recognition and evade inflammasome activation. Proceedings of the National Academy of Sciences of the United States of America *110*, E4345-4354.

Pasetti, M.F., Simon, J.K., Sztein, M.B., and Levine, M.M. (2011). Immunology of gut mucosal vaccines. Immunological reviews *239*, 125-148.

Phalipon, A., Michetti, P., Kaufmann, M., Cavaillon, J.M., Huerre, M., Kraehenbuhl, J.P., and Sansonetti, P.J. (1994). Protection against invasion of the mouse pulmonary epithelium by a monoclonal IgA directed against Shigella flexneri lipopolysaccharide. Annals of the New York Academy of Sciences *730*, 356-358.

Phalipon, A., and Sansonetti, P.J. (2007). Shigella's ways of manipulating the host intestinal innate and adaptive immune system: a tool box for survival? Immunology and cell biology *85*, 119-129.

Philpott, D.J., Yamaoka, S., Israel, A., and Sansonetti, P.J. (2000). Invasive Shigella flexneri activates NF-kappa B through a lipopolysaccharide-dependent innate intracellular response and leads to IL-8 expression in epithelial cells. J Immunol *165*, 903-914.

Raqib, R., Sarker, P., Bergman, P., Ara, G., Lindh, M., Sack, D.A., Nasirul Islam, K.M., Gudmundsson, G.H., Andersson, J., and Agerberth, B. (2006). Improved outcome in shigellosis associated with butyrate induction of an endogenous peptide antibiotic. Proceedings of the National Academy of Sciences of the United States of America *103*, 9178-9183.

Rasolofo-Razanamparany, V., Cassel-Beraud, A.M., Roux, J., Sansonetti, P.J., and Phalipon, A. (2001). Predominance of serotype-specific mucosal antibody response in Shigella flexneri-infected humans living in an area of endemicity. Infection and immunity *69*, 5230-5234.

Salgado-Pabon, W., Celli, S., Arena, E.T., Nothelfer, K., Roux, P., Sellge, G., Frigimelica, E., Bousso, P., Sansonetti, P.J., and Phalipon, A. (2013). Shigella impairs T lymphocyte dynamics in vivo. Proceedings of the National Academy of Sciences of the United States of America *110*, 4458-4463.

Samandari, T., Kotloff, K.L., Losonsky, G.A., Picking, W.D., Sansonetti, P.J., Levine, M.M., and Sztein, M.B. (2000). Production of IFN-gamma and IL-10 to Shigella invasins by mononuclear cells from volunteers orally inoculated with a Shiga toxin-deleted Shigella dysenteriae type 1 strain. J Immunol *164*, 2221-2232.

Sansonetti, P.J., and Phalipon, A. (1999). M cells as ports of entry for enteroinvasive pathogens: mechanisms of interaction, consequences for the disease process. Seminars in immunology *11*, 193-203.

Sansonetti, P.J., Tran Van Nhieu, G., and Egile, C. (1999). Rupture of the intestinal epithelial barrier and mucosal invasion by Shigella flexneri. Clinical infectious diseases : an official publication of the Infectious Diseases Society of America *28*, 466-475.

Sarker, P., Ahmed, S., Tiash, S., Rekha, R.S., Stromberg, R., Andersson, J., Bergman, P., Gudmundsson, G.H., Agerberth, B., and Raqib, R. (2011). Phenylbutyrate counteracts Shigella mediated downregulation of cathelicidin in rabbit lung and intestinal epithelia: a potential therapeutic strategy. PloS one *6*, e20637.

Sellge, G., Magalhaes, J.G., Konradt, C., Fritz, J.H., Salgado-Pabon, W., Eberl, G., Bandeira, A., Di Santo, J.P., Sansonetti, P.J., and Phalipon, A. (2010). Th17 cells are the dominant T cell subtype primed by Shigella flexneri mediating protective immunity. J Immunol *184*, 2076-2085.

Shata, M.T., and Hone, D.M. (2001). Vaccination with a Shigella DNA vaccine vector induces antigen-specific CD8(+) T cells and antiviral protective immunity. Journal of virology *75*, 9665-9670.

Shim, D.H., Ryu, S., and Kweon, M.N. (2010). Defensins play a crucial role in protecting mice against oral Shigella flexneri infection. Biochemical and biophysical research communications *401*, 554-560.

Simon, J.K., Wahid, R., Maciel, M., Jr., Picking, W.L., Kotloff, K.L., Levine, M.M., and Sztein, M.B. (2009). Antigen-specific B memory cell responses to lipopolysaccharide

(LPS) and invasion plasmid antigen (Ipa) B elicited in volunteers vaccinated with live-attenuated Shigella flexneri 2a vaccine candidates. Vaccine *27*, 565-572.

Sizemore, D.R., Branstrom, A.A., and Sadoff, J.C. (1995). Attenuated Shigella as a DNA delivery vehicle for DNA-mediated immunization. Science *270*, 299-302.

Sperandio, B., Fischer, N., Joncquel Chevalier-Curt, M., Rossez, Y., Roux, P., Robbe Masselot, C., and Sansonetti, P.J. (2013). Virulent Shigella flexneri affects the secretion, expression, and glycosylation of gel-forming mucins in mucus-producing cells. Infection and immunity.

Sperandio, B., Regnault, B., Guo, J., Zhang, Z., Stanley, S.L., Jr., Sansonetti, P.J., and Pedron, T. (2008). Virulent Shigella flexneri subverts the host innate immune response through manipulation of antimicrobial peptide gene expression. The Journal of experimental medicine *205*, 1121-1132.

Sudha, P.S., Devaraj, H., and Devaraj, N. (2001). Adherence of Shigella dysenteriae 1 to human colonic mucin. Current microbiology *42*, 381-387.

Tattoli, I., Sorbara, M.T., Vuckovic, D., Ling, A., Soares, F., Carneiro, L.A., Yang, C., Emili, A., Philpott, D.J., and Girardin, S.E. (2012). Amino acid starvation induced by invasive bacterial pathogens triggers an innate host defense program. Cell host and microbe *11*, 563-575.

Treiner, E., Duban, L., Bahram, S., Radosavljevic, M., Wanner, V., Tilloy, F., Affaticati, P., Gilfillan, S., and Lantz, O. (2003). Selection of evolutionarily conserved mucosal-associated invariant T cells by MR1. Nature *422*, 164-169.

Wassef, J.S., Keren, D.F., and Mailloux, J.L. (1989). Role of M cells in initial antigen uptake and in ulcer formation in the rabbit intestinal loop model of shigellosis. Infection and immunity *57*, 858-863.

Way, S.S., Borczuk, A.C., Dominitz, R., and Goldberg, M.B. (1998). An essential role for gamma interferon in innate resistance to Shigella flexneri infection. Infection and immunity *66*, 1342-1348.

Way, S.S., Borczuk, A.C., and Goldberg, M.B. (1999). Thymic independence of adaptive immunity to the intracellular pathogen Shigella flexneri serotype 2a. Infection and immunity *67*, 3970-3979.

Way, S.S., and Goldberg, M.B. (1998). Clearance of Shigella flexneri infection occurs through a nitric oxide-independent mechanism. Infection and immunity *66*, 3012-3016.

Weinrauch, Y., Drujan, D., Shapiro, S.D., Weiss, J., and Zychlinsky, A. (2002). Neutrophil elastase targets virulence factors of enterobacteria. Nature *417*, 91-94.

Yang, J., Zhao, Y., Shi, J., and Shao, F. (2013). Human NAIP and mouse NAIP1 recognize bacterial type III secretion needle protein for inflammasome activation. Proceedings of the National Academy of Sciences of the United States of America *110*, 14408-14413.

Zychlinsky, A., Thirumalai, K., Arondel, J., Cantey, J.R., Aliprantis, A.O., and Sansonetti, P.J. (1996). In vivo apoptosis in Shigella flexneri infections. Infection and immunity *64*, 5357-5365.

12

Dissecting *Shigella flexneri* Type III Secretion System Tip Complex Maturation

Olivia Arizmendi[1] and Nicholas E. Dickenson[2]

[1]Department of Pharmaceutical Chemistry, The University of Kansas, Lawrence, Kansas 66047, USA, olivia.arizmendi@ku.edu
[2]Department of Chemistry and Biochemistry, Utah State University, Logan, Utah 84322, USA, nick.dickenson@usu.edu

Abstract
The type III secretion system (T3SS) is used by many Gram negative bacteria as a means for communicating with target eukaryotic cells. In the most thoroughly characterized systems, the T3SS is used to subvert normal cellular functions as a weapon in the bacterium's virulence arsenal. The T3SS machinery consists of two main structural components. The first is a basal body which is a complex assemblage of proteins that spans the bacterial periplasm and is anchored in the bacterial inner and outer membranes. The second is an exposed needle structure that is involved in sensing host cell contact and delivering the effector proteins responsible for subverting host functions directly into the eukaryotic cell's membrane and cytoplasm. The secreted effector proteins vary greatly with respect to function from one pathogen to another and these effectors ultimately determine the outcome of the pathogen-host interaction. In contrast, there is notable structural and functional homology among the type III secretion apparatuses (T3SAs) studied thus far. Despite such conservation, the mechanisms by which different T3SSs sense the environment and mature is only now being understood. The identification of protein complexes at the tip of the exposed needle that are responsible for secretion control and determination of the structures for some needle and needle tip complex proteins has provided a starting point for dissecting the conserved and divergent aspects of type III secretion induction. This chapter discusses the structures and mechanisms of maturation of the T3SA tip complex, focusing on the model system, *Shigella flexneri*.

Bacterial protein secretion
Bacteria have developed numerous systems for delivering proteins across their membrane(s). The movement of proteins out of the cell, whether into the periplasm or into the extracellular milieu, is essential for bacterial survival. Secreted proteins are needed for cell wall biosynthesis, biogenesis/maintenance of the outer membrane, transport,

modification of the bacterial environment (*e.g.* hydrolysis of macromolecules as food sources, detoxification or the release of toxins that provide competitive advantages), establishment of microbial communities (*e.g.* attachment to biotic or abiotic surfaces and communication with other bacteria), and conjugal transfer of genetic material to name a few of the important roles played by these exported macromolecules (Deane et al., 2006a; Heras et al., 2014; Jessen et al., 2014; Konovalova et al., 2014; Morita et al., 2014). There has long been an appreciation of the importance of protein secretion and this is underscored by the fact that nearly one fifth of the proteins produced by bacteria are destined to reside at least partially outside of the cell (Pugsley, 1993). The majority of these proteins are secreted by the general secretory pathway (GSP) which entails transport into or across the cytoplasmic membrane via a secretion signal (called Sec). The Sec signal is located at the N terminus of the protein (Pugsley, 1993) where it is recognized by the Sec machinery (Desvaux et al., 2004) and is cleaved during the translocation across the cytoplasmic membrane. The Sec signal can routinely be identified using bioinformatics approaches based on the placement of positively charged and nonpolar residues. In Gram negative bacteria, Sec-dependent export results in either integration of the protein into the cytoplasmic membrane or export into the periplasm. In parallel with the Sec-dependent secretion pathway is the twin-arginine translocator (Tat) system, which also results in the export of cargo proteins into the periplasm of Gram negative bacteria (De Buck et al., 2008; Robinson et al., 2011).

Because of the double membrane system of Gram negative bacteria and other restraints such as the thick, hydrophobic cell wall of *mycobacterium*, secretion of proteins into the extracellular milieu often requires steps that go beyond the Tat and Sec-dependent pathways. These additional pathways show conservation between the organisms that employ them and are designated as the type I to type IX secretion systems (Chagnot et al., 2013). In many cases, Sec-dependent export of proteins into the periplasm is critical for the assembly of these secretion systems as they often require complex machinery that resides in the inner membrane, the outer membrane, or both. In the case of the autotransporters (also called type V secretion systems), the first step of secretion involves the Sec pathway and the second step involves self-insertion of a translocation domain into the outer membrane followed by threading of the cargo through the large β-barrel structure of the translocation domain to a position on the surface of the bacterium (Gawarzewski et al., 2013). From this position, the cargo can remain as a surface protein or be auto-catalytically cleaved to allow release into the environment. Other multistep processes that allow for the secretion of proteins from the periplasm to the cell exterior include those involved in the assembly of pili such as the chaperone-usher pathway (type II secretion pathway) which may also be involved in the formation of pili or in toxin secretion (Ayers et al., 2010). The recently identified type VIII secretion system, also referred to as the extracellular nucleation-precipitation pathway, is responsible for transport of prepilins across the outer membrane where they assemble into fimbriae at the bacterial surface (Evans and Chapman, 2014). Many of the details of the most recently named secretion system (type IX) remain unknown, however, it has been found to represent a unique system in members of the *Bacteroidetes* phylum where it may secrete a diverse set of proteins ranging from components responsible for gliding motility to proteases required for virulence (McBride and Zhu, 2013).

Many organisms, including Gram negative bacteria have also evolved mechanisms that allow the transport of proteins from the cytoplasm to the extracellular environment in a single step. For example, the type I secretion system which is also known as the ATP-binding cassette (ABC) transport system is often involved in the extracellular release of hydrolases (Kanonenberg et al., 2013) while the type IV secretion system represents a diverse group of secretion machines that can be involved in the transfer of genetic material into other bacteria or the transport of DNA and/or proteins into targeted eukaryotic cells (Guglielmini et al., 2014). The type VI secretion system was first discovered in *Vibrio cholera* and *Pseudomonas aeruginosa* (Mougous et al., 2006; Pukatzki et al., 2006), but has since been identified in numerous proteobacteria including both plant and animal pathogens where it has been shown to play an important role in pathogenesis as well as in providing a defense mechanism against other microorganisms (Basler et al., 2013). The type VII secretion system, also known as the ESX secretion system in mycobacteria, provides an avenue of transport across the exceptionally impermeable cell wall of mycobacteria and is believed to have homologous systems in some Gram positive bacteria as well (Abdallah et al., 2007). Finally, the type III secretion system is related to the flagellar biosynthetic pathway and appears to be dedicated to injecting proteins into target eukaryotic cells (Buttner, 2012). The role of the type III secretion system is highly coordinated and controls secretion of proteins across both bacterial membranes and the cytoplasmic membrane of a designated host to modify the normal behavior of the host cell. Due to this particular role, most of the identified type III secretion systems have been shown to be linked to the virulence of the bacterium, and are regarded as highly structurally conserved between different species and across phyla (Demers et al., 2013). Structural homology has in turn provided the flexibility of establishing the core aspects of the pathogenesis-structure interface in one organism and further establishing parallels in similar pathogens (Reis and Horn, 2010). This has certainly been the case with *Shigella flexneri*, which, as a non-motile pathogen requiring mild biosafety precautions and a single T3SS encoded in a virulence plasmid, has provided a valuable model for studying T3SSs of other bacterial systems, including those of the highly epidemiologically relevant *E. coli* and *S. enterica,* as well as other high profile pathogens including *Yersinia* and *Burkholderia* species. The remainder of this chapter will be dedicated to examining the nature of the *Shigella* T3SS with a focus on the extracellular tip complex and its maturation in response to extracellular signaling and induction of the secretion process.

T3SS pathogenicity basis *of Shigella* spp.

Shigella spp. virulence and its ability to invade the human colonic epithelium is conferred by a 220-kb virulence plasmid (pINV), which is also present in the close relative enteroinvasive *E. coli* (EIEC) (Lan et al., 2003). Loss of this megaplasmid by either organism renders the bacteria avirulent as it has been shown that epithelial cell entry, rupture of the endocytic vacuole, induction of pyroptosis, and several more virulence traits are all under the control of the plasmid's ipa-mxi-spa loci (Schuch et al., 1999; Watarai et al., 1995; Xiong et al., 2006). This entry region encodes the structural components of the T3SS through VirF, an AraC type transcriptional regulator that is activated at 37°C (Prosseda et al., 2004), enabling the transcription of VirB, another transcriptional regulator responsible for controlling production of the T3SA components (Kane and Dorman, 2012). Expression and assembly of the apparatus components, however, is only the first step in pathogenicity, as this nanomachine appears to remain dormant prior to activation by contact with specific host cell membrane components. Recently, in fact, a novel *S. flexneri* fluorescent reporter system was developed that takes advantage of this tight transcriptional control to follow the on/off regulation of the T3SA, finding that T3SA activity peaks during initial cell entry and cell-to-cell spread (Campbell-Valois et al., 2014). Activation is marked by the expression of early effectors encoded both in the virulence plasmid and the bacterial chromosome as well as by the secretion of effectors through the T3SA, all of which are necessary for the virulence of the pathogen.

Because of this, there has recently been a great deal of effort toward understanding the activation of the T3SS related to molecular signaling, structural intermediates, and mechanisms involved in maturation of the T3SA tip complex. Again, while the focus of this chapter is the T3SA tip complex in *S. flexneri*, conservation among Gram negative T3SSs suggests that these findings can serve as a platform for work in many other systems as well.

Type III secretion apparatus

The type III secretion apparatus is a molecular complex that bears structural resemblance to flagella and is commonly described as a syringe and needle. The T3SA is composed of 25 or more different protein types, the majority of which are dedicated to forming the basal body of the needle complex that anchors the apparatus to both the inner and outer bacterial membranes (Burkinshaw and Strynadka, 2014). This basal structure possesses a cytoplasmic bulb that contributes to controlling the hierarchy of secretion resultant in apparatus assembly and controlled release of effector proteins (Barison et al., 2013). Spa47 is part of this cytoplasmic bulb and is a member of an ATPase family sharing similarity with the β subunit of F_0F_1 ATPases. This ATPase is believed to provide at least a portion of the catalytic power of the T3SS, perhaps at the point of loading the system with cargo after separating the effector protein targeted for secretion from its cognate chaperone. Energy to power secretion may also be provided by this ATPase, however, the transport of protons across the cytoplasmic membrane has also been implicated with this process and the precise driving forces and mode of regulation remain unclear (Ohgita et al., 2013). It has been shown however, that additional levels of regulation are achieved by a series of proteins located at the cytoplasmic face of the T3SA that act as substrate switches as well as the use of several protein chaperones to time the specific points of assembly of this apparatus. In *Shigella* spp., MxiC interacts with Spa47 to avoid premature secretion of effectors (but not of translocators or structural components), presumably, acting as a T3SA substrate that would activate secretion only once a translocon is properly assembled in the host cell membrane (Botteaux et al., 2009; Martinez-Argudo and Blocker, 2010). Similarly, the class II chaperone, IpgC, has been shown to bind to the translocators IpaB and IpaC, thus preventing their toxicity in the bacterial cytosol and signaling to the bacteria to proceed with the release of effectors once IpgC is freed (Birket et al., 2007; Page et al., 1999).

The remaining "structural" components of the needle complex basal body consist of: a) The inner membrane rings embedded in the cytoplasmic membrane; b) The periplasmic or inner rod that spans the periplasm; and c) The outer membrane ring that connects the base with the extracellular portions of the T3SA. Like the T3SA tip complex, the precise architecture and roles of the individual basal body components have just recently begun to be understood and while the details are beyond the scope of this chapter, another within this book and several thorough reviews are available that describe the structural and functional roles of the basal body components (Barison et al., 2013; Cornelis, 2006; Erhardt et al., 2010; Yip and Strynadka, 2006). Extending outward from the base and beyond the outer membrane of the pathogen, the T3SA possesses an external needle with an associated "tip complex". These are the structures exposed to the extracellular environment where they would be expected to have a role in sensing cues needed for T3SS induction and apparatus maturation. The needle itself is typically 45 to 60 nm in length, with an outer diameter of approximately 7 nm and a 2 to 3 nm diameter inner channel through which T3SS components are secreted (Enninga and Rosenshine, 2009; Schraidt and Marlovits, 2011). Based on early electron microscopy and more recently cryoelectron microscopy reconstructions of the *Shigella* needle, the MxiH monomers that make up the needle have been found to pack in a precise helical fashion to give rise to an extended cylinder with an architecture of 5.6 subunits per turn and a 24 Å helical pitch (Cordes et al., 2003; Deane et al., 2006b; Fujii et al., 2012). This packing is not unlike that seen for the FliC filament of Gram negative flagella which is also composed of a polymer of a single protein type. High-resolution structures for the MxiH needle monomers have been reported, identifying a helix-turn-helix motif (Deane et al., 2006a; Fujii et al., 2012) and allowing coordination of the high-resolution monomer structure with the lower resolution needle EM reconstructions, providing valuable insight into the mechanism of polymerization and needle assembly (Deane et al., 2006a; Fujii et al., 2012). Furthermore, several interacting residues between MxiH and IpaD have been established through mutational analyses, showing that the needle architecture provides the crucial base from which the tip complex is assembled, ultimately resulting in an invasive phenotype (Zhang et al., 2007).

Biochemistry of the tip protein IpaD

IpaD is a hydrophilic 37 kDa protein that is secreted by the T3SS and resides at the tip of the MxiH needle as a high molecular weight complex. IpaD is required for virulence and proper secretion of translocators and effectors, as null mutants both fail to invade cells and display a hypersecretion phenotype, in which several bacterial proteins, including the translocators IpaB and IpaC, are continuously secreted onto the extracellular environment (Espina et al., 2006b). Recombinant IpaD is highly soluble and stable under aqueous conditions, allowing its structure to be studied by complementary techniques such as Fourier transform infrared spectroscopy (FTIR) and circular dichroism spectroscopy (CD). These studies provided early insight suggesting that IpaD is a mainly alpha-helical protein with a core consisting of coiled-coils. Thermodynamic studies of recombinant IpaD using scanning calorimetry further suggested that IpaD has two distinct domains capable of independently interacting with the protein core (Espina et al., 2006a), which was later confirmed with a 3.0Å crystal structure (Johnson et al., 2007). The structure of IpaD is shown in Figure 12.1A, with the central coiled-coil in green, the small N-terminal domain in blue, and the distal domain in pink.

While solving the IpaD structure provided valuable insight into its role in controlling type three secretion, it is the body of work directed toward understanding IpaD's role in T3SA maturation and T3SS secretion control that has highlighted the importance of IpaD in the *Shigella* T3SS and ultimately virulence. Early in these studies, antibodies against the N-terminus of IpaD were shown to block pore formation in erythrocytes, pointing to the fact that IpaD must be exposed in the T3SA and that accessibility of specific regions are required for continued formation of the tip complex and ultimately a virulence phenotype (Picking et al., 2005). This has been supported by a series of transmission electron microscopy studies which began by using immuno-gold labeling techniques to indirectly locate IpaD at the exposed tip of immature needles (Espina et al., 2006b). More recently, three-dimensional electron density maps resulting from transmission electron microscopy reconstructions have revealed that IpaD not only resides at the tip of nascent MxiH needles, but that it resides as a homo-pentamer, providing a septor-like structure that is likely responsible for proper secretion control (Epler et al., 2012) (Figure 12.1B-D). The same study found that deletion of the distal region of IpaD (Δ192-267) maintained a pentameric tip complex, but one that formed an "open" structure incapable of controlling subsequent traslocator secretion, resulting in a non-invasive phenotype (Epler et al., 2012). These results not only provide structural insight into the ability of IpaD to control secretion, but tie together previous *in vitro* biophysical studies which used FRET and NMR to characterize conformational changes within IpaD following interaction with the bile salt deoxycholate (DOC) (Dickenson et al., 2011). Results from these studies as well as from later crystal structures of both DOC bound and apo forms of IpaD identified the DOC binding site of IpaD near the N-terminal domain between helices α3 and α7 (Barta et al., 2012b) (green in Figure 12.1A). Somewhat unexpectedly, however, this interaction results in a switch in the orientation of one of the long central helices (helix α7) promoting significant conformational changes at the IpaD distal domain. This evidence suggests that IpaD serves not only as a structural scaffold for the T3SA tip complex, but that it is also an environmental sensor for small molecules such as the bile salt DOC (found at high concentration within the small intestine of its human host).

Maturation of the tip complex

It has been shown that IpaD is central to maintaining secretion control of the *Shigella* T3SS and likely acts as a sensor for environmental small molecules, however, virulence

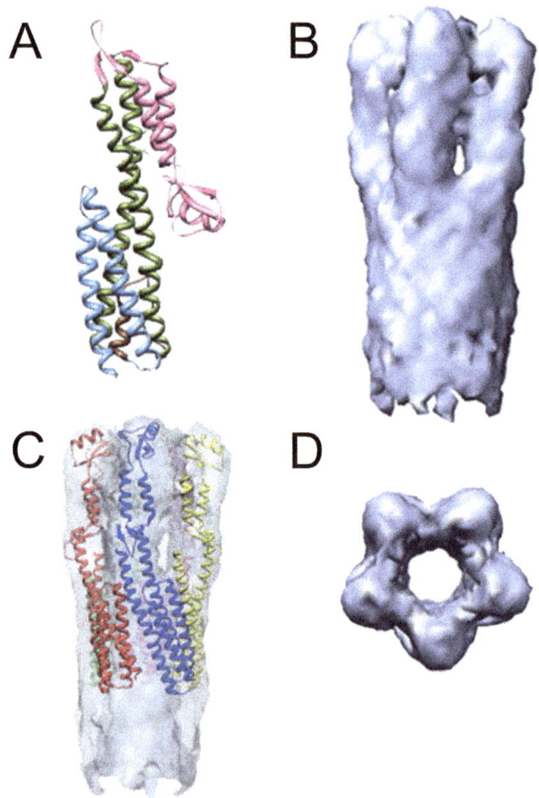

Figure 12.1. Structure of the IpaD tip complex of *S. flexneri*. (A) X-ray crystal structure of IpaD from *S. flexneri* (PDB ID 2J0O). The distal domain is shown in pink, the N-terminal domain in blue, the central coiled-coil in green, and the small 2' helix is shown in brown. A 3-D TEM electron density map generated from wild-type *S. flexneri* needle tip complexes highlights the pentameric IpaD structure (B), which becomes more apparent when the crystal structure is modeled into the EM density map (C). This required that the IpaD distal domain (pink) be rotated upward from the central coiled coil (green) approximately 165 degrees, potentially providing means for controlling the channel diameter of the tip complex (D). This figure was modified from reference (Epler et al., 2012) with permission from the publisher.

of the organism relies on a series of events resulting in the step-wise maturation of the T3SA and ultimately induction of the T3SS (Marteyn et al., 2012; Parsot, 2009). Through our work and the findings of many other labs, we propose an updated model for *S. flexneri* T3SS tip complex maturation (Figure 12.2). The first step in type III secretion induction includes the DOC-induced recruitment of the hydrophobic translocator protein, IpaB, to the tip of the needle where it forms a defined oligomeric structure with the awaiting pentamer of the tip protein IpaD. It is hypothesized that this IpaD/IpaB tip structure contains a distal homotetramer if IpaB that is capable of sensing host cell contact, recruiting the second translocator, IpaC, ultimately forming the translocon pore within the target cell cytoplasmic membrane (Blocker et al., 1999; Epler et al., 2009). The details of the structural contribution of IpaC to the translocon have not been determined, but it is clear that IpaC recruitment is important in completing the translocon pore and providing an open passageway between the bacterial and host cytoplasm. Following full secretion induction, effector proteins are injected directly through the conduit formed by the T3SA basal body, needle, and translocon into the target cell cytoplasm (Bahrani et al., 1997; Parsot, 2009). Interestingly, secretion induction can be activated artificially by exposure of *Shigella* cultures to small amphipathic dyes such as Congo red or Evans blue (Bahrani et al., 1997). The mechanism for secretion induction by these small molecules is not clear; however, it has provided a valuable means for *Shigella* phenotype characterization and has led to a series of works investigating the specific triggers and mechanisms of *Shigella* T3SA maturation and secretion induction. A summary of these studies and their implications in this model for *Shigella* T3SA maturation will be discussed.

IpaB represents the first translocator protein mobilized to the maturing *Shigella* T3SA tip complex. It is a hydrophobic 62 kDa protein that is not found as a part of the nascent needle tip complex with IpaD and is not required for localization of IpaD at the needle tip (Olive et al., 2007). Conversely, though it is hypersecreted through the needle, IpaB cannot localize to the needle tip in the absence of IpaD, indicating that it is required for IpaB recruitment and subsequent formation of a functional translocon (Olive et al., 2007). Several recent studies involving IpaB have provided insight into its role in the complex series of events that ultimately lead to formation of the translocon and an open channel between the pathogen and host cell cytoplasm. First, exposure of the bacteria to small molecules such as DOC recruits and maintains IpaB at the needle tip, which has been observed by both immunofluorescence techniques and high-resolution immunogold localization detected by transmission electron microscopy (Olive et al., 2007) (Figure 12.3). This recruitment can be thought of as a priming step preparing the bacteria for contact with its host cell target. In fact, exposure of *S. flexneri* cultures to DOC not only brings IpaB to the needle tip, but also increases its ability to invade cultured cells by several hundred percent, suggesting that bile salt exposure during infection may be a critical event for *Shigella* virulence (Pope et al., 1995). The precise mechanism for this enhancement is not well understood, though it appears that DOC may play a multifaceted role in the virulence of the organism. First, IpaB has been shown to bind cholesterol and CD44 during invasion (Hayward et al., 2005; Skoudy et al., 2000), and it is proposed that its presence at the tip of the needle may serve *Shigella* as an environmental sensor via host cell membrane recognition (Epler et al., 2009). Additionally, DOC exposure has been implicated in IscA (VirG) mediated host cell adherence which also leads to increased invasiveness of the organism (Brotcke Zumsteg et al., 2014). While the two mechanisms and signaling pathways appear not to be directly linked, *S. flexneri* lacking needle structures (Δspa33) lose adherence sensitivity to DOC exposure, suggesting that the T3SA plays a critical role in this signaling cascade (Brotcke Zumsteg et al., 2014).

The size and hydrophobic nature of IpaB has made *in vitro* biochemical analysis slow to come. However, detergent-based purification techniques (Dickenson et al., 2013b) and the use of limited proteolysis to solve the structure of the stable IpaB core (Barta et al., 2012a) have guided several important follow-up studies. Together, these works have elucidated key points in our understanding of how IpaB is able to interact with IpaD in the context of the maturation of the tip complex: first, a report of differential oligomeric states by *in vitro* interactions of IpaB with select detergents showed that although monomeric IpaB was able to interact with liposomes, it was only when it was found forming homo-tetramers that it was capable of triggering formation of pores that efficiently released small molecules from within (Dickenson et al., 2013b), suggesting that this tetrameric state may represent the scaffold for the translocon formed at the tip of the mature *Shigella* T3SA. Secondly, several soluble N-terminal peptides of IpaB were generated, allowing the first direct observations of interaction between IpaD and IpaB (Dickenson et al., 2013a). These results showed that the N-terminal peptides of IpaB are capable of strong interactions with IpaD and that this interaction was dependent on the presence of DOC. Furthermore, inter- and intra-molecular FRET studies found that the interaction was dependent on the ability of the IpaD distal domain to move away from the central coiled-coil where it has been speculated to serve as a self-chaperone protecting against premature interaction with IpaB (Dickenson et al., 2013a). These findings help to bring

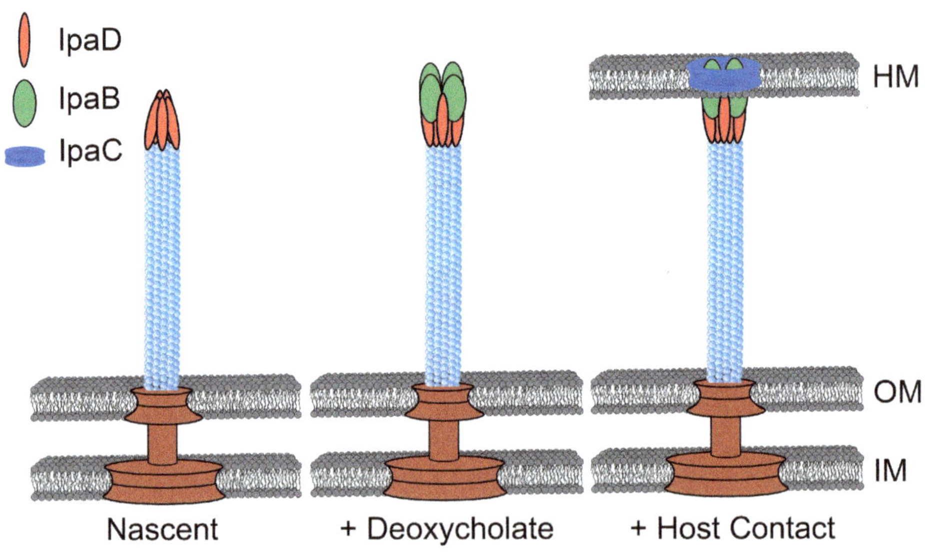

Figure 12.2. Proposed step-wise model for maturation of the *S. flexneri* type III secretion system tip complex. The nascent complex consists of the basal body spanning the inner (IM) and outer (OM) membranes of the bacteria, an exposed hollow needle, and a pentameric structure of IpaD which is responsible for controlling secretion at this step. Exposure to the bile salt deoxycholate results in a conformational change within the individual units of the IpaD pentamer, allowing IpaB to mobilize from the bacterial cytoplasm and needle channel to the tip complex where it binds IpaD and likely resides as a tetramer. The IpaB complex is now responsible for sensing contact with host cell membrane (HM) components including cholesterol and sphingomyelin. Though the mechanisms and stoichiometry remain unclear, this recruits IpaC to the maturing complex where it completes the translocon pore spanning the host membrane and induces secretion of effectors into the host cell cytoplasm.

together the previously discussed results that found DOC-induced conformational changes in IpaD near the distal domain of the protein structure (Barta et al., 2012b; Dickenson et al., 2011). This also reinforces the evidence found with TEM reconstructions of the IpaD needle tip identifying the distal domains of the homo-pentameric complex in an upward extended conformation, which appears to be responsible for control of effector secretion (Epler et al., 2012) (Figure 12.1A and 1C). From these and the body of other works, we are beginning to understand the triggers and mechanisms controlling *Shigella* type three secretion and the steps involved in the recruitment of IpaB to the maturing tip complex. It is known, however, that this only represents an intermediate form of the mature complex and that the recruitment of the second hydrophobic traslocator, IpaC, is required for completion of the translocon and induction of full effector secretion into the host cell.

IpaC is a potent effector in the *Shigella* T3SS arsenal, responsible for directing actin rearrangement via Cdc42 activation (Terry et al., 2008; Tran Van Nhieu et al., 1999), ultimately orchestrating the macropinocytotic uptake of the pathogen by the non-phagocytic host cell (Mounier et al., 2009). IpaC has also been shown to play an important structural role in the *Shigella* T3SA and specifically in formation of the mature translocon. Both IpaB and IpaC are found in isolated red blood cell membranes following incubation with virulent *Shigella* and while organisms lacking IpaC are still capable of

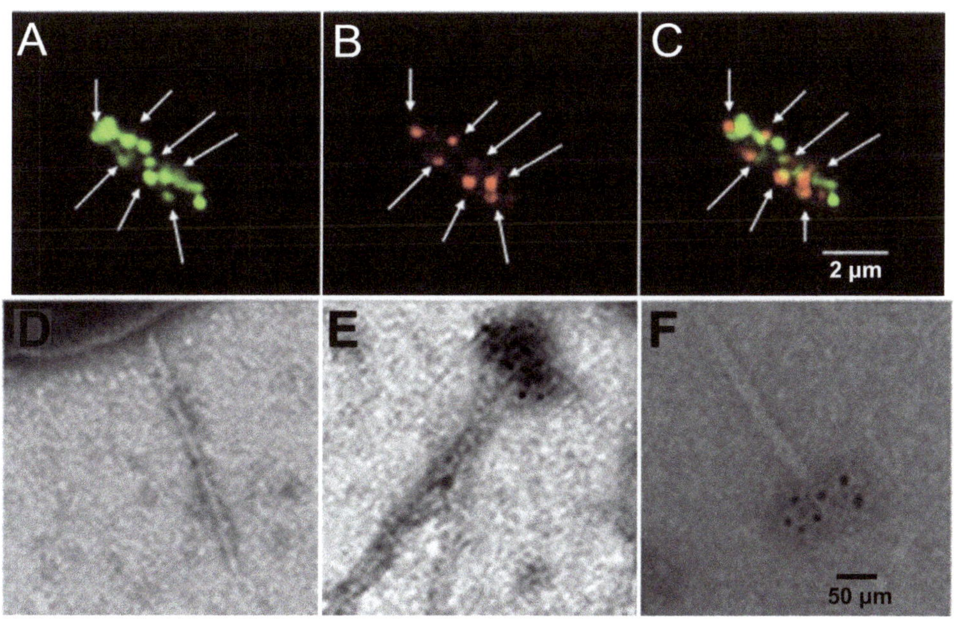

Figure 12.3. *S. flexneri* exposure to deoxycholate results in IpaB localization to the type III secretion apparatus needle tip. Immunofluorescence images of non-permeabilized *S. flexneri* show both IpaD (A) and IpaB (B) co-localized to punctate regions on the bacterial surface (C) following deoxycholate (DOC) exposure. Identical conditions lacking DOC found that IpaD was surface localized, but that IpaB was not (data not shown), indicating that DOC is a key trigger in this step of maturation. High-resolution TEM imaging of needles using 5 nm gold-conjugated secondary IgGs and anti-IpaB antibodies confirmed that IpaB is not present at the needle tip prior to DOC exposure (D), yet is readily detected following incubation with DOC (E). Anti-IpaD antibodies detected IpaD at the needle tip under all conditions tested including post-DOC incubation (F). This figure was modified from reference (Olive et al., 2007) with permission from the publisher.

incorporating IpaB into the host membrane, their ability to form functional pores (as measured by an erythrocyte hemolysis assay) is diminished approximately 90%, suggesting both are involved in the formation of the translocon (Blocker et al., 1999). Immunofluorescence and high-resolution electron microscopy have shown that IpaC localizes to the IpaB primed needle tip in the presence of defined liposomes containing sphingomyelin and cholesterol, implicating host cell lipid raft components in the signaling pathway (Epler et al., 2009) (Figure 12.4). The same study also found that *S. flexneri* exposure to these components not only recruits IpaC to the needle tip, but also initiates full secretion. This makes sense with the model suggesting that contact with a suitable host cell recruits IpaC, representing the last step in *Shigella* T3SA maturation and completion of the translocon spanning the host cell plasma membrane. From here, secreted effectors have a clear path into the host cell where they are able to subvert normal processes to enhance virulence and survival. The mechanistic details of IpaC's role in translocon formation remain unclear, but are an area of great interest as IpaC may represent the final component in *Shigella* T3SA maturation prior to host cell infection. The link between virulence and the formation of the translocon has made it a highly anticipated target for therapeutic agents against not only *Shigella* infection, but the many other Gram negative pathogens also expressing T3SSs as essential virulence factors.

Concluding remarks and future directions

The T3SA is relatively well conserved and can be roughly divided into families based on the structural features of the exposed portions of the apparatus. In contrast, the effector proteins that are ultimately responsible for subverting normal host cell functions vary

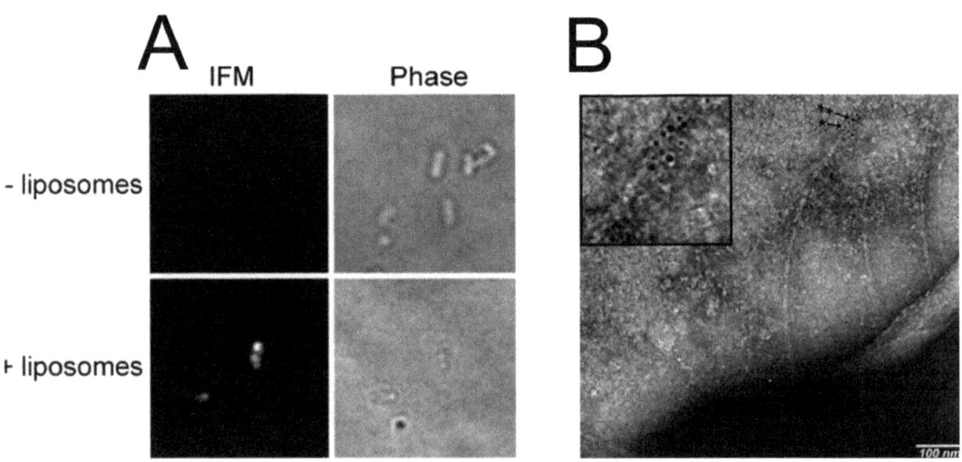

Figure 12.4. Defined liposomes containing cholesterol and sphingomyelin recruit IpaC to the *S. flexneri* type III secretion apparatus tip complex. The presence of surface exposed IpaC on *S. flexneri was* probed with anti-IpaC antibodies in the presence and absence of liposomes containing DOPC, sphingomyelin, and cholesterol (A). Bacteria are easily identified in the phase contrast channel while only those incubated with liposomes result in fluorescence detection of IpaC on the bacterial surface. (B) High-resolution transmission electron microscopy of *S. flexneri* T3SA needles after incubation with liposomes containing DOPC, sphingomyelin and cholesterol. Anti-IpaB and anti-IpaC antibodies identify the proteins at the T3SA needle tip with 10 nm gold particles representing IpaB and 5 nm particles IpaC, suggesting that IpaC resides at the needle tip distal to IpaB following liposome exposure. This figure was modified from reference (Epler et al., 2009) with permission from the publisher.

greatly and tend to be highly pathogen specific with regard to their functions and roles in virulence. Thus, dissecting the triggers and mechanisms involved in maturation of the tip complex of *Shigella flexneri* has been and should continue to prove useful as a model organism for other bacteria containing T3SSs. However, the apparent completeness of the *Shigella* model can make it tempting to believe that the mysteries of the T3SS have been exposed and that our understanding of the system is more complete than it truly is. While a combination of both routine and creatively executed biochemical and biophysical analyses have opened the doors to understanding the importance of T3SSs in these organisms and continue to provide new insight, advancements in high-resolution microscopy and spectroscopy techniques have propelled the field by providing spatial and temporal information that was previously impossible. Advancements in these fields will remain at the forefront of T3SS studies as we continue to unravel the molecular structures, interactions, and mechanisms of the T3SS with an ultimate goal of broad spectrum therapeutic agents against this important class of pathogens.

References

Abdallah, A.M., Gey van Pittius, N.C., Champion, P.A., Cox, J., Luirink, J., Vandenbroucke-Grauls, C.M., Appelmelk, B.J., and Bitter, W. (2007). Type VII secretion--mycobacteria show the way. Nature reviews Microbiology 5, 883-891.

Ayers, M., Howell, P.L., and Burrows, L.L. (2010). Architecture of the type II secretion and type IV pilus machineries. Future Microbiol 5, 1203-1218.

Bahrani, F.K., Sansonetti, P.J., and Parsot, C. (1997). Secretion of Ipa proteins by Shigella flexneri: Inducer molecules and kinetics of activation. Infect Immun 65, 4005-4010.

Barison, N., Gupta, R., and Kolbe, M. (2013). A sophisticated multi-step secretion mechanism: how the type 3 secretion system is regulated. Cell Microbiol 15, 1809-1817.

Barta, M.L., Dickenson, N.E., Patil, M., Keightley, A., Wyckoff, G.J., Picking, W.D., Picking, W.L., and Geisbrecht, B.V. (2012a). The Structures of Coiled-Coil Domains from Type III Secretion System Translocators Reveal Homology to Pore-Forming Toxins. J Mol Biol 417, 395-405.

Barta, M.L., Guragain, M., Adam, P., Dickenson, N.E., Patil, M., Geisbrecht, B.V., Picking, W.L., and Picking, W.D. (2012b). Identification of the bile salt binding site on IpaD from Shigella flexneri and the influence of ligand binding on IpaD structure. Proteins 80, 935-945.

Basler, M., Ho, B.T., and Mekalanos, J.J. (2013). Tit-for-tat: type VI secretion system counterattack during bacterial cell-cell interactions. Cell 152, 884-894.

Birket, S.E., Harrington, A.T., Espina, M., Smith, N.D., Terry, C.M., Darboe, N., Markham, A.P., Middaugh, C.R., Picking, W.L., and Picking, W.D. (2007). Preparation and characterization of translocator/chaperone complexes and their component proteins from Shigella flexneri. Biochemistry 46, 8128-8137.

Blocker, A., Gounon, P., Larquet, E., Niebuhr, K., Cabiaux, V., Parsot, C., and Sansonetti, P. (1999). The tripartite type III secreton of Shigella flexneri inserts IpaB and IpaC into host membranes. J Cell Biol 147, 683-693.

Botteaux, A., Sory, M.P., Biskri, L., Parsot, C., and Allaoui, A. (2009). MxiC is secreted by and controls the substrate specificity of the Shigella flexneri type III secretion apparatus. Mol Microbiol 71, 449-460.

Brotcke Zumsteg, A., Goosmann, C., Brinkmann, V., Morona, R., and Zychlinsky, A. (2014). IcsA is a Shigella flexneri adhesin regulated by the type III secretion system and required for pathogenesis. Cell Host Microbe 15, 435-445.

Burkinshaw, B.J., and Strynadka, N.C.J. (2014). Assembly and structure of the T3SS. Biochimica et biophysica acta 1844, 1649-1663.
Buttner, D. (2012). Protein Export According to Schedule: Architecture, Assembly, and Regulation of Type III Secretion Systems from Plant- and Animal-Pathogenic Bacteria. Microbiol Mol Biol Rev 76, 262-310.
Campbell-Valois, F.X., Schnupf, P., Nigro, G., Sachse, M., Sansonetti, P.J., and Parsot, C. (2014). A Fluorescent Reporter Reveals On/Off Regulation of the Shigella Type III Secretion Apparatus during Entry and Cell-to-Cell Spread. Cell Host Microbe 15, 177-189.
Chagnot, C., Zorgani, M.A., Astruc, T., and Desvaux, M. (2013). Proteinaceous determinants of surface colonization in bacteria: bacterial adhesion and biofilm formation from a protein secretion perspective. Front Microbiol 4.
Cordes, F.S., Komoriya, K., Larquet, E., Yang, S.X., Egelman, E.H., Blocker, A., and Lea, S.M. (2003). Helical structure of the needle of the type III secretion system of Shigella flexneri. J Biol Chem 278, 17103-17107.
Cornelis, G.R. (2006). The type III secretion injectisome. Nat Rev Microbiol 4, 811-825.
De Buck, E., Lammertyn, E., and Anne, J. (2008). The importance of the twin-arginine translocation pathway for bacterial virulence. Trends Microbiol 16, 442-453.
Deane, J.E., Roversi, P., Cordes, F.S., Johnson, S., Kenjale, R., Daniell, S., Booy, F., Picking, W.D., Picking, W.L., Blocker, A.J., et al. (2006a). Molecular model of a type III secretion system needle: Implications for host-cell sensing. Proc Natl Acad Sci U S A 103, 12529-12533.
Deane, J.E., Roversi, P., Cordes, F.S., Johnson, S., Kenjale, R., Daniell, S., Booy, F., Picking, W.D., Picking, W.L., Blocker, A.J., et al. (2006b). Molecular model of a type III secretion system needle: Implications for host-cell sensing. Proc Natl Acad Sci U S A 103, 12529-12533.
Demers, J.P., Sgourakis, N.G., Gupta, R., Loquet, A., Giller, K., Riedel, D., Laube, B., Kolbe, M., Baker, D., Becker, S., et al. (2013). The Common Structural Architecture of Shigella flexneri and Salmonella typhimurium Type Three Secretion Needles. PLoS Pathog 9.
Desvaux, M., Parham, N.J., Scott-Tucker, A., and Henderson, I.R. (2004). The general secretory pathway: a general misnomer? Trends Microbiol 12, 306-309.
Dickenson, N.E., Arizmendi, O., Patil, M.K., Toth, R.T., Middaugh, C.R., Picking, W.D., and Picking, W.L. (2013a). N-Terminus of IpaB Provides a Potential Anchor to the Shigella Type III Secretion System Tip Complex Protein IpaD. Biochemistry 52, 8790-8799.
Dickenson, N.E., Choudhari, S.P., Adam, P.R., Kramer, R.M., Joshi, S.B., Middaugh, C.R., Picking, W.L., and Picking, W.D. (2013b). Oligomeric states of the Shigella translocator protein IpaB provide structural insights into formation of the type III secretion translocon. Protein Sci 22, 614-627.
Dickenson, N.E., Zhang, L.L., Epler, C.R., Adam, P.R., Picking, W.L., and Picking, W.D. (2011). Conformational Changes in IpaD from Shigella flexneri upon Binding Bile Salts Provide Insight into the Second Step of Type III Secretion. Biochemistry 50, 172-180.
Enninga, J., and Rosenshine, I. (2009). Imaging the assembly, structure and activity of type III secretion systems. Cell Microbiol 11, 1462-1470.
Epler, C.R., Dickenson, N.E., Bullitt, E., and Picking, W.L. (2012). Ultrastructural Analysis of IpaD at the Tip of the Nascent MxiH Type III Secretion Apparatus of Shigella flexneri. J Mol Biol 420, 29-39.
Epler, C.R., Dickenson, N.E., Olive, A.J., Picking, W.L., and Picking, W.D. (2009). Liposomes Recruit IpaC to the Shigella flexneri Type III Secretion Apparatus Needle as a Final Step in Secretion Induction. Infect Immun 77, 2754-2761.

Erhardt, M., Namba, K., and Hughes, K.T. (2010). Bacterial Nanomachines: The Flagellum and Type III Injectisome. Cold Spring Harbor Perspect Biol 2.

Espina, M., Ausar, S.F., Middaugh, C.R., Picking, W.D., and Picking, W.L. (2006a). Spectroscopic and calorimetric analyses of invasion plasmid antigen D (IpaD) from Shigella flexneri reveal the presence of two structural domains. Biochemistry 45, 9219-9227.

Espina, M., Olive, A.J., Kenjale, R., Moore, D.S., Ausar, S.F., Kaminski, R.W., Oaks, E.V., Middaugh, C.R., Picking, W.D., and Picking, W.L. (2006b). IpaD localizes to the tip of the type III secretion system needle of Shigella flexneri. Infect Immun 74, 4391-4400.

Evans, M.L., and Chapman, M.R. (2014). Curli biogenesis: Order out of disorder. Biochim Biophys Acta 1844, 1551-1558.

Fujii, T., Cheung, M., Blanco, A., Kato, T., Blocker, A.J., and Namba, K. (2012). Structure of a type III secretion needle at 7-angstrom resolution provides insights into its assembly and signaling mechanisms. Proc Natl Acad Sci U S A 109, 4461-4466.

Gawarzewski, I., Smits, S.H.J., Schmitt, L., and Jose, J. (2013). Structural comparison of the transport units of type V secretion systems. Biol Chem 394, 1385-1398.

Guglielmini, J., Neron, B., Abby, S.S., Garcillan-Barcia, M.P., la Cruz, F.d., and Rocha, E.P.C. (2014). Key components of the eight classes of type IV secretion systems involved in bacterial conjugation or protein secretion. Nucleic acids research 42, 5715-5727.

Hayward, R.D., Cain, R.J., McGhie, E.J., Phillips, N., Garner, M.J., and Koronakis, V. (2005). Cholesterol binding by the bacterial type III translocon is essential for virulence effector delivery into mammalian cells. Molecular microbiology 56, 590-603.

Heras, B., Totsika, M., Peters, K.M., Paxman, J.J., Gee, C.L., Jarrott, R.J., Perugini, M.A., Whitten, A.E., and Schembri, M.A. (2014). The antigen 43 structure reveals a molecular Velcro-like mechanism of autotransporter-mediated bacterial clumping. Proc Natl Acad Sci U S A 111, 457-462.

Jessen, D.L., Osei-Owusu, P., Toosky, M., Roughead, W., Bradley, D.S., and Nilles, M.L. (2014). Type III Secretion Needle Proteins Induce Cell Signaling and Cytokine Secretion via Toll-Like Receptors. Infect Immun 82, 2300-2309.

Johnson, S., Roversi, P., Espina, M., Olive, A., Deane, J.E., Birket, S., Field, T., Picking, W.D., Blocker, A.J., Galyov, E.E., et al. (2007). Self-chaperoning of the type III secretion system needle tip proteins IpaD and BipD. J Biol Chem 282, 4035-4044.

Kane, K.A., and Dorman, C.J. (2012). VirB-Mediated Positive Feedback Control of the Virulence Gene Regulatory Cascade of Shigella flexneri. J Bacteriol 194, 5264-5273.

Kanonenberg, K., Schwarz, C.K.W., and Schmitt, L. (2013). Type I secretion systems - a story of appendices. Res Microbiol 164, 596-604.

Konovalova, A., Sogaard-Andersen, L., and Kroos, L. (2014). Regulated proteolysis in bacterial development. Fems Microbiol Rev 38, 493-522.

Lan, R.T., Stevenson, G., and Reeves, P.R. (2003). Comparison of two major forms of the Shigella virulence plasmid pINV: Positive selection is a major force driving the divergence. Infect Immun 71, 6298-6306.

Marteyn, B., Gazi, A., and Sansonetti, P. (2012). Shigella: a model of virulence regulation in vivo. Gut microbes 3, 104-120.

Martinez-Argudo, I., and Blocker, A.J. (2010). The Shigella T3SS needle transmits a signal for MxiC release, which controls secretion of effectors. Mol Microbiol 78, 1365-1378.

McBride, M.J., and Zhu, Y. (2013). Gliding motility and Por secretion system genes are widespread among members of the phylum bacteroidetes. J Bacteriol 195, 270-278.

Morita, Y., Tomida, J., and Kawamura, Y. (2014). Responses of Pseudomonas aeruginosa to antimicrobials. Front Microbiol 4.

Mougous, J.D., Cuff, M.E., Raunser, S., Shen, A., Zhou, M., Gifford, C.A., Goodman, A.L., Joachimiak, G., Ordonez, C.L., Lory, S., et al. (2006). A virulence locus of Pseudomonas aeruginosa encodes a protein secretion apparatus. Science 312, 1526-1530.

Mounier, J., Popoff, M.R., Enninga, J., Frame, M.C., Sansonetti, P.J., and Van Nhieu, G.T. (2009). The IpaC Carboxyterminal Effector Domain Mediates Src-Dependent Actin Polymerization during Shigella Invasion of Epithelial Cells. PLoS Pathog 5.

Ohgita, T., Hayashi, N., Hama, S., Tsuchiya, H., Gotoh, N., and Kogure, K. (2013). A novel effector secretion mechanism based on proton-motive force-dependent type III secretion apparatus rotation. Faseb J 27, 2862-2872.

Olive, A.J., Kenjale, R., Espina, M., Moore, D.S., Picking, W.L., and Picking, W.D. (2007). Bile salts stimulate recruitment of IpaB to the Shigella flexneli surface, where it colocalizes with IpaD at the tip of the type III secretion needle. Infect Immun 75, 2626-2629.

Page, A.L., Ohayon, H., Sansonetti, P.J., and Parsot, C. (1999). The secreted IpaB and IpaC invasins and their cytoplasmic chaperone IpgC are required for intercellular dissemination of Shigella flexneri. Cell Microbiol 1, 183-193.

Parsot, C. (2009). Shigella type III secretion effectors: how, where, when, for what purposes? Current opinion in microbiology 12, 110-116.

Picking, W.L., Nishioka, H., Hearn, P.D., Baxter, M.A., Harrington, A.T., Blocker, A., and Picking, W.D. (2005). IpaD of Shigella flexneri is independently required for regulation of Ipa protein secretion and efficient insertion of IpaB and IpaC into host membranes. Infect Immun 73, 1432-1440.

Pope, L.M., Reed, K.E., and Payne, S.M. (1995). Increased protein secretion and adherence to HeLa-cells by Shigella spp following growth in the presence of bile-salts. Infect Immun 63, 3642-3648.

Prosseda, G., Falconi, M., Giangrossi, M., Gualerzi, C.O., Micheli, G., and Colonna, B. (2004). The virF promoter in Shigella: more than just a curved DNA stretch. Mol Microbiol 51, 523-537.

Pugsley, A.P. (1993). THE COMPLETE GENERAL SECRETORY PATHWAY IN GRAM-NEGATIVE BACTERIA. Microbiol Rev 57, 50-108.

Pukatzki, S., Ma, A.T., Sturtevant, D., Krastins, B., Sarracino, D., Nelson, W.C., Heidelberg, J.F., and Mekalanos, J.J. (2006). Identification of a conserved bacterial protein secretion system in Vibrio cholerae using the Dictyostelium host model system. Proc Natl Acad Sci U S A 103, 1528-1533.

Reis, R.S.D., and Horn, F. (2010). Enteropathogenic Escherichia coli, Samonella, Shigella and Yersinia: cellular aspects of host-bacteria interactions in enteric diseases. Gut pathogens 2, 8.

Robinson, C., Matos, C., Beck, D., Ren, C., Lawrence, J., Vasisht, N., and Mendel, S. (2011). Transport and proofreading of proteins by the twin-arginine translocation (Tat) system in bacteria. Biochim Biophys Acta-Biomembr 1808, 876-884.

Schraidt, O., and Marlovits, T.C. (2011). Three-Dimensional Model of Salmonella's Needle Complex at Subnanometer Resolution. Science 331, 1192-1195.

Schuch, R., Sandlin, R.C., and Maurelli, A.T. (1999). A system for identifying post-invasion functions of invasion genes: requirements for the Mxi-Spa type III secretion pathway of Shigella flexneri in intercellular dissemination. Mol Microbiol 34, 675-689.

Skoudy, A., Mounier, J., Aruffo, A., Ohayon, H., Gounon, P., Sansonetti, P., and Tran Van Nhieu, G. (2000). CD44 binds to the Shigella IpaB protein and participates in bacterial invasion of epithelial cells. Cellular microbiology 2, 19-33.

Terry, C.M., Picking, W.L., Birket, S.E., Flentie, K., Hoffman, B.M., Barker, J.R., and Picking, W.D. (2008). The C-terminus of IpaC is required for effector activities related to Shigella invasion of host cells. Microbial pathogenesis 45, 282-289.
Tran Van Nhieu, G., Caron, E., Hall, A., and Sansonetti, P.J. (1999). IpaC induces actin polymerization and filopodia formation during Shigella entry into epithelial cells. The EMBO journal 18, 3249-3262.
Watarai, M., Tobe, T., Yoshikawa, M., and Sasakawa, C. (1995). Contact of Shigella with host-cells triggers release of Ipa invasins and is an essential function of invasiveness. Embo J 14, 2461-2470.
Xiong, Z.H., Tang, X.D., Yang, F., Zhang, X.B., Yang, J., Chen, L.H., Nie, H., Yan, Y.L., Jiang, Y., Wang, J., et al. (2006). Comparison of the virulence plasmid genomes of two strains of Shigella which lost the ability to bind Congo red. Sci China Ser C-Life Sci 49, 141-148.
Yip, C.K., and Strynadka, N.C.J. (2006). New structural insights into the bacterial type III secretion system. Trends BiochemSci 31, 223-230.
Zhang, L., Wang, Y., Olive, A.J., Smith, N.D., Picking, W.D., De Guzman, R.N., and Picking, W.L. (2007). Identification of the MxiH needle protein residues responsible for anchoring invasion plasmid antigen D to the type III secretion needle tip. J Biol Chem 282, 32144-32151.

13

Molecular Approaches to Identifying Type 3 Secreted Effectors: a *Shigella* Case Study

Analise Z. Reeves, Sonia C. Costa and Cammie F. Lesser

Department of Medicine, Division of Infectious Diseases, Massachusetts General Hospital, Cambridge, MA, USA
Department of Microbiology and Immunobiology, Harvard Medical School, Boston, MA, USA, clesser@mgh.harvard.edu

Abstract
Type 3 secretion systems are specialized machines common to plant and animal bacterial pathogens as well as insect endosymbionts that act to deliver tens of proteins directly from bacteria into eukaryotic cells. While protein components of these complex machines are highly conserved, each bacterium delivers its own unique repertoire of proteins, commonly referred to as effectors, into host cells. The function of effectors can be difficult to identify experimentally, as they often act in a functionally redundant manner to mediate specific steps in pathogenesis such that loss of expression of an individual protein does not result in a detectable phenotype. In addition, despite the growing availability of annotated bacterial genome sequences, effectors are challenging to identify via bioinformatic approaches as they have poorly defined secretion sequences and commonly do not share homology with other proteins. Extensive efforts over the past 30-40 years have been extended towards identifying these secreted virulence proteins, and discerning how they act to usurp host cell processes to promote bacterial survival and spread. Here, we provide a historical review of the means by which the currently known ~30 *Shigella* effectors were discovered to illustrate the variety of experimental approaches including genetic, biochemical, functional and bioinformatic based assays that have been used to identify these important virulence factors.

Initial characterization of the *Shigella* virulence plasmid

The *Shigella* type 3 secretion system and all but five of its currently known secreted effectors are encoded on a large ~220 kB plasmid, often referred to as the *Shigella* virulence plasmid. All four pathogenic *Shigella* species, *S. flexneri, S. sonnei, S. dysenteriae* and *S. boydii* as well as enteroinvasive *E. coli* carry this plasmid. A role for this plasmid in virulence was discovered in the early 1980s, at least a decade before the discovery of type 3 secretion. The first clue that a plasmid plays a role in *Shigella* virulence was the observation that strains often irreversibly and spontaneously transition from a virulent to an avirulent phenotype, suggesting that an element, like a plasmid, is lost. Consistent with these observations in 1982, Sansonetti and colleagues reported that strains lacking the large ~220 kB plasmid are unable to invade epithelial cells or, when injected into the conjunctival sac of a guinea pig eye, elicit inflammation (conjunctivitis) (the Sereny test) (Sereny, 1955). Remarkably, the introduction of this plasmid into avirulent strains restores the ability of the strains to invade epithelial cells and cause disease (Sansonetti et al., 1982), thus strongly demonstrating an association of the plasmid with virulence.

Subsequently, in 1983, Sansonetti and colleagues established that the introduction of this large plasmid into laboratory strains of *E. coli* is sufficient to generate strains that invade and replicate inside non-phagocytic epithelial cells (Sansonetti et al., 1983), suggesting that the plasmid encodes a fully functional virulence system. In a complementary set of experiments, Hale and colleagues demonstrated that anucleate *Shigella* mini-cells that contain the virulence plasmid are also capable of invading HeLa cells (Hale et al., 1983). Furthermore, in 1985, Maurelli and colleagues showed that the introduction of a cosmid carrying a 37 kB fragment of virulence plasmid DNA into plasmid-cured strains of *Shigella* is sufficient to generate invasive strains (Maurelli et al., 1985). As these strains are invasive, but not virulent, this 37 kB region of DNA is often referred to as the *Shigella* entry region (SER) (Figure 13.1).

Identification of the Ipa (Invasion Plasmid Antigen) secreted proteins
Once the large plasmid was identified to be an important *Shigella* virulence factor, efforts became focused on finding plasmid-encoded proteins. Towards this end, in 1983, Hale and colleagues screened for proteins produced when metabolically active anucleate *S. flexneri*-derived minicells containing the virulence plasmid were grown in the presence of [^{35}S] methionine. Fourteen polypeptides were observed, the majority of which were associated with the outer membrane of the bacterial minicells (Hale et al., 1983). Subsequently, in 1985, by comparing the 2-dimensional gel electrophoresis profiles of [^{35}S] methionine labeled proteins produced by anucleate minicells derived from virulent and avirulent *Shigella*, the investigators identified 7 discrete spots, polypeptides they designated a-g, which are specific to virulent *Shigella* derived minicells (Hale et al., 1985).

In parallel with these studies, other groups were isolating convalescent antibodies from the serum of humans and monkeys recovering from infection with *Shigella.* Interestingly, these antibodies recognized four of the seven proteins, polypeptides a, b, c and d (Hale et al., 1985; Oaks et al., 1986). These four immunogenic polypeptides were found to be produced by virulence-plasmid cured *Shigella* strains that carry the 37 kB SER cosmid (Maurelli et al., 1985), thus suggesting that they are involved in host cell invasion. The antigenicity of these proteins provided the first clues that these proteins were exposed on the surface of and/or secreted by *Shigella* at some point during an infection. These proteins were later identified to be IpaB, IpaC and IpaD, the components of the *Shigella* translocon apparatus and IpaA, a type 3 secreted effector.

While analyzing 2D gel electrophoresis profiles of *Shigella* minicells, Hale and colleagues not only investigated the ability of the patient-derived convalescent serum to recognize the proteins, but also generated rabbit polyclonal antibodies against fractions of the three most abundant proteins visible on the gels, polypeptides b, c, and d (Hale et al., 1985). In 1987, Buysse and colleagues used phage display technology to identify DNA fragments that encode polypeptides recognized by the rabbit-derived antibodies (Buysse et al.,

1987). In agreement with the earlier studies demonstrating that polypeptides a-d were encoded in the SER, restriction mapping experiments discovered that the corresponding genes encoding these polypeptides were adjacent to one another within this region. These genes were designated *ipaB, ipaC*, and *ipaD* for invasive plasmid antigens b, c and d. However, curiously, the screen with the serum generated against polypeptide b also led to the identification of a second polypeptide, unrelated to IpaB. The investigators demonstrated that this serum contained a mix of two antibodies, each of which recognizes an ~58kD protein. Since the rabbit polyclonal antibodies were raised against enriched fractions as opposed to extensively purified proteins, it turns out that two antigenic polypeptides of similar molecular weight were present in the 2D gel spot used as a source to generate the "b serum". The second protein recognized by the b serum was designated IpaH.

Although the genes encoding IpaA, B, C, and D were rapidly cloned and sequenced in 1988 and determined to be encoded within a single operon (Sasakawa et al., 1989; Venkatesan et al., 1988), the identification of the gene corresponding to IpaH turned out to be much more complicated. In 1990, while restriction mapping the DNA fragments identified by phage display technology to encode proteins recognized by the IpaH antibody, Hartman and colleagues discovered that phage carrying 5 unique fragments of DNA encoded proteins recognized by the antibody. Using Southern blot analyses, they established that five closely related IpaH homologs are encoded on the virulence plasmid (Hartman et al., 1990). The homologs were designated IpaH1.4, IpaH2.4, IpaH4.5, IpaH7.8 and IpaH9.8 based on the sizes of unique equal intensity bands observed when an *ipaH* probe was used to probe *Hind*III restriction digested virulence plasmid DNA. Interestingly, future studies would later confirm that 4-5 additional *ipaH* alleles are encoded on the *Shigella* chromosome (Ashida et al., 2007). Curiously, although the IpaH proteins were identified through analyses of the rabbit-derived polyclonal antibodies, convalescent serum from *Shigella* infected monkeys also recognized the IpaH proteins, suggesting that the IpaH proteins, like IpaB, C, D and A, are found on the surface and/or secreted from *Shigella* during the course of an infection. Thus, the first type 3 secreted proteins to be discovered, IpaA and the IpaH homologs, were originally identified due to their antigenicity.

Genetic screens in search of *Shigella* virulence determinants
In parallel with the efforts to characterize proteins on the *Shigella* large plasmid based on antigenicity, Sasakawa and colleagues conducted a series of complementary transposon (Tn5)-based genetic screens to identify loci on the large virulence plasmid involved in virulence. The strains carrying transposon insertions were screened in four phenotypic assays: (1) invasion of HeLa cells (Inv), (2) induction of conjunctivitis (Sereny test) (Ser), (3) ability to bind the dye Congo Red (Pcr) and (4) colony size (Igr = inhibition of growth). The latter two phenotypes were tested given that earlier studies had established that in contrast to avirulent *Shigella* strains, when grown on solid media, virulent strains bind the dye Congo red and are impaired for growth, resulting in decreased colony size (Maurelli et al., 1984; Sasakawa et al., 1986a).

In parallel, the investigators utilized a clever *Sal*I restriction enzyme digestion strategy to map the locations of Tn5 insertions in the virulence plasmid. Basically, digestion of purified virulence plasmid DNA with *Sal*I results in 23 distinct fragments visible when analyzed on a 0.8% agarose gel, fragments A, B, C .....T, R', S' and T' ranging in size from 32.6 to 1.5 kb (Sasakawa et al., 1986a). Since the Tn5 transposon itself also contains a *Sal*I site, the restriction digestion based analyses also resulted in the mapping of the relative location of the Tn5 insertion within one of the 23 fragments. Their initial transposon screens identified transposons in 6 *Sal*I fragments (F, G, B, P, H and D) that resulted in loss of virulence (Sasakawa et al., 1986b) (Figure 13.1). Insertions into 5 fragments (F, B, P, H and D) exhibited loss of invasion (Inv^-,), conjunctivitis (Ser^-) and Congo red binding (Pcr^-) and exhibited increased bacterial growth rates (Igr^-), while insertions in fragment G resulted in bacteria that invaded cells, bound Congo red, and grew at rates similar to wild type *Shigella,* but did not elicit conjunctivitis (Inv^+, Ser^-, Pcr^+, Igr^+).

Insertions into fragments B, P, H and D that confer the (Inv^-, Ser^-, Pcr^-, Igr^-) phenotype were determined to map to a contiguous 31 kB region of the virulence plasmid (Figure 13.1) (Sasakawa et al., 1986b), a fragment contained within that previously identified to promote *Shigella* entry into host cells, the *Shigella* entry region (SER) (Maurelli et al., 1985). As discussed in detail below, it would later be discovered that these transposon insertions rendered strains avirulent by disrupting expression of components of the type 3 secretion machinery.

Furthermore, several years later, Tn5 insertions were identified that resulted in a new phenotype. These strains were unable to invade HeLa cells or cause disease, and thus, were completely avirulent. However, like wild type *Shigella*, they bound Congo red and exhibited impaired growth (Inv^-, Ser^-, Pcr^+, Igr^+). These transposon insertions all mapped to region 2, a 5.1kb section of DNA contained within *Sal*I fragments B and P (Figure 13.1). Western blot analyses using convalescent serum of infected humans determined that region 2 carries the genes that encode IpaB, IpaC, IpaD and IpaA (Sasakawa et al., 1989; Sasakawa et al., 1988). While insertions into IpaB, IpaC or IpaD resulted in Inv^-, Ser^-, Pcr^+, Igr^+ strains, insertions into IpaA did not confer a detectable phenotype in the screening assays. It would later be determined that IpaB, IpaC and IpaD are components of the translocon apparatus, while IpaA is a secreted effector.

Analysis of insertions into fragments F and B eventually led to the identification of VirF and VirB (vir for virulence), the main transcriptional regulators that control expression of components of the *Shigella* type 3 secretion system and many effectors (Adler et al., 1989; Sakai et al., 1988). Mapping of the Tn5 insertions in fragment G led to the identification of two proteins involved at post-invasion steps in *Shigella* virulence, VirG (IcsA) (Makino et al., 1986) and VirA (Uchiya et al., 1995). VirG is not a type 3 secreted protein; however, it is delivered into the cytosol of host cells, as it is a type V autotransporter (Goldberg et al., 1993). VirG plays an essential role in mediating the formation of actin tails that propel *Shigella* through the host cell cytosol and promote cell-to-cell spread. VirA is a type 3 secreted effector that like VirG is involved in cell-to-cell spread (Uchiya et al., 1995), and potentially host cell invasion (Bergounioux et al., 2012). The gene encoding VirA is about one-third the size of that encoding VirG, likely explaining why the former gene was identified almost 10 years later, when the size of the Tn5 library was expanded. Of note, VirA was one of the first secreted effectors to be formally demonstrated to depend on the presence of an intact type 3 secretion system to be secreted, as it was identified around the first time that mutations in components of the secretion apparatus were being defined (Uchiya et al., 1995).

Sequencing of the *Shigella* entry region (SER) led to the identification of the two additional effectors, IcsB and IpgD.

Concurrent with the genetic and biochemical approaches described above, given the established role for the large plasmid in virulence, several groups focused efforts on completing the sequence of virulence plasmid DNA within the *Shigella* entry region. For example, in 1992, Allaoui and colleagues identified three open reading frames (ORFs) directly adjacent to IpaB. The first two were designated *ipgA* and *ipgB* for invasion plasmid genes A and B (Allaoui et al., 1992). These genes were not further analyzed at this point, although IpgB would later be determined to be a type 3 secreted protein while IpgA would be identified as a type 3 secretion chaperone. Rather efforts at this time focused on establishing a potential role for the third open reading frame named IcsB for intercellular-intracellular spread. The investigators observed that strains no longer expressing IcsB could still invade HeLa cells at levels equivalent to wild type *Shigella,* but like strains carrying insertions in fragment G, were markedly impaired in cell-to-cell spread and completely defective in the Sereny test.

Years later, subsequent studies with the deletion strain led to the discovery that after invading and escaping into the cytosol of infected cells, the *icsB* deletion strain is sequestered within double membrane-bound vacuoles via autophagy (Ogawa et al., 2005). Nevertheless, notably, this was the first effector whose loss of expression resulted in a detectable phenotype in *Shigella* pathogenesis. Furthermore, although the investigators did not observe any IcsB to be secreted from *Shigella,* IcsB was observed in the periplasmic fraction. At the time, this was not too surprising, as conditions to induce activation of the type 3 secretion system had not yet been discovered. Indeed, when conditions that induce type 3 secretion *in vitro* were identified, IcsB was confirmed as a type 3 secreted protein.

While characterizing the expression of IcsB, the investigators observed that the introduction of a fragment of DNA carrying the *icsB* promoter in the opposite orientation upstream of a *lacZ* reporter resulted in ß-galactosidase activity (Allaoui et al., 1992), suggesting that this region encoded bidirectional promoters. Sequencing of the DNA located downstream of this newly identified promoter led to the identification of three additional new genes on the *Shigella* plasmid, *ipgD*, *ipgE* and *ipgF* (Allaoui et al., 1993a). In order to investigate the potential roles of these genes in *Shigella* pathogenesis, Allaoui and colleagues generated strains deleted for the genes encoding either IpgD or IpgF by integrating an antibiotic resistance cassette within the gene's open reading frame. Neither deletion strain exhibited a phenotype as assessed by their ability to bind Congo red, invade or spread between HeLa cells, or cause disease in the Sereny test. The investigators next compared the secreted protein profiles of the two deletion strains with wild type *Shigella.* Notably, a 58 kD band was absent in the secreted fraction of the Δ*ipgD* strain. As the gene encoding IpgD is predicted to encode a 60 kDa protein, the investigators correctly proposed that IpgD was the missing protein. No differences were observed between the secretion profiles of wild type and Δ*ipgF Shigella.* In later studies IpgE was demonstrated to be the chaperone that mediates IpgD secretion (Niebuhr et al., 2000).

There now exists a wealth of data demonstrating that targeted deletion of most type 3 effectors individually does not elicit marked impairment in host cell invasion, cell-to-cell spread or Congo red binding. These observations support the current hypothesis that many type 3 secreted bacterial effectors from *Shigella* and other pathogens act in a functionally redundant manner to mediate major steps in pathogenesis. Thus, in retrospect, it is perhaps not so surprising that Sasakawa and colleagues did not identify more effectors in their transposon screens, particularly as subtle phenotypes might have been missed in these large-scale screens. However, this was clearly far from investigator's expectations when beginning to identify effectors, as exemplified by a statement in the discussion of a paper from Allaoui and colleagues after discovering IpgD in 1993. "It thus appears that 2 secreted proteins, IpaA and possibly IpgD, are not involved in the virulent phenotype of *S. flexneri*. This is very surprising, especially considering the positions of the *ipaA* and *ipgD* genes in the invasion region of the virulence plasmid and the presence of the encoded proteins among the few proteins that are secreted into the culture supernatant (Allaoui et al., 1992)."

Discovery of *Shigella* type 3 secretion system

In the early 1990s, extensive work from numerous groups studying Gram-negative pathogens of both plants and animals led to the discovery of type 3 secretion (Van Gijsegem et al., 1993). Some of the *Shigella* based findings that played a role in this exciting advance arose from analyses of genes present in the *mxi* (membrane expression of invasion plasmid antigens) and *spa* (surface presentation of Ipa antigens) loci (Allaoui et al., 1993b; Andrews and Maurelli, 1992; Sasakawa et al., 1993; Venkatesan et al., 1992) located within the 37kB SER adjacent to the IpaBCDA operon in *SalI* fragments P, H and D. Deletions of genes in these regions were determined to result in strains that no longer secreted or displayed IpaA, IpaB, IpaC and IpaD on their surface. These results led to the hypothesis and eventual discovery that these proteins form a secretion machine that mediates secretion of at least IpaA, IpaB, IpaC and IpaD. In addition, as genes within the *mxi* and *spa* loci were identified and sequenced they were found to encode proteins that share extensive homology with proteins from *Yersinia* and *Salmonella* also hypothesized to be involved in what was designated the type 3 secretion system, the first complex transkingdom protein delivery system to be discovered.

Soon afterwards, several groups identified conditions that greatly enhanced secretion of Ipa proteins, a major advance that facilitated the identification of secreted effectors. In the mid 1990s, Menard as well as Watari and colleagues demonstrated that the secretion of Ipa proteins was greatly enhanced when *Shigella* were exposed to epithelial cells (Menard et al., 1994a; Watarai et al., 1995). In addition, Menard reported that strains lacking IpaB or IpaD exhibited enhanced levels of secreted Ipa proteins, providing the first evidence that IpaB and IpaD could act to modulate secretion (Menard et al., 1994a). Lastly, in 1995, Parsot and colleagues observed that the same proteins were secreted when wild type bacteria were grown in the presence of the dye Congo red, thus developing an *in vitro* assay to efficiently monitor type 3 secreted proteins from intact wild type *Shigella* (Parsot et al., 1995), an assay refined in 1997 by Bahrani and colleagues (Watarai et al., 1995). These were important technical achievements that led to the discovery that numerous proteins, in addition to the Ipa proteins, were likely substrates of the newly discovered type 3 secretion system (Parsot et al., 1995).

Sequencing and annotation of the entire *Shigella* virulence plasmid combined with N-terminal sequencing of secreted proteins identified a new group of type 3 effectors

In 2000, in a tour-de-force, using two complementary approaches, Buchrieser and colleagues identified multiple new probable type 3 effectors (Buchrieser et al., 2000). First, the investigators revealed the entire annotated 220kB sequence of the virulence plasmid from *Shigella flexneri* serotype 5 strain M90T. The sequence of the 37kB entry region was consistent with the previous smaller scale sequencing projects. In addition, many single genes and small operons were found to be present throughout the virulence plasmid. While plasmid maintenance and housekeeping gene sequences demonstrated a G+C content similar to those encoded on the *Shigella* chromosome, many ORFs of unknown function exhibited a G+C content similar to that observed with genes present within the operons in the *Shigella* entry region, suggesting that the genes were acquired by horizontal transfer and potentially encoded virulence factors including previously unknown type 3 secreted proteins.

Once the virulence plasmid sequence was determined, the investigators were in position to identify proteins as type 3 substrates. To do this, they took advantage of *Shigella* strains lacking either IpaB or IpaD which were previously observed to hyper-secrete proteins (Menard et al., 1993; Parsot et al., 1995). The investigators collected supernatants from a *ΔipaBCDA* strain and determined the N-terminal sequences of 14 secreted proteins. Included among the sequenced proteins were six previously characterized proteins. Two, VirA and IpaH9.8, were already established type 3 secretion substrates (Demers et al., 1998; Uchiya et al., 1995), providing proof-of-concept for this approach. Four additional secreted proteins were identified, IpgB1, MxiC, MxiL and Spa32. The gene encoding IpgB1 is present adjacent to that encoding IcsB, a known *Shigella* effector, and had been previously identified when sequencing genes present in the *Shigella* entry region, but not further studied. Likewise, the genes encoding MxiC, MxiL and Spa32 are located in the *mxi/spa* operons, alongside genes that encode protein components or regulators of the type 3 secretion apparatus. Interestingly, future studies established that MxiC and Spa32 play roles in the control of secretion of type 3 effectors while a role for MxiL remains unknown (Botteaux et al., 2009; Magdalena et al., 2002).

The remaining nine proteins identified to be present in the secreted supernatant corresponded to genes present on the virulence plasmid that had not previously been sequenced. As predicted earlier by the Southern blot analyses of Hartman and colleagues, alleles encoding five IpaH proteins were identified, IpaH1.4, IpaH2.4, IpaH4.5, IpaH7.8 and IpaH9.8 (Menard et al., 1993; Parsot et al., 1995), three (IpaH4.5, IpaH7.8 and IpaH9.8) of which were secreted by the *ΔipaABCD* strain. All five *ipaH* genes share a very high degree of homology but yet the sequence of each uniquely accounted for the hybridizing bands observed via Southern blot analyses with *HinDIII d*igested virulence plasmid DNA. Curiously, the G-C content of the almost invariant C-termini were similar to genes encoded on the *Shigella* chromosome while the G-C content of their variable N-termini resembled those of the genes present in the *Shigella* entry region. The predicted N-terminal sequences of six additional plasmid encoded ORFs accounted for the remaining proteins secreted by the *ΔipaABCD* strain. These proteins were subsequently designated OspB, OspC1, OspD1, OspE1/2, OspF and OspG for outer *Shigella* proteins. As suggested by their "1" designations, some of the putative effectors were the founding members of effector families, based on homology. For example, two homologs of OspC1 and OspD1 were identified. The amino termini of the proteins differ significantly, but homology extends over the remaining portion of the proteins. These homologs were later demonstrated to also be type 3 secreted substrates. Thus, by the completion of these studies, on the order of 25 probable effectors encoded on the *Shigella* virulence plasmid were identified.

Chaperone based studies lead to the identification of new effectors

In 1993, it was recognized that effectors require chaperones in order to be recognized as secreted substrates (Wattiau et al., 1994; Wattiau and Cornelis, 1993). While this did not turn out to be the case for all effectors, it opened up new avenues to identify effectors. Type 3 secretion chaperones are small acidic proteins that fall into two classes. Class I chaperones mediate the secretion of effectors, while class II chaperones promote the secretion of components of the translocon apparatus. *Shigella* has one class II chaperone, IpgC, which is required for the secretion of IpaB and IpaC (Menard et al.,

1994b), and three class I chaperones. IpgE and IpgA are class IA chaperones that are each devoted to the delivery of a single effector into host cells, IpgD and IcsB, respectively (Niebuhr et al., 2000; Ogawa et al., 2003). The genes encoding IpgE and IpgA are adjacent to their cognate effectors. In contrast, Spa15 is a multi-cargo class IB chaperone. Spa15 is not encoded in the vicinity of any of its cognate effectors, but rather alongside proteins components of the *Shigella* type 3 secretion apparatus in the *spa* operon (Figure 13.1).

In 2002, Page and colleagues utilized two complementary approaches to identify proteins that interact with Spa15 (Page et al., 2002). First, they screened for proteins that directly bind Spa15 using a yeast two-hybrid (Y2H) screen. The investigators screened for interactions between Spa15 fused to the GAL4 binding domain and a library of random fragments of a 65,000 base pair region of the *Shigella* virulence plasmid fused to the GAL4 activation domain. Three proteins, OspC3, IpaA and IpgB1 were identified to interact with Spa15. To complement the yeast studies, the investigators screened for proteins that co-immunoprecipitate with Spa15 present in the cytosol. Consistent with the Y2H results, Spa15 co-immunoprecipitated with IpaA, IpgB1 and OspC3 and/or OspC2

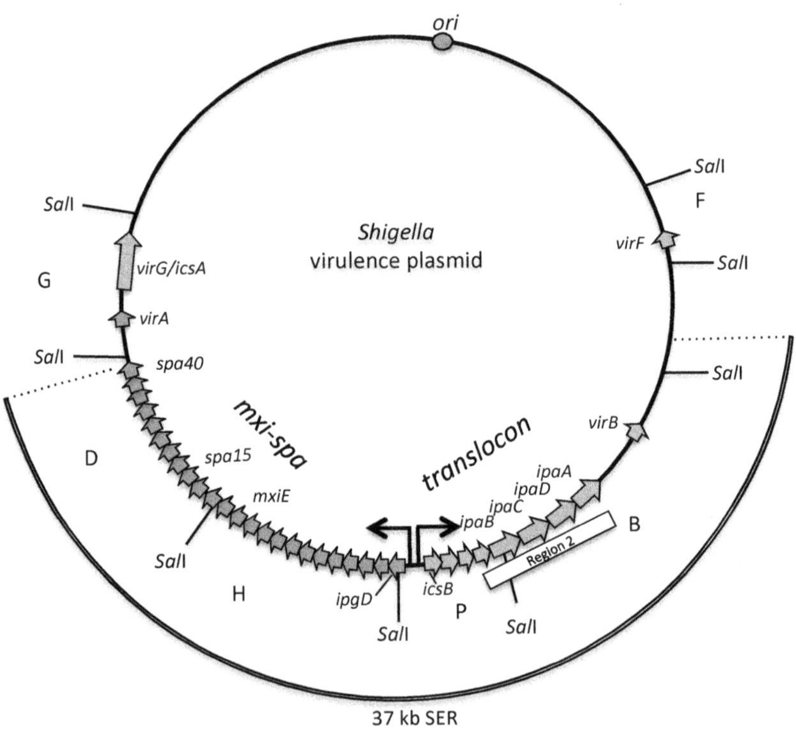

Figure 13.1. Depiction of the *Shigella* virulence plasmid The plasmid shown is based on the annotated sequence of virulence plasmid pWR100 from *Shigella flexneri* M90T (serotype 5), as the sequence of MSY6000 the virulence plasmid used to map the transposons is not publically available. The *mxi-spa* and *ipa* operons are present in opposite orientations on these two plasmids (Jin et al., 2002). The letters G, D, H, P, B, and F label 6 of the 23 fragments of the virulence plasmid DNA generated upon *Sal*I restriction digestion. The *mxi-spa* region contains the genes and proteins required to form the secretion apparatus. The *Shigella* Entry Region (SER) identified to encompass 37 kB through cosmid analysis and transposon screens is denoted. The plasmid map is not drawn to scale.

(as OspC2 and OspC3 are >90% identical, it was impossible to differentiate between the two based on N-terminal sequence). Lastly, to confirm that interactions with Spa15 played a role in protein secretion, the investigators compared the patterns of proteins secreted by Δ*ipaB* and Δ*ipaB*Δ*spa15* strains. As predicted, IpgB1 and OspC2/3 along with a third protein, OspB (a protein encoded on the virulence plasmid outside of the region used to generate the Y2H library and hence missed by the Y2H screen) were only secreted in the presence of Spa15. Curiously, IpaA is still secreted in the absence of Spa15. A finding that led the investigators to determine that IpaA is secreted in both a Spa15-dependent and a Spa15-indepent manner.

In 2009, Schmitz and colleagues utilized another protein interaction screen, the newly developed yeast Protein Interaction Platform (PIP) assay to screen for proteins that interact with Spa15. This approach, in contrast to the Y2H, enables investigators to assay and visually screen for interacting proteins in real-time in live cells (Schmitz et al., 2009) (Figure 13.2). In this assay, one protein is fused to μNS, a reoviral scaffolding protein, while the second is fused to a fluorescent protein such as GFP. Each of the genes encoding the fusion proteins is expressed under the control of an inducible promoter. When fused to μNS, proteins are displayed on platforms in yeast (or mammalian cells). These foci are not visible, unless a second fluorescently labeled protein is recruited to the platform. As proof of concept Schmitz and colleagues screened a collection of 21 putative and known *Shigella* effectors with Spa15, IpgE and IpgA using both the Y2H and PIP assays. In this case, the investigators screened for interactions with full length ORFs as opposed to the sheared fragments of virulence plasmid DNA used in the Y2H screen conducted by Page et. al. Interactions were assayed for when each protein was fused to μNS, GFP or the GAL4AD (activation domain) or the GAL4BD (binding domain).

As expected, when assayed via PIP, IpgE and IpgA interacted with their cognate effectors, IpgD and IcsB, respectively. Similarly, as observed by Page and colleagues, Spa15 interacted with IpaA, IpgB1, OspC2, OspC3 and OspB. In addition, Spa15 was observed to bind OspC1, OspD1, OspD2 and IpgB2. Notably, the PIP assay markedly outperformed the Y2H in detecting binary protein interactions. PIP not only identified more interaction pairs overall, but also recognized the pairs regardless of which was fused to μNS or GFP. Several reasons likely account for this increased detection of chaperone-effector interactions (Figure 13.2). First, given that proteins are expressed in an inducible manner in the PIP assay and that fluorescent foci can be observed 4-6 hours post-induction of the fusion proteins, PIP was able to capture interactions involving effectors, like IpgB2, IpgD and IcsB, whose expression is toxic to yeast (Slagowski et al., 2008). Such proteins cannot be studied in the Y2H assay where growth is the readout. Second, for a positive readout the Y2H assay relies on reconstitution of a split transcription factor reporter protein. To detect a positive interaction, the two split domains of the transcription factor must be oriented in the correct three-dimensional manner when juxtaposed by the interacting proteins. This type of precise geometry appears to be much less relevant for a positive readout from PIP. And lastly, while the Y2H assay measures binary protein interactions, in the PIP assay, multiple "bait" proteins fused to μNS are present on a display platform such that transient and/or weak interactions might be stabilized due to the high local concentration of the bait protein.

Schmitz and colleagues proceeded to demonstrate that the efficient secretion of all nine of these effectors requires Spa15. Interestingly, none of the remaining 11 effectors on the virulence plasmid tested bound or required Spa15 for their efficient secretion. Currently no additional *Shigella* type 3 chaperones are known, nor are any candidate proteins identified by protein homology chaperones. Thus, it appears that at least half of the known *Shigella* effectors do not require a chaperone to be recognized as a secreted substrate. Together the studies from Page and Schmitz highlight how studies focused on studying chaperone-effector interactions can potentially lead to identification of previously unknown effectors.

Functional genomic approaches to identify effectors

Technologies have advanced significantly since the discovery of the *Shigella* virulence plasmid in the early 1980s. Currently bacterial genomes are being sequenced and annotated at an astonishing rate. Comparative genomics can be used to assign putative functions to the majority of ORFs, particularly housekeeping genes. However, the identification of effectors still remains challenging given that their secretion signal sequences are poorly defined and many lack homology with other effectors. Given that effectors often act to target conserved eukaryotic proteins and processes, in 2001 Lesser and Miller demonstrated that effectors from *Salmonella* and *Yersinia* often inhibit growth when expressed in the model eukaryote, the yeast *Saccharomyces cerevisiae* (Lesser

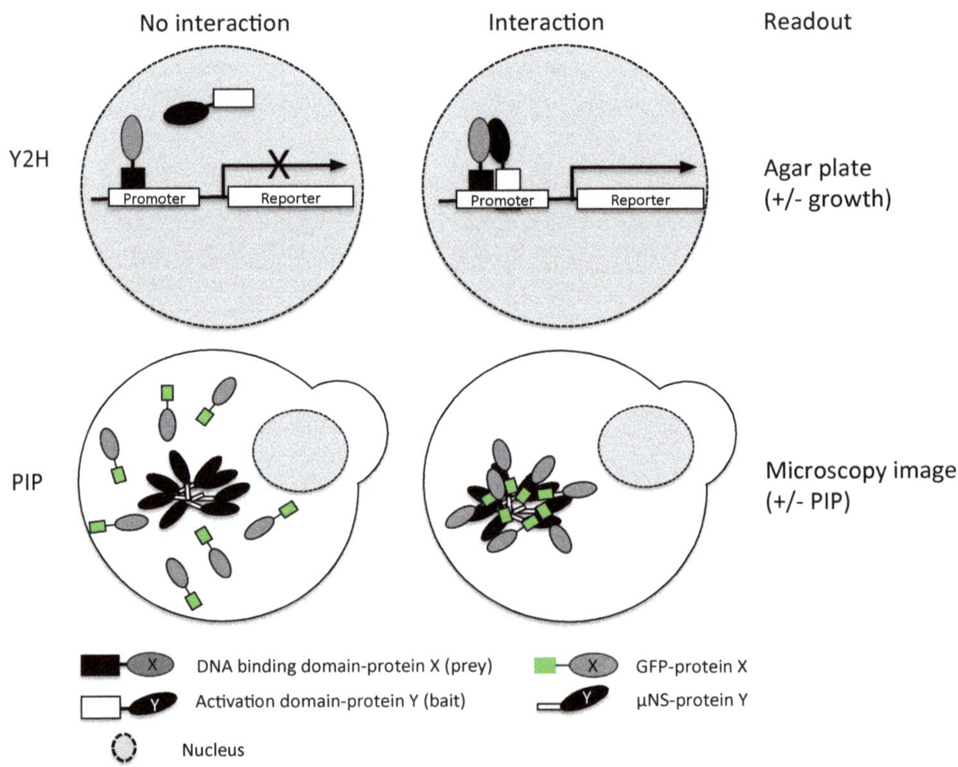

Figure 13.2. Schematic comparison of the Yeast 2 Hybrid (Y2H) and Protein Interaction Platform (PIP) assays. In the Y2H assay, candidate proteins are fused to either the DNA binding or activation domain of a transcription factor. If the two proteins interact, correctly juxtapose the transcription factor activation and binding domains and localize to the nucleus transcription of a reporter gene is activated and a positive read out is generated. In the PIP assay, each candidate protein is fused to either GFP or µNS. The protein fused to µNS is subsequently displayed on intracellular platforms. If the protein fused to GFP is recruited to the platforms, foci are observed via fluorescence microscopy.

and Miller, 2001). Based on these observations, in 2008, Slagowski and colleagues developed a high throughput yeast growth assay to screen libraries of potentially hundreds of proteins whose expression results in growth inhibition and used this assay to quantify growth inhibition when proteins encoded on the *Shigella* virulence plasmid were expressed in yeast (Slagowski et al., 2008).

Interestingly, the investigators observed that almost half of the known effectors tested inhibit growth, to variable degrees, when expressed in yeast. Expression of low-levels of IpgB2, IpgD, VirA and IcsB completely inhibited growth, while similar levels of expression of OspC1, OspD3, OspF and IpgB1 moderately impaired growth. Additional effector proteins inhibit growth when stressors are added to the media at low doses. Expression of 0/21 non-translocated proteins inhibited growth. The investigators subsequently identified a new virulence plasmid encoded effector, IpaJ, based on the severe growth inhibition observed with its expression. Similar approaches have been utilized by multiple other groups to identify candidate secreted effectors from both plant and animal pathogens including *Legionella* (Campodonico et al., 2005; Heidtman et al., 2009), *Chlamydia* (Sisko et al., 2006), *Pseudomonas* (Munkvold et al., 2008) and *Anaplasma* (Sukumaran et al., 2011). A major advantage of this system is that one can study candidate effector proteins from genetically intractable organisms, given that it only requires access to DNA. In addition, the "awesome power of yeast genetics" provides a powerful platform for the subsequent identification of the conserved eukaryotic processes and/or proteins targeted by effectors (Kramer et al., 2007; Siggers and Lesser, 2008).

Bioinformatic approaches to effector identification

The chromosomes and plasmids of bacterial pathogens are currently being rapidly sequenced and annotated opening up new avenues for bioinformatic approaches to identify type 3 secreted effectors. While each pathogen encodes its own unique set of effectors, effectors from one species often share homology with effectors from another. Thus, it is possible to identify putative effectors from bacteria using homology searches and, as more bioinformatic data is accumulated on known effectors, new ones will likely be rapidly and accurately identified based on sequence similarity. Although the majority of *Shigella* effectors were identified prior to the advent of bioinformatic databases, we now know that many effectors share homology with effectors from other bacteria. In fact, the *Shigella* effector OspZ was identified through homology searches. While its ORF is encoded on the *Shigella* virulence plasmid, it was not recognized or demonstrated to be a secreted substrate until 2008 when Zurawski and colleagues recognized that this protein shared extensive homology with the enteropathogenic *Escherichia coli* type 3 effector NleE (Zurawski et al., 2008).

Another bioinformatic-based approach to identify potential effectors that is being explored is based on screening for secretion signal or chaperone binding domain sequences common to type 3 effectors. However, no clear consensus sequence is apparent when examining the N-terminal ~20 residues of type 3 effectors, the region that encodes the secretion signal sequence. Despite any clear homology, several groups have recently developed machine learning algorithm-based approaches to identify potential effectors (for review see (McDermott et al., 2011)). Although none of the groups have tested the ability of the algorithms to identify *Shigella* effectors, several web-based interfaces are available to screen lists of proteins for those predicted to be type 3 secreted substrates: BEAN (Bacterial Effector Analyzer) (http://protein.cau.edu.cn:8080/bean/) (Dong et al., 2013), EffectiveT3 (http://www.effectors.org) (Arnold et al., 2009; Jehl et al., 2011), Modlab T3SS effector prediction (http://gecco.org.chemie.uni-frankfurt.de/

T3SS_prediction/T3SS_prediction) (Lower and Schneider, 2009) and SIEVE (http://www.sysbep.org/sieve/) (Arnold et al., 2009). While none of these algorithms have incorporated *Shigella* proteins in their training sets, they are each able to correctly designate the majority of known *Shigella* virulence plasmid effectors while minimally predicting non-secreted virulence plasmid proteins as being secreted substrates (Table 13.1).

With the exception of the Bean algorithm, the machine learning based approaches discussed above attempt to identify effectors based on their N-terminal ~20 residues. However, as described above many effectors require chaperones for their efficient secretion. In 2006, based on analyses of chaperone-effector co-crystals, Lilic et al., in agreement with earlier work from Birtalan et. al., suggested that effectors encode a conserved structural motif, the ß-motif, which serves as a chaperone-binding domain. This region is characterized by three hydrophobic residues (ψ) separated by 1-4 variable residues ($\psi x_{2\text{-}4}\psi x_{1\text{-}2}\psi$) (Birtalan et al., 2002; Lilic et al., 2006). More recently, Costa and colleagues established that nine Spa15-dependent *Shigella* and four *Salmonella* effectors bind a conserved sequence, the conserved chaperone binding domain (CCBD) sequence $[(LMIF)_1xxx(IV)_5xx(IV)_8x(Ns)_{10}]$(Costa et al., 2012). Notably, in the case of *Salmonella* SipA, one effector whose crystal structure in complex with its class IB chaperone, InvB, has been solved, it is clear that the CCBD domain overlaps with the β-strand motif. While it is not possible to perform pattern search searches for the poorly defined ß-motif ($\psi x_{2\text{-}4}\psi x_{1\text{-}2}\psi$), it is possible to screen annotated genomes for open reading frames that encode the CCBD sequence.

Using pattern-search algorithms for the CCBD sequence, Costa and colleagues confirmed that the CCBD sequence is common only to Spa15 effectors. Although this approach did not result in the identification of any new *Shigella* effectors, the investigators did identify the first two putative effectors of *Sodalis glossinidius,* a tsetse fly endosymbiont that encodes two Spa15 homologs. Taken together, these examples demonstrate how bioinformatic pattern search algorithms for the CCBD sequence could potentially provide a useful means for identifying probable type 3 effectors from bacteria that encode class IB chaperones.

Table 13.1: Comparison of correct detection of known *Shigella* type 3 secreted effectors via predictive machine learning algorithms.

Algorithm	Type 3 effectors	Non-secreted proteins
BEAN	15/24 (63%)	40/45 (89%)
Effective T3(selective paramenters)(p=.0001)	9/24 (38%)	40/45 (89%)
Effective T3 (sensitive parameters) (p=.05)	15/24 (63%)	38/45 (84%)
ModLab	20/24 (83%)	42/45 (93%)
SIEVE (p=.05)	18/24 (75%)	25/45 (56%)
SIEVE (p=.02)	15/24 (63%)	36/45 (80%)

Concluding remarks

Many years ago, long before the discovery that bacterial pathogens encode complex transkingdom protein delivery systems for the direct delivery of bacterial proteins into eukaryotic host cells, the large ~220 kB plasmid of *Shigella* was recognized to be an important, essential virulence factor. Subsequently, over the following 35 years, a variety of genetic and biochemical approaches led to the identification of over 30 type 3 secreted *Shigella* effectors, a list that will surely continue to grow, particularly as many ORFs on the virulence plasmid remain to be characterized and additional secreted effectors are likely encoded on the *Shigella* chromosome. Many of the experimental approaches reviewed in this chapter have been replaced by newer more refined assays, but the strategies remain the same. For example, mass spectrometry based approaches have the potential to assist in identifying low abundant and/or low molecular weight effectors previously missed. Similarly, refined phenotypic genetic screens focused on detecting alterations in specific host cellular pathways as opposed to *Shigella* invasion and cell-to-cell spread have the potential to provide means to identify new effectors. And, lastly bioinformatic approaches will only improve as the sequences of more effectors are revealed and analyzed. Nevertheless, the identification of type 3 effectors only serves as a starting point for dissecting the means by which pathogens like *Shigella* manipulate host cells to promote their own survival and spread. The greater challenge has been and will continue to be the identification of host cell processes and proteins targeted by type 3 effectors during the course of an infection. It is these studies that will provide novel insights regarding the different and common strategies used by bacterial pathogens to cause disease that have the potential to lead to the identification of novel targets for the development of new and effective antimicrobial agents.

References

Adler, B., Sasakawa, C., Tobe, T., Makino, S., Komatsu, K., and Yoshikawa, M. (1989). A dual transcriptional activation system for the 230 kb plasmid genes coding for virulence-associated antigens of Shigella flexneri. Molecular microbiology *3*, 627-635.

Allaoui, A., Menard, R., Sansonetti, P.J., and Parsot, C. (1993a). Characterization of the Shigella flexneri ipgD and ipgF genes, which are located in the proximal part of the mxi locus. Infection and immunity *61*, 1707-1714.

Allaoui, A., Mounier, J., Prevost, M.C., Sansonetti, P.J., and Parsot, C. (1992). icsB: a Shigella flexneri virulence gene necessary for the lysis of protrusions during intercellular spread. Mol Microbiol *6*, 1605-1616.

Allaoui, A., Sansonetti, P.J., and Parsot, C. (1993b). MxiD, an outer membrane protein necessary for the secretion of the Shigella flexneri Ipa invasins. Mol Microbiol *7*, 59-68.

Andrews, G.P., and Maurelli, A.T. (1992). mxiA of Shigella flexneri 2a, which facilitates export of invasion plasmid antigens, encodes a homolog of the low-calcium-response protein, LcrD, of Yersinia pestis. Infection and immunity *60*, 3287-3295.

Arnold, R., Brandmaier, S., Kleine, F., Tischler, P., Heinz, E., Behrens, S., Niinikoski, A., Mewes, H.W., Horn, M., and Rattei, T. (2009). Sequence-based prediction of type III secreted proteins. PLoS Pathog *5*, e1000376.

Ashida, H., Toyotome, T., Nagai, T., and Sasakawa, C. (2007). Shigella chromosomal IpaH proteins are secreted via the type III secretion system and act as effectors. Mol Microbiol *63*, 680-693.

Bergounioux, J., Elisee, R., Prunier, A.L., Donnadieu, F., Sperandio, B., Sansonetti, P., and Arbibe, L. (2012). Calpain activation by the Shigella flexneri effector VirA regulates key steps in the formation and life of the bacterium's epithelial niche. Cell host and microbe *11*, 240-252.

Birtalan, S.C., Phillips, R.M., and Ghosh, P. (2002). Three-dimensional secretion signals in chaperone-effector complexes of bacterial pathogens. Molecular cell *9*, 971-980.

Botteaux, A., Sory, M.P., Biskri, L., Parsot, C., and Allaoui, A. (2009). MxiC is secreted by and controls the substrate specificity of the Shigella flexneri type III secretion apparatus. Mol Microbiol *71*, 449-460.

Buchrieser, C., Glaser, P., Rusniok, C., Nedjari, H., D'Hauteville, H., Kunst, F., Sansonetti, P., and Parsot, C. (2000). The virulence plasmid pWR100 and the repertoire of proteins secreted by the type III secretion apparatus of Shigella flexneri. Mol Microbiol *38*, 760-771.

Buysse, J.M., Stover, C.K., Oaks, E.V., Venkatesan, M., and Kopecko, D.J. (1987). Molecular cloning of invasion plasmid antigen (ipa) genes from Shigella flexneri: analysis of ipa gene products and genetic mapping. Journal of bacteriology *169*, 2561-2569.

Campodonico, E.M., Chesnel, L., and Roy, C.R. (2005). A yeast genetic system for the identification and characterization of substrate proteins transferred into host cells by the Legionella pneumophila Dot/Icm system. Mol Microbiol *56*, 918-933.

Costa, S.C., Schmitz, A.M., Jahufar, F.F., Boyd, J.D., Cho, M.Y., Glicksman, M.A., and Lesser, C.F. (2012). A new means to identify type 3 secreted effectors: functionally interchangeable class IB chaperones recognize a conserved sequence. mBio *3*.

Demers, B., Sansonetti, P.J., and Parsot, C. (1998). Induction of type III secretion in Shigella flexneri is associated with differential control of transcription of genes encoding secreted proteins. The EMBO journal *17*, 2894-2903.

Dong, X., Zhang, Y.J., and Zhang, Z. (2013). Using weakly conserved motifs hidden in secretion signals to identify type-III effectors from bacterial pathogen genomes. PloS one *8*, e56632.

Goldberg, M.B., Barzu, O., Parsot, C., and Sansonetti, P.J. (1993). Unipolar localization and ATPase activity of IcsA, a Shigella flexneri protein involved in intracellular movement. Journal of bacteriology *175*, 2189-2196.

Hale, T.L., Oaks, E.V., and Formal, S.B. (1985). Identification and antigenic characterization of virulence-associated, plasmid-coded proteins of *Shigella* spp. and enteroinvasive Escherichia coli. Infection and immunity *50*, 620-629.

Hale, T.L., Sansonetti, P.J., Schad, P.A., Austin, S., and Formal, S.B. (1983). Characterization of virulence plasmids and plasmid-associated outer membrane proteins in Shigella flexneri, Shigella sonnei, and Escherichia coli. Infection and immunity *40*, 340-350.

Hartman, A.B., Venkatesan, M., Oaks, E.V., and Buysse, J.M. (1990). Sequence and molecular characterization of a multicopy invasion plasmid antigen gene, ipaH, of Shigella flexneri. Journal of bacteriology *172*, 1905-1915.

Heidtman, M., Chen, E.J., Moy, M.Y., and Isberg, R.R. (2009). Large-scale identification of Legionella pneumophila Dot/Icm substrates that modulate host cell vesicle trafficking pathways. Cellular microbiology *11*, 230-248.

Jehl, M.A., Arnold, R., and Rattei, T. (2011). Effective--a database of predicted secreted bacterial proteins. Nucleic acids research *39*, D591-595.

Jin, Q., Yuan, Z., Xu, J., Wang, Y., Shen, Y., Lu, W., Wang, J., Liu, H., Yang, J., Yang, F., *et al.* (2002). Genome sequence of Shigella flexneri 2a: insights into pathogenicity through comparison with genomes of Escherichia coli K12 and O157. Nucleic acids research *30*, 4432-4441.

Kramer, R.W., Slagowski, N.L., Eze, N.A., Giddings, K.S., Morrison, M.F., Siggers, K.A., Starnbach, M.N., and Lesser, C.F. (2007). Yeast functional genomic screens lead to identification of a role for a bacterial effector in innate immunity regulation. PLoS Pathog *3*, e21.

Lesser, C.F., and Miller, S.I. (2001). Expression of microbial virulence proteins in Saccharomyces cerevisiae models mammalian infection. The EMBO journal *20*, 1840-1849.

Lilic, M., Vujanac, M., and Stebbins, C.E. (2006). A common structural motif in the binding of virulence factors to bacterial secretion chaperones. Molecular cell *21*, 653-664.

Lower, M., and Schneider, G. (2009). Prediction of type III secretion signals in genomes of gram-negative bacteria. PloS one *4*, e5917.

Magdalena, J., Hachani, A., Chamekh, M., Jouihri, N., Gounon, P., Blocker, A., and Allaoui, A. (2002). Spa32 regulates a switch in substrate specificity of the type III secreton of Shigella flexneri from needle components to Ipa proteins. Journal of bacteriology *184*, 3433-3441.

Makino, S., Sasakawa, C., Kamata, K., Kurata, T., and Yoshikawa, M. (1986). A genetic determinant required for continuous reinfection of adjacent cells on large plasmid in S. flexneri 2a. Cell *46*, 551-555.

Maurelli, A.T., Baudry, B., d'Hauteville, H., Hale, T.L., and Sansonetti, P.J. (1985). Cloning of plasmid DNA sequences involved in invasion of HeLa cells by Shigella flexneri. Infection and immunity *49*, 164-171.

Maurelli, A.T., Blackmon, B., and Curtiss, R., 3rd (1984). Loss of pigmentation in Shigella flexneri 2a is correlated with loss of virulence and virulence-associated plasmid. Infection and immunity *43*, 397-401.

McDermott, J.E., Corrigan, A., Peterson, E., Oehmen, C., Niemann, G., Cambronne, E.D., Sharp, D., Adkins, J.N., Samudrala, R., and Heffron, F. (2011). Computational prediction of type III and IV secreted effectors in gram-negative bacteria. Infection and immunity *79*, 23-32.

Menard, R., Sansonetti, P., and Parsot, C. (1994a). The secretion of the Shigella flexneri Ipa invasins is activated by epithelial cells and controlled by IpaB and IpaD. The EMBO journal *13*, 5293-5302.

Menard, R., Sansonetti, P., Parsot, C., and Vasselon, T. (1994b). Extracellular association and cytoplasmic partitioning of the IpaB and IpaC invasins of S. flexneri. Cell *79*, 515-525.

Menard, R., Sansonetti, P.J., and Parsot, C. (1993). Nonpolar mutagenesis of the ipa genes defines IpaB, IpaC, and IpaD as effectors of Shigella flexneri entry into epithelial cells. J Bacteriol *175*, 5899-5906.

Munkvold, K.R., Martin, M.E., Bronstein, P.A., and Collmer, A. (2008). A survey of the Pseudomonas syringae pv. tomato DC3000 type III secretion system effector repertoire reveals several effectors that are deleterious when expressed in Saccharomyces cerevisiae. Molecular plant-microbe interactions : MPMI *21*, 490-502.

Niebuhr, K., Jouihri, N., Allaoui, A., Gounon, P., Sansonetti, P.J., and Parsot, C. (2000). IpgD, a protein secreted by the type III secretion machinery of Shigella flexneri, is chaperoned by IpgE and implicated in entry focus formation. Mol Microbiol *38*, 8-19.

Oaks, E.V., Hale, T.L., and Formal, S.B. (1986). Serum immune response to Shigella protein antigens in rhesus monkeys and humans infected with Shigella spp. Infection and immunity *53*, 57-63.

Ogawa, M., Suzuki, T., Tatsuno, I., Abe, H., and Sasakawa, C. (2003). IcsB, secreted via the type III secretion system, is chaperoned by IpgA and required at the post-invasion stage of Shigella pathogenicity. Mol Microbiol *48*, 913-931.

Ogawa, M., Yoshimori, T., Suzuki, T., Sagara, H., Mizushima, N., and Sasakawa, C. (2005). Escape of intracellular Shigella from autophagy. Science *307*, 727-731.

Page, A.L., Sansonetti, P., and Parsot, C. (2002). Spa15 of Shigella flexneri, a third type of chaperone in the type III secretion pathway. Mol Microbiol *43*, 1533-1542.

Parsot, C., Menard, R., Gounon, P., and Sansonetti, P.J. (1995). Enhanced secretion through the Shigella flexneri Mxi-Spa translocon leads to assembly of extracellular proteins into macromolecular structures. Mol Microbiol *16*, 291-300.

Sakai, T., Sasakawa, C., and Yoshikawa, M. (1988). Expression of four virulence antigens of Shigella flexneri is positively regulated at the transcriptional level by the 30 kiloDalton virF protein. Mol Microbiol *2*, 589-597.

Sansonetti, P.J., Hale, T.L., Dammin, G.J., Kapfer, C., Collins, H.H., Jr., and Formal, S.B. (1983). Alterations in the pathogenicity of Escherichia coli K-12 after transfer of plasmid and chromosomal genes from Shigella flexneri. Infection and immunity *39*, 1392-1402.

Sansonetti, P.J., Kopecko, D.J., and Formal, S.B. (1982). Involvement of a plasmid in the invasive ability of Shigella flexneri. Infection and immunity *35*, 852-860.

Sasakawa, C., Adler, B., Tobe, T., Okada, N., Nagai, S., Komatsu, K., and Yoshikawa, M. (1989). Functional organization and nucleotide sequence of virulence Region-2 on the large virulence plasmid in Shigella flexneri 2a. Molecular microbiology *3*, 1191-1201.

Sasakawa, C., Kamata, K., Sakai, T., Makino, S., Yamada, M., Okada, N., and Yoshikawa, M. (1988). Virulence-associated genetic regions comprising 31 kilobases of the 230-kilobase plasmid in Shigella flexneri 2a. Journal of bacteriology *170*, 2480-2484.

Sasakawa, C., Kamata, K., Sakai, T., Murayama, S.Y., Makino, S., and Yoshikawa, M. (1986a). Molecular alteration of the 140-megadalton plasmid associated with loss of virulence and Congo red binding activity in Shigella flexneri. Infection and immunity *51*, 470-475.

Sasakawa, C., Komatsu, K., Tobe, T., Suzuki, T., and Yoshikawa, M. (1993). Eight genes in region 5 that form an operon are essential for invasion of epithelial cells by Shigella flexneri 2a. Journal of bacteriology *175*, 2334-2346.

Sasakawa, C., Makino, S., Kamata, K., and Yoshikawa, M. (1986b). Isolation, characterization, and mapping of Tn5 insertions into the 140-megadalton invasion plasmid defective in the mouse Sereny test in Shigella flexneri 2a. Infection and immunity *54*, 32-36.

Schmitz, A.M., Morrison, M.F., Agunwamba, A.O., Nibert, M.L., and Lesser, C.F. (2009). Protein interaction platforms: visualization of interacting proteins in yeast. Nat Methods *6*, 500-502.

Sereny, B. (1955). Experimental shigella keratoconjunctivitis; a preliminary report. Acta microbiologica Academiae Scientiarum Hungaricae *2*, 293-296.

Siggers, K.A., and Lesser, C.F. (2008). The Yeast Saccharomyces cerevisiae: a versatile model system for the identification and characterization of bacterial virulence proteins. Cell Host Microbe *4*, 8-15.

Sisko, J.L., Spaeth, K., Kumar, Y., and Valdivia, R.H. (2006). Multifunctional analysis of Chlamydia-specific genes in a yeast expression system. Mol Microbiol *60*, 51-66.

Slagowski, N.L., Kramer, R.W., Morrison, M.F., LaBaer, J., and Lesser, C.F. (2008). A functional genomic yeast screen to identify pathogenic bacterial proteins. PLoS Pathog *4*, e9.

Sukumaran, B., Mastronunzio, J.E., Narasimhan, S., Fankhauser, S., Uchil, P.D., Levy, R., Graham, M., Colpitts, T.M., Lesser, C.F., and Fikrig, E. (2011). Anaplasma phagocytophilum AptA modulates Erk1/2 signalling. Cellular microbiology *13*, 47-61.

Uchiya, K., Tobe, T., Komatsu, K., Suzuki, T., Watarai, M., Fukuda, I., Yoshikawa, M., and Sasakawa, C. (1995). Identification of a novel virulence gene, virA, on the large plasmid of Shigella, involved in invasion and intercellular spreading. Mol Microbiol *17*, 241-250.

Van Gijsegem, F., Genin, S., and Boucher, C. (1993). Conservation of secretion pathways for pathogenicity determinants of plant and animal bacteria. Trends in microbiology *1*, 175-180.

Venkatesan, M.M., Buysse, J.M., and Kopecko, D.J. (1988). Characterization of invasion plasmid antigen genes (ipaBCD) from Shigella flexneri. Proceedings of the National Academy of Sciences of the United States of America *85*, 9317-9321.

Venkatesan, M.M., Buysse, J.M., and Oaks, E.V. (1992). Surface presentation of Shigella flexneri invasion plasmid antigens requires the products of the spa locus. Journal of bacteriology *174*, 1990-2001.

Watarai, M., Tobe, T., Yoshikawa, M., and Sasakawa, C. (1995). Contact of Shigella with host cells triggers release of Ipa invasins and is an essential function of invasiveness. The EMBO journal *14*, 2461-2470.

Wattiau, P., Bernier, B., Deslee, P., Michiels, T., and Cornelis, G.R. (1994). Individual chaperones required for Yop secretion by Yersinia. Proceedings of the National Academy of Sciences of the United States of America *91*, 10493-10497.

Wattiau, P., and Cornelis, G.R. (1993). SycE, a chaperone-like protein of Yersinia enterocolitica involved in Ohe secretion of YopE. Mol Microbiol *8*, 123-131.

Zurawski, D.V., Mumy, K.L., Badea, L., Prentice, J.A., Hartland, E.L., McCormick, B.A., and Maurelli, A.T. (2008). The NleE/OspZ family of effector proteins is required for polymorphonuclear transepithelial migration, a characteristic shared by enteropathogenic Escherichia coli and Shigella flexneri infections. Infection and immunity *76*, 369-379.

Index

A

ABC transporter 30-31, 34, 244
Abl 217
Actin-based motility 67, 169-171, 178-179, 181, 183, 219
Adherent junction 200
Aerobactin 32, 39, 41, 52
Akt 161
Antimicrobial peptide 194, 201, 231
Anti-sense RNA 111, 114, 115
Antivirulence genes (AVG) 49-54, 56-59
Apoptosis 56, 146, 170, 196, 233
ArcA 36-37, 76, 90
ArcB 36
ARF1p/2p 163
Arg/Abl 154-255
Arp2/3 153, 157-158, 180-181, 183, 199
ATG5 183-184, 198, 232
Autophagy 198, 221-222, 232
Autophagy protein 5 *See* ATG5
Autotransporter 172, 174, 177

B

Bacterial effector analyzer 269
Bacterioferritin 31

C

CadA 49, 53-56
Cadherin 183, 196-197
Calcium 156-157
Caspase 139, 143-146, 162, 201, 233
Catenin 196
CCL20 231, 234
CD4 235-236
CD44 13, 156, 162, 249
CD8 235
Cdc42 153-154, 157-158, 180, 198, 251
Cir 32
Claudin 195, 197
Cofilin 181-182, 199
Connexin 162, 200
Conserved chaperone binding domain (CCBD) 270
Coprogen 32
Correlative light electron microscopy (CLEM) 212, 214, 221
Cortactin 155, 198
CpxR 76, 87-88
Cra 76
Crk 154-155, 198
CsrA 76, 90, 118, 123
CsrB 118, 122
CsrC 122
Curli 53

D

Danger-associated molecular patterns (DAMP) 201
DegP 174
Deoxycholate (DOC) 248-249, 251
Dimethyldodecylamine-N-oxide (LDAO) 141
DnaK 178
Dock180 159
Dynamin-2 222

E

Enteroinvasive *E. coli* (EIEC) 52, 54, 56, 66-67, 171, 245
ELMO 159
Enterobactin 31
Enteroendocrine cell 193-194
Epsin-1 222
ERK 202, 204, 217, 234
ERM 162
EvgA 120-122
Exb 30
Extracellular ATP 162
Ezrin 154, 162, 199

F

F-actin 154, 160, 180, 198-199, 221
Fec 33
Feo 33-34, 37, 39
Ferrioxamine B 32
Filamin 199
Filopodia 155
FIS 86
Flagellar system 53, 58
FlAsH 212-216, 223
Fluorescence recovery after photobleaching (FRAP) 213, 217

Fluorescence resonance energy transfer (FRET) 213, 216, 218-219, 223, 250
Formin 153, 182-183
FtsQ 177-178
FtsZ 177
Fumarate and nitrate reduction (FNR) 36-37, 88, 90, 92
Fur 27, 34-36, 76, 89, 121

G

Galectin-3 219, 222
GAP 163
GEF 154, 158-159, 161-162
Global enteric multicenter study (GEMS) 9
Goblet cell 193-194, 230

H

Heme 27, 30, 32-33, 38, 126
Hemolytic uremic syndrome 11
Hepoxilin A3 (HXA3 202-203)
Hfq 34, 112, 118
H-NS 65, 69-73, 76-79, 81-82, 91-93

I

IcsA/VirG 58, 67, 73, 77-78, 82, 119, 124-125, 169-170, 172-176, 178-184, 198, 232, 249, 262
IcsB 81, 83, 92, 162, 172, 198-199, 221, 232, 262-263, 266-267, 269,
IcsP 81-82, 89, 174-176
IFN-gamma 233-234, 236
IHF 76, 86
IKK (IκB kinase) 203
IL-10 236
IL-17 236
IL-18 152, 201
IL-1beta 140, 144, 152, 197, 201
IL-8 152, 204, 231-232
Inflammasome 162, 201
Inositol 156-157, 160-161, 180, 236
Integrin 13, 155, 160-161
Invasion plasmid antigens (Ipa) 12-13
IpaA 13-14, 67, 157, 159, 160, 198, 261, 264, 266-267
IpaB 12, 39, 139-146, 155-156, 162, 198, 213, 232, 246, 249, 250-251, 260-261, 264-265
IpaC 12, 39, 139-140, 156-158, 198, 213-214, 217, 232, 246, 249, 251-252, 260-261, 261, 265
IpaD 67, 155-156, 213-214, 214, 247, 249-251, 260-261, 261, 264-265
IPAF 139, 143-146
IpaH 14, 67, 203, 261, 265
IpaJ 163, 269
Ipa-mxi-spa 81
IpgA 13, 263, 266-267
IpgB 13, 159, 161-162, 198, 217, 263, 265-266, 269
IpgC 13, 68, 74, 83-84, 92, 140, 213, 246, 265
IpgD 13, 161, 198, 236, 262-264, 266-267, 269
IpgE 263, 266-267
IpgF 263
Iron transport system 27-28, 30, 29
IroN 32

L

Lactoferrin 38-39
Lamina propria 193-194
Lipopolysaccharide *See* LPS
LPS 11, 39, 50, 175, 201, 229, 231-235, 237

M

M cell 57, 140, 151-152, 197, 229-230, 233
Mad2L2 141
MaG 125
MAIT cell 236
MAPK 14, 201-202
Membrane expression of invasion plasmid antigens 264
Mena 181-182
MicF 72
MLCK 196, 200
MreB 177
Mucosa-associated invariant T cell *See* MAIT cell
MxiC 265
MxiE 13, 68, 74-75, 81, 83-86, 94
MxiH 233
MxiL 265
Myosin 159, 182, 189, 200, 222
Myosin light chain kinase *See* MLCK

N

Nad 49, 56-57
Natural killer cell (NK cell) 233-234

Necrotic death 153
Neutrophil extracellular trap (NET) 231-232
NF-κB 162, 201-204, 223, 231
Nicotinic acid 56
NlrC4 162
n-Octyl-oligo-oxyethylene (OPOE) 141
N-WASP 153, 179, 180-182, 199

O

Occludin 195
OmpA 124, 127-128
OmpC 90
OmpF 72
OmpR 90
OmpT 49, 58, 175
Optical reconstruction microscopy (STORM) 212, 221
Osp 13, 14, 81-84, 202-204, 234, 265-267, 269
OxyR 36

P

Paneth cell 193-194
Paracellualr transport 194-195
ParB 79-80
Pathogen-associated molecular patterns (PAMP) 201
PBP 30-31, 34
Periplasmic binding protein *See* PBP
PhoP 88-89
Photoactivated localization microscopy (PALM) 212, 215-216, 221
PI3 kinase 161
Pic 16
pINV 65, 67-69, 71, 73, 75, 83, 85-86, 89-90, 91-92, 94, 245
PIP 267-268
Plastin 199
PMN 54, 152, 197
Polymorphorneuclear leukocyte *See* PMN
PopB 140-141, 146
Profilin 181-182, 199
Protein interaction platform *See* PIP
Pyroptosis 57, 139-140, 143, 145, 201

R

Rab 163, 218
Rac 153-154, 157-158, 198, 217, 218
Reactive arthritis 12
Rho 153-155, 160, 180, 198, 217
Rhodotorulic acid 32
Ribo-regulator 109-111, 113, 116, 118, 122, 124, 128, 129, 130
Riboswitch 111, 114-117
RNA mediated regulation 125
RNA thermometer 37, 111, 113, 115-116, 124, 126-130
RnaG 77, 124-125
RNA-mediated regulation 109-110
Rns 77-78
RyfA 118, 123, 130
RyhB 34-35, 89-90, 118-121, 130

S

Salmochelin 32, 39
Scar/WAVE 153
SepA 15
Septin 221-222
SER 260, 2620264
Sfh 68-69, 71-72
ShET 15-16, 54-55
ShiA 119-120
Shiga toxin 15, 52
Shigella entry region *See* SER
Shigella pathogenicity island 51-52, 68
Shu 125-126
ShuA 27, 37, 127
Siderocalin 39
Siderophore 27-28, 30-32, 56
SipA 162
SipB 140-141, 146
SipC 157, 162
Sit 34, 37, 39-41
Skp 174
Small RNA 111-112, 114, 118, 121-123, 125, 128-130
SodB 34, 118-119, 122
SopA 174
SopB 161-162
SopE 161-162, 215
SoxS 36
Spa 264
Spa15 83-84, 266-267
Spa32 88, 265
Spa33 88, 249
Spa47 246
SpeG 57
Spermidine 57
SptP 215
Src 154-155, 217

Stimulated emission depletion 212
StpA 69, 71-72
Structured illumination microscopy 212
Superoxide dismutase *See* SodB
SurA 174
Surface presentation of Ipa antigens –see Spa

T

T3SS 12, 14, 51-53, 57, 66, 88, 95, 121, 139, 142, 144, 146, 153, 155-157, 172, 198, 212-214, 216-217, 223, 230, 236, 243, 246, 248-249, 251-253, 259, 263-264, 269-270
Talin 160-161
Tectonin domain-containing protein 199
TH17 231, 236
Tight junction 14, 194-197, 200-201
TNF-alpha 201, 232, 236
Toca-1 180-182
TonB 30, 32-33, 37, 39
TRAF6 203
Transcellular 194
Transcription-based secretion reporter 214
Transferrin 38
Tricellulin 200-201, 222
Type 3 secretion system (T3SS/TTSS) *See* T3SS

V

Vasodilator-stimulated phosphoprotein 181-182, 199
Vinculin 14, 160, 181-182
VirA 13, 163, 198, 265, 269
VirB 13, 69, 73, 75-76, 78-83, 85, 87, 89-90, 92, 121-123,
VirF 35, 69, 73, 75-78, 86, 90, 92, 122-123, 213, 245
VirK 67, 89

W

WASP/WAVE family 154
WASP-interacting protein 179
WAVE 161

X

Xenosiderophores 32

Y

YidC 177-178
YopB 140-141, 146

Z

Zonula occludens proteins 196-197

www.ingramcontent.com/pod-product-compliance
Ingram Content Group UK Ltd.
Pitfield, Milton Keynes, MK11 3LW, UK
UKHW022034041026
15591UKWH00012B/669

9 781910 190197